# Abiogenesis

" Natural Processes for The Origin of Life "

Edited by Paul F. Kisak

# Contents

**1 Abiogenesis**     1
   1.1 Early geophysical conditions .................................... 2
     1.1.1 The earliest biological evidence for life on Earth ............... 3
   1.2 Conceptual history ............................................ 4
     1.2.1 Spontaneous generation ................................. 4
     1.2.2 Etymology ........................................... 4
     1.2.3 Louis Pasteur and Charles Darwin ......................... 5
     1.2.4 "Primordial soup" hypothesis ............................. 5
     1.2.5 Proteinoid microspheres ................................ 6
   1.3 Current models .............................................. 7
   1.4 Chemical origin of organic molecules ............................. 8
     1.4.1 Chemical synthesis .................................... 8
     1.4.2 Autocatalysis ......................................... 9
     1.4.3 Information theory .................................... 10
     1.4.4 Homochirality ........................................ 10
   1.5 Self-enclosement, reproduction, duplication and the RNA world ......... 10
     1.5.1 Protocells ........................................... 10
     1.5.2 RNA world .......................................... 11
     1.5.3 RNA synthesis and replication ........................... 12
     1.5.4 Pre-RNA world ...................................... 13
   1.6 Origin of biological metabolism .................................. 13
     1.6.1 Iron–sulfur world ..................................... 13
     1.6.2 Zn-world hypothesis ................................... 14
     1.6.3 Deep sea vent hypothesis ............................... 14
     1.6.4 Thermosynthesis ...................................... 16
   1.7 Other models of abiogenesis .................................... 16
     1.7.1 Clay hypothesis ...................................... 17
     1.7.2 Gold's "deep-hot biosphere" model ........................ 17
     1.7.3 Panspermia .......................................... 17

| | 1.7.4 | Extraterrestrial organic molecules | 18 |
| | 1.7.5 | Lipid world | 19 |
| | 1.7.6 | Polyphosphates | 20 |
| | 1.7.7 | PAH world hypothesis | 20 |
| | 1.7.8 | Radioactive beach hypothesis | 20 |
| | 1.7.9 | Thermodynamic dissipation | 20 |
| | 1.7.10 | Multiple genesis | 21 |
| | 1.7.11 | Fluctuating hydrothermal pools on volcanic islands | 22 |
| 1.8 | See also | | 22 |
| 1.9 | Notes | | 22 |
| 1.10 | References | | 22 |
| 1.11 | Bibliography | | 35 |
| 1.12 | Further reading | | 38 |
| 1.13 | External links | | 39 |
| | 1.13.1 | Video resources | 40 |

## 2 Spontaneous generation — 41

- 2.1 Description — 41
- 2.2 Pre-Aristotelian philosophers — 42
- 2.3 Aristotle — 42
- 2.4 Classical writers after Aristotle — 43
- 2.5 Adoption in Christianity — 43
- 2.6 Scientific method — 44
- 2.7 See also — 46
- 2.8 References — 46

## 3 Self-organization — 48

- 3.1 Overview — 48
  - 3.1.1 Principles of self-organization — 48
- 3.2 History of the idea — 49
  - 3.2.1 Developing views — 49
- 3.3 Examples — 50
  - 3.3.1 Physics — 50
  - 3.3.2 Chemistry — 51
  - 3.3.3 Biology — 51
  - 3.3.4 Computer Science — 52
  - 3.3.5 Cybernetics — 52
  - 3.3.6 Human society — 53
  - 3.3.7 Psychology and education — 54

| | 3.3.8 | Traffic flow | 56 |
|---|---|---|---|
| | 3.3.9 | Methodology | 56 |
| 3.4 | Criticism | | 57 |
| 3.5 | See also | | 57 |
| 3.6 | References | | 57 |
| 3.7 | Further reading | | 60 |
| 3.8 | External links | | 61 |
| | 3.8.1 | Dissertations and theses on self-organization | 62 |

## 4 Self-replication     63

| 4.1 | Overview | | 63 |
|---|---|---|---|
| | 4.1.1 | Theory | 63 |
| | 4.1.2 | Classes of self-replication | 63 |
| | 4.1.3 | A self-replicating computer program | 64 |
| | 4.1.4 | Self-replicating tiling | 64 |
| | 4.1.5 | Applications | 64 |
| 4.2 | Mechanical self-replication | | 65 |
| 4.3 | Fields involving study of self-replication | | 65 |
| 4.4 | Self-replication in industry | | 65 |
| | 4.4.1 | Space exploration and manufacturing | 65 |
| | 4.4.2 | Molecular manufacturing | 66 |
| 4.5 | See also | | 66 |
| 4.6 | References | | 66 |

## 5 Orthogenesis     68

| 5.1 | Definition | | 68 |
|---|---|---|---|
| 5.2 | Origins | | 69 |
| | 5.2.1 | Theories | 69 |
| | 5.2.2 | Teleological | 70 |
| 5.3 | Collapse of the hypothesis | | 70 |
| 5.4 | Modern co-opted usage | | 72 |
| 5.5 | See also | | 72 |
| 5.6 | References | | 72 |
| 5.7 | Further reading | | 73 |

## 6 Primordial soup     75

| 6.1 | A reducing atmosphere | 75 |
|---|---|---|
| 6.2 | Monomer formation | 75 |
| 6.3 | Further transformation | 75 |

|     |       |                                     |    |
| --- | ----- | ----------------------------------- | -- |
|     | 6.4   | See also                            | 76 |
|     | 6.5   | References                          | 76 |

## 7 Miller–Urey experiment — 77

|     |     |                          |    |
| --- | --- | ------------------------ | -- |
|     | 7.1 | Experiment               | 77 |
|     | 7.2 | Chemistry of experiment  | 78 |
|     | 7.3 | Other experiments        | 78 |
|     | 7.4 | Earth's early atmosphere | 78 |
|     | 7.5 | Extraterrestrial sources | 79 |
|     | 7.6 | Recent related studies   | 79 |
|     | 7.7 | Amino acids identified   | 80 |
|     | 7.8 | References               | 80 |
|     | 7.9 | External links           | 81 |

## 8 Biogenic substance — 82

|     |     |            |    |
| --- | --- | ---------- | -- |
|     | 8.1 | Examples   | 82 |
|     | 8.2 | References | 82 |

## 9 Biotic material — 83

|     |     |            |    |
| --- | --- | ---------- | -- |
|     | 9.1 | References | 84 |

## 10 Common descent — 86

|     |        |                                       |    |
| --- | ------ | ------------------------------------- | -- |
|     | 10.1   | History                               | 86 |
|     | 10.2   | Evidence of universal common descent  | 88 |
|     | 10.2.1 | Common biochemistry and genetic code  | 88 |
|     | 10.2.2 | Phylogenetic trees                    | 88 |
|     | 10.3   | Illustrations of common descent       | 88 |
|     | 10.3.1 | Artificial selection                  | 88 |
|     | 10.3.2 | Natural selection                     | 89 |
|     | 10.4   | See also                              | 90 |
|     | 10.5   | References                            | 90 |
|     | 10.6   | Bibliography                          | 90 |
|     | 10.7   | External links                        | 91 |

## 11 Last universal ancestor — 92

|     |      |                     |    |
| --- | ---- | ------------------- | -- |
|     | 11.1 | Features            | 92 |
|     | 11.2 | Hypotheses          | 94 |
|     | 11.3 | Location of the root| 95 |
|     | 11.4 | See also            | 95 |
|     | 11.5 | References          | 95 |

# 12 Proteinoid — 97

- 12.1 History — 97
- 12.2 Polymerization — 97
- 12.3 Legacy — 98
- 12.4 See also — 98
- 12.5 References — 98
- 12.6 Further reading — 98

# 13 Autocatalysis — 99

- 13.1 Chemical reactions — 99
  - 13.1.1 Chemical equilibrium — 99
  - 13.1.2 Far from equilibrium — 99
  - 13.1.3 Autocatalytic reactions — 100
- 13.2 Creation of order — 100
  - 13.2.1 Background — 100
  - 13.2.2 Temporal order — 101
  - 13.2.3 Spatial order — 102
- 13.3 Biological example — 102
- 13.4 Phase transitions — 103
- 13.5 Asymmetric autocatalysis — 103
- 13.6 Role in origin of life — 103
- 13.7 Examples of autocatalytic reactions — 103
- 13.8 See also — 104
- 13.9 References — 104
- 13.10 External links — 104

# 14 Homochirality — 105

- 14.1 Mirror-symmetry breaking — 105
- 14.2 Chiral amplification — 105
- 14.3 Chiral transmission — 106
- 14.4 Optical resolution in racemic amino acids — 106
- 14.5 History — 107
- 14.6 See also — 107
- 14.7 References — 107
- 14.8 External links — 108

# 15 Protocell — 109

- 15.1 Overview — 109
- 15.2 Selectivity for compartmentalization — 110

|   |   | 15.2.1 Energy gradient . . . . . . . . . . . . . . . . . . . . . . . . . . . . . . . . . . | 111 |
|---|---|---|---|

- 15.3 Vesicles and micelles . . . . . . . . . . . . . . . . . . . . . . . . . . . . . . . . . . . . . . . 111
    - 15.3.1 Geothermal ponds and clay . . . . . . . . . . . . . . . . . . . . . . . . . . . . . . 111
    - 15.3.2 Montmorillonite bubbles . . . . . . . . . . . . . . . . . . . . . . . . . . . . . . . 112
    - 15.3.3 Membrane transport . . . . . . . . . . . . . . . . . . . . . . . . . . . . . . . . . 112
- 15.4 Artificial models . . . . . . . . . . . . . . . . . . . . . . . . . . . . . . . . . . . . . . . . . 112
    - 15.4.1 Langmuir-Blodgett deposition . . . . . . . . . . . . . . . . . . . . . . . . . . . . 112
    - 15.4.2 Jeewanu . . . . . . . . . . . . . . . . . . . . . . . . . . . . . . . . . . . . . . . . 113
- 15.5 Ethics and controversy . . . . . . . . . . . . . . . . . . . . . . . . . . . . . . . . . . . . . . 113
- 15.6 See also . . . . . . . . . . . . . . . . . . . . . . . . . . . . . . . . . . . . . . . . . . . . . . 113
- 15.7 References . . . . . . . . . . . . . . . . . . . . . . . . . . . . . . . . . . . . . . . . . . . . 113
- 15.8 External links . . . . . . . . . . . . . . . . . . . . . . . . . . . . . . . . . . . . . . . . . . 115

## 16 Iron–sulfur world hypothesis 116
- 16.1 Origin of life . . . . . . . . . . . . . . . . . . . . . . . . . . . . . . . . . . . . . . . . . . . 116
    - 16.1.1 Pioneer organism . . . . . . . . . . . . . . . . . . . . . . . . . . . . . . . . . . . 116
    - 16.1.2 Nutrient conversions . . . . . . . . . . . . . . . . . . . . . . . . . . . . . . . . . 116
    - 16.1.3 Synthetic reactions . . . . . . . . . . . . . . . . . . . . . . . . . . . . . . . . . . 116
- 16.2 Early evolution . . . . . . . . . . . . . . . . . . . . . . . . . . . . . . . . . . . . . . . . . . 117
    - 16.2.1 Cellularization . . . . . . . . . . . . . . . . . . . . . . . . . . . . . . . . . . . . 117
    - 16.2.2 Proto-ecological systems . . . . . . . . . . . . . . . . . . . . . . . . . . . . . . . 117
- 16.3 References . . . . . . . . . . . . . . . . . . . . . . . . . . . . . . . . . . . . . . . . . . . . 118

## 17 Panspermia 120
- 17.1 History . . . . . . . . . . . . . . . . . . . . . . . . . . . . . . . . . . . . . . . . . . . . . . 120
- 17.2 Proposed mechanisms . . . . . . . . . . . . . . . . . . . . . . . . . . . . . . . . . . . . . . 122
    - 17.2.1 Radiopanspermia . . . . . . . . . . . . . . . . . . . . . . . . . . . . . . . . . . . 122
    - 17.2.2 Lithopanspermia . . . . . . . . . . . . . . . . . . . . . . . . . . . . . . . . . . . 122
    - 17.2.3 Accidental panspermia . . . . . . . . . . . . . . . . . . . . . . . . . . . . . . . . 123
    - 17.2.4 Directed panspermia . . . . . . . . . . . . . . . . . . . . . . . . . . . . . . . . . 123
    - 17.2.5 Pseudo-panspermia . . . . . . . . . . . . . . . . . . . . . . . . . . . . . . . . . . 124
- 17.3 Extraterrestrial life . . . . . . . . . . . . . . . . . . . . . . . . . . . . . . . . . . . . . . . . 125
    - 17.3.1 Hypotheses on extraterrestrial sources of illnesses . . . . . . . . . . . . . . . . . . 126
    - 17.3.2 Case studies . . . . . . . . . . . . . . . . . . . . . . . . . . . . . . . . . . . . . . 127
    - 17.3.3 Hoaxes . . . . . . . . . . . . . . . . . . . . . . . . . . . . . . . . . . . . . . . . 128
- 17.4 Extremophiles . . . . . . . . . . . . . . . . . . . . . . . . . . . . . . . . . . . . . . . . . . 128
    - 17.4.1 Research in outer space . . . . . . . . . . . . . . . . . . . . . . . . . . . . . . . . 129
- 17.5 Criticism . . . . . . . . . . . . . . . . . . . . . . . . . . . . . . . . . . . . . . . . . . . . . 131
- 17.6 Science fiction . . . . . . . . . . . . . . . . . . . . . . . . . . . . . . . . . . . . . . . . . . 131

## Contents

- 17.7 See also . . . . . . . . . . . . . . . . . . . . . . . . . . . . . . . . . . . . . . . . . . . . . 132
- 17.8 References . . . . . . . . . . . . . . . . . . . . . . . . . . . . . . . . . . . . . . . . . . 132
- 17.9 Further reading . . . . . . . . . . . . . . . . . . . . . . . . . . . . . . . . . . . . . . . 139
- 17.10 External links . . . . . . . . . . . . . . . . . . . . . . . . . . . . . . . . . . . . . . . . 139

**18 List of interstellar and circumstellar molecules**   **140**
- 18.1 Detection . . . . . . . . . . . . . . . . . . . . . . . . . . . . . . . . . . . . . . . . . . 140
  - 18.1.1 History . . . . . . . . . . . . . . . . . . . . . . . . . . . . . . . . . . . . . . 140
- 18.2 Molecules . . . . . . . . . . . . . . . . . . . . . . . . . . . . . . . . . . . . . . . . . . 141
  - 18.2.1 Diatomic (43) . . . . . . . . . . . . . . . . . . . . . . . . . . . . . . . . . . 142
  - 18.2.2 Triatomic (43) . . . . . . . . . . . . . . . . . . . . . . . . . . . . . . . . . . 142
  - 18.2.3 Four atoms (27) . . . . . . . . . . . . . . . . . . . . . . . . . . . . . . . . . 142
  - 18.2.4 Five atoms (19) . . . . . . . . . . . . . . . . . . . . . . . . . . . . . . . . . 142
  - 18.2.5 Six atoms (16) . . . . . . . . . . . . . . . . . . . . . . . . . . . . . . . . . . 142
  - 18.2.6 Seven atoms (10) . . . . . . . . . . . . . . . . . . . . . . . . . . . . . . . . 142
  - 18.2.7 Eight atoms (11) . . . . . . . . . . . . . . . . . . . . . . . . . . . . . . . . 142
  - 18.2.8 Nine atoms (10) . . . . . . . . . . . . . . . . . . . . . . . . . . . . . . . . . 142
  - 18.2.9 Ten or more atoms (15) . . . . . . . . . . . . . . . . . . . . . . . . . . . . . 143
- 18.3 Deuterated molecules (20) . . . . . . . . . . . . . . . . . . . . . . . . . . . . . . . . . 143
- 18.4 Unconfirmed (13) . . . . . . . . . . . . . . . . . . . . . . . . . . . . . . . . . . . . . 143
- 18.5 See also . . . . . . . . . . . . . . . . . . . . . . . . . . . . . . . . . . . . . . . . . . 143
- 18.6 References . . . . . . . . . . . . . . . . . . . . . . . . . . . . . . . . . . . . . . . . . 144
- 18.7 External links . . . . . . . . . . . . . . . . . . . . . . . . . . . . . . . . . . . . . . . 152

**19 Gard model**   **153**
- 19.1 The GARD model . . . . . . . . . . . . . . . . . . . . . . . . . . . . . . . . . . . . . 153
- 19.2 Selection in GARD . . . . . . . . . . . . . . . . . . . . . . . . . . . . . . . . . . . . . 153
- 19.3 GARD and Quasispecies . . . . . . . . . . . . . . . . . . . . . . . . . . . . . . . . . . 153
- 19.4 See also . . . . . . . . . . . . . . . . . . . . . . . . . . . . . . . . . . . . . . . . . . 153
- 19.5 References . . . . . . . . . . . . . . . . . . . . . . . . . . . . . . . . . . . . . . . . . 154
- 19.6 External links . . . . . . . . . . . . . . . . . . . . . . . . . . . . . . . . . . . . . . . 154

**20 PAH world hypothesis**   **155**
- 20.1 Background . . . . . . . . . . . . . . . . . . . . . . . . . . . . . . . . . . . . . . . . 155
- 20.2 Polycyclic aromatic hydrocarbons . . . . . . . . . . . . . . . . . . . . . . . . . . . . . 155
- 20.3 Attachment of nucleobases to PAH scaffolding . . . . . . . . . . . . . . . . . . . . . . 156
- 20.4 Attachment of oligomeric backbone . . . . . . . . . . . . . . . . . . . . . . . . . . . . 156
- 20.5 Detachment of the RNA-like strands . . . . . . . . . . . . . . . . . . . . . . . . . . . 156
- 20.6 Formation of ribozyme-like structures . . . . . . . . . . . . . . . . . . . . . . . . . . 156

| | | |
|---|---|---|
| 20.7 | See also | 157 |
| 20.8 | References | 157 |
| 20.9 | External links | 158 |

# 21 Albert von Kölliker — 159

- 21.1 Biography . . . 159
- 21.2 Works . . . 159
- 21.3 Heterogenesis . . . 160
- 21.4 Notes . . . 160
- 21.5 References . . . 161
- 21.6 Further reading . . . 161
- 21.7 Text and image sources, contributors, and licenses . . . 162
  - 21.7.1 Text . . . 162
  - 21.7.2 Images . . . 166
  - 21.7.3 Content license . . . 170

# Chapter 1

# Abiogenesis

**Abiogenesis**

*Precambrian stromatolites in the Siyeh Formation, Glacier National Park. In 2002, a paper in the scientific journal* Nature *suggested that these 3.5 Ga (billion years) old geological formations contain fossilized cyanobacteria microbes. This suggests they are evidence of one of the earliest known life forms on Earth.*

or **biopoiesis** [5] or **OoL (Origins of Life )**,[6] is the natural process of life arising from non-living matter, such as simpleorganic compounds .[7][8][9][10] It is thought to have occurred on Earth between 3.8 and 4.1[11] billion years ago. Abiogenesis is studied through a combination of laboratory experiments and extrapolation of the genetic information contained within modern organisms , and aims to make reasonable conjectures about how certain pre-life chemical reactions might have given rise to the Earth's living systems.[12]

The study of abiogenesis involves three main types of considerations: the geophysical, the chemical, and the biological,[13] with more recent approaches attempting a synthesis of all three.[14] Many approaches investigate how self-replicating molecules, or their components, came into existence. It is generally thought that current life on Earth is descended from an RNA world,[15] although RNA-based life may not have been the first life to have existed.[16][17] The classic Miller–Urey experiment and similar experiments demonstrated that most amino acids, the basic chemical constituents of the proteins used in all living organisms, can be synthesized from inorganic compounds under conditions intended to replicate those of the early Earth. Various external sources of energy that may have triggered this organic molecule synthesis have been proposed, including lightning and radiation. Other approaches ("metabolism-first" hypotheses) focus on understanding how catalysis in chemical systems on the early Earth might have provided the precursor molecules necessary for self-replication.[18] Complex organic molecules have been found in the Solar System and in interstellar space, and these molecules may have provided starting material for the development of life on Earth.[19][20][21][22]

The panspermia hypothesis alternatively suggests that microscopic life was distributed to the early Earth by meteoroids, asteroids and other small Solar System bodies and that life may exist throughout the Universe.[23] It is speculated that the biochemistry of life may have begun shortly after the Big Bang, 13.8 billion years ago, during a habitable epoch when the age of the universe was only 10 to 17 million years.[24][25] The panspermia hypothesis therefore answers questions of where, not how, life came to be; it only postulates that life may have originated in a locale outside the Earth.

Nonetheless, Earth remains the only place in the Universe known to harbor life,[26][27] and fossil evidence from the Earth supplies most studies of abiogenesis. The age of the Earth is about 4.54 billion years;[28][29][30] the earliest undisputed evidence of life on Earth dates from at least 3.5 billion years ago,[31][32][33] and possibly as early as the Eoarchean Era, after a geological crust started to solidify following the earlier molten Hadean Eon. Microbial mat fossils have been found in 3.48 billion-year-old sandstone in Western Australia.[34][35][36] Other early physical evidence of biogenic substances includes graphite discovered in 3.7 billion-year-old metasedimentary rocks in southwestern Greenland,[37] as well as "remains of biotic life" found in 4.1 billion-year-old rocks in Western Australia.[38][39] According to one of the researchers, "If life arose relatively quickly on Earth ... then it could be common in the universe."[38]

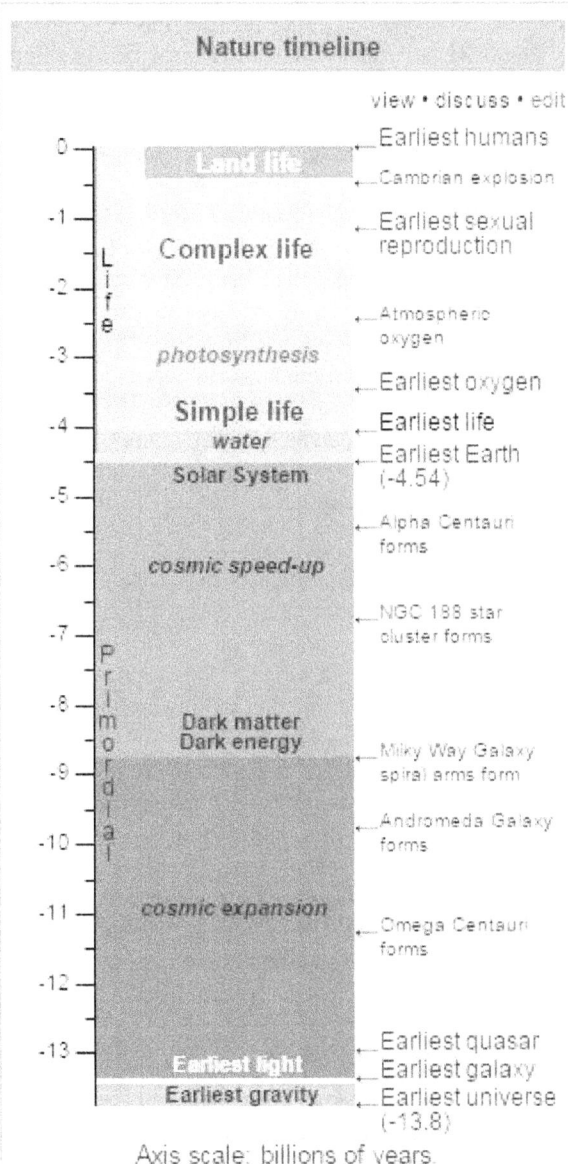

The Hadean Earth is thought to have had a secondary atmosphere, formed through degassing of the rocks that accumulated from planetesimal impactors. At first, it was thought that the Earth's atmosphere consisted of hydrogen compounds—methane, ammonia and water vapour—and that life began under such reducing conditions, which are conducive to the formation of organic molecules. During its formation, the Earth lost a significant part of its initial mass, with a nucleus of the heavier rocky elements of the protoplanetary disk remaining.[41] According to later models, suggested by study of ancient minerals, the atmosphere in the late Hadean period consisted largely of nitrogen and carbon dioxide, with smaller amounts of carbon monoxide, hydrogen, and sulfur compounds.[42] As Earth lacked the gravity to hold any molecular hydrogen, this component of the atmosphere would have been rapidly lost during the Hadean period, along with the bulk of the original inert gases. The solution of carbon dioxide in water is thought to have made the seas slightly acidic, giving it a pH of about 5.5. The atmosphere at the time has been characterized as a "gigantic, productive outdoor chemical laboratory."[43] It may have been similar to the mixture of gases released today by volcanoes, which still support some abiotic chemistry.[43]

Oceans may have appeared first in the Hadean Eon, as soon as two hundred million years (200 Ma) after the Earth was formed, in a hot 100 °C (212 °F) reducing environment, and the pH of about 5.8 rose rapidly towards neutral.[44] This has been supported by the dating of 4.404 Ga-old zircon crystals from metamorphosed quartzite of Mount Narryer in Western Australia, which are evidence that oceans and continental crust existed within 150 Ma of Earth's formation.[45] Despite the likely increased vulcanism and existence of many smaller tectonic "platelets," it has been suggested that between 4.4 and 4.3 Ga (billion year), the Earth was a water world, with little if any continental crust, an extremely turbulent atmosphere and a hydrosphere subject to intense ultraviolet (UV) light, from a T Tauri stage Sun, cosmic radiation and continued bolide impacts.[46]

The Hadean environment would have been highly hazardous to modern life. Frequent collisions with large objects, up to 500 kilometres (310 mi) in diameter, would have been sufficient to sterilise the planet and vaporise the ocean within a few months of impact, with hot steam mixed with rock vapour becoming high altitude clouds that would completely cover the planet. After a few months, the height of these clouds would have begun to decrease but the cloud base would still have been elevated for about the next thousand years. After that, it would have begun to rain at low altitude. For another two thousand years, rains would slowly have drawn down the height of the clouds, returning the oceans to their original depth only 3,000 years after the impact event.[47]

### 1.1.1 The earliest biological evidence for life on Earth

The earliest life on Earth existed before 3.5 billion years ago,[31][32][33] during the Eoarchean Era when sufficient crust had solidified following the molten Hadean Eon. Physical evidence has been found in biogenic graphite in 3.7 billion-year-old metasedimentary rocks from southwestern Greenland[37] and microbial mat fossils found in 3.48 billion-year-old sandstone from Western Australia.[34][36] Evidence of early life in rocks from Akilia Island, near the Isua supracrustal belt in southwestern Greenland, dating to 3.7 billion years ago have shown biogenic carbon isotopes.[48] At Strelley Pool, in the Pilbarra region of Western Australia, compelling evidence of early life has been found in pyrite-bearing sandstone in a fossilized beach, that showed rounded tubular cells that oxidised sulfur by photosynthesis in the absence of oxygen.[49] More recently, geochemists have found evidence that life likely existed on Earth at least 4.1 billion years ago — 300 million years earlier than previous research suggested.[38][39][50]

In the earlier period between 3.8 and 4.1 Ga, changes in the orbits of the giant planets may have caused a heavy bombardment by asteroids and comets[51] that pockmarked the Moon and the other inner planets (Mercury, Mars, and presumably Earth and Venus). This would likely have repeatedly sterilized the planet, had life appeared before that time.[43] Geologically, the Hadean Earth would have been far more active than at any other time in its history. Studies of meteorites suggests that radioactive isotopes such as aluminium-26 with a half-life of $7.17 \times 10^5$ years, and potassium-40 with a half-life of $1.250 \times 10^9$ years, isotopes mainly produced in supernovae, were much more common.[52] Coupled with internal heating as a result of gravitational sorting between the core and the mantle, there would have been a great deal of mantle convection, with the probable result of many more smaller and much more active tectonic plates than now exist.

The time periods between such devastating environmental events give time windows for the possible origin of life in the early environments. A study by Kevin A. Maher and David J. Stevenson shows that if the deep marine hydrothermal setting provides a suitable site for the origin of life, then abiogenesis could have happened as early as 4.0 to 4.2 Ga, whereas if it occurred at the surface of the Earth, abiogenesis could only have occurred between 3.7 and 4.0 Ga.[53] Somewhat related, in July 2016, scientists reported identifying a set of 355 genes from the Last Universal Common Ancestor (LUCA) of all organisms living on Earth.[54]

## 1.2 Conceptual history

### 1.2.1 Spontaneous generation

Main article: Spontaneous generation

Belief in spontaneous generation of certain forms of life from non-living matter goes back to Aristotle and ancient Greek philosophy and continued to have support in Western scholarship until the 19th century.[55] This belief was paired with a belief in heterogenesis, i.e., that one form of life derived from a different form (e.g., bees from flowers).[56] Classical notions of spontaneous generation held that certain complex, living organisms are generated by decaying organic substances. According to Aristotle, it was a readily observable truth that aphids arise from the dew that falls on plants, flies from putrid matter, mice from dirty hay, crocodiles from rotting logs at the bottom of bodies of water, and so on.[57] In the 17th century, people began to question such assumptions. In 1646, Sir Thomas Browne published his *Pseudodoxia Epidemica* (subtitled *Enquiries into Very many Received Tenets, and commonly Presumed Truths*), which was an attack on false beliefs and "vulgar errors." His contemporary, Alexander Ross, erroneously refuted him, stating: "To question this [Ed.: i.e., spontaneous generation], is to question Reason, Sense, and Experience: If he doubts of this, let him go to *Ægypt*, and there he will finde the fields swarming with mice begot of the mud of *Nylus*, to the great calamity of the Inhabitants."[58][59]

In 1665, Robert Hooke published the first drawings of a microorganism. Hooke was followed in 1676 by Antonie van Leeuwenhoek, who drew and described microorganisms that are now thought to have been protozoa and bacteria.[60] Many felt the existence of microorganisms was evidence in support of spontaneous generation, since microorganisms seemed too simplistic for sexual reproduction, and asexual reproduction through cell division had not yet been observed. Van Leeuwenhoek took issue with the ideas common at the time that fleas and lice could spontaneously result from putrefaction, and that frogs could likewise arise from slime. Using a broad range of experiments ranging from sealed and open meat incubation and the close study of insect reproduction he became, by the 1680s, convinced that spontaneous generation was incorrect.[61]

The first experimental evidence against spontaneous generation came in 1668 when Francesco Redi showed that no maggots appeared in meat when flies were prevented from laying eggs. It was gradually shown that, at least in the case of all the higher and readily visible organisms, the previous sentiment regarding spontaneous generation was false. The alternative seemed to be biogenesis: that every living thing came from a pre-existing living thing (*omne vivum ex ovo*, Latin for "every living thing from an egg").

In 1768, Lazzaro Spallanzani demonstrated that microbes were present in the air, and could be killed by boiling. In 1861, Louis Pasteur performed a series of experiments that demonstrated that organisms such as bacteria and fungi do not spontaneously appear in sterile, nutrient-rich media, but could only appear by invasion from without.

The belief that self-ordering by spontaneous generation was impossible begged for an alternative. By the middle of the 19th century, the theory of biogenesis had accumulated so much evidential support, due to the work of Pasteur and others, that the alternative theory of spontaneous generation had been effectively disproven. John Desmond Bernal, a pioneer in X-ray crystallography, suggested that earlier theories such as spontaneous generation were based upon an explanation that life was continuously created as a result of chance events.[62]

### 1.2.2 Etymology

Main article: Biogenesis

The term *biogenesis* is usually credited to either Henry Charlton Bastian or to Thomas Henry Huxley.[63] Bastian used the term around 1869 in an unpublished exchange with John Tyndall to mean "life-origination or commencement". In 1870, Huxley, as new president of the British Association for the Advancement of Science, delivered an address entitled *Biogenesis and Abiogenesis*.[64] In it he introduced the term *biogenesis* (with an opposite meaning to Bastian's) as well as *abiogenesis*:

> And thus the hypothesis that living matter always arises by the agency of pre-existing living matter, took definite shape; and had, henceforward, a right to be considered and a claim to be refuted, in each particular case, before the production of living matter in any other way could be admitted by careful reasoners. It will be necessary for me to refer to this hypothesis so frequently, that, to save circumlocution, I shall call it the hypothesis of *Biogenesis*; and I shall term the contrary doctrine–that living matter may be produced by not living matter–the hypothesis of *Abiogenesis*.[64]

Subsequently, in the preface to Bastian's 1871 book, *The Modes of Origin of Lowest Organisms*,[65] Bastian referred to the possible confusion with Huxley's usage and explicitly renounced his own meaning:

> A word of explanation seems necessary with

## 1.2. CONCEPTUAL HISTORY

regard to the introduction of the new term *Archebiosis*. I had originally, in unpublished writings, adopted the word *Biogenesis* to express the same meaning—viz., life-origination or commencement. But in the mean time the word *Biogenesis* has been made use of, quite independently, by a distinguished biologist [Huxley], who wished to make it bear a totally different meaning. He also introduced the word *Abiogenesis*. I have been informed, however, on the best authority, that neither of these words can—with any regard to the language from which they are derived—be supposed to bear the meanings which have of late been publicly assigned to them. Wishing to avoid all needless confusion, I therefore renounced the use of the word *Biogenesis*, and being, for the reason just given, unable to adopt the other term, I was compelled to introduce a new word, in order to designate the process by which living matter is supposed to come into being, independently of pre-existing living matter.[66]

### 1.2.3 Louis Pasteur and Charles Darwin

*Charles Darwin in 1879*

Louis Pasteur remarked, about a finding of his in 1864 which he considered definitive, "Never will the doctrine of spontaneous generation recover from the mortal blow struck by this simple experiment."[67][68] One alternative was that life's origins on Earth had come from somewhere else in the Universe. Periodically resurrected (see Panspermia, above) Bernal said that this approach "is equivalent in the last resort to asserting the operation of metaphysical, spiritual entities... it turns on the argument of creation by design by a creator or demiurge."[69] Such a theory, Bernal said, was unscientific. A theory popular around the same time was that life was the result of an inner "life force", which in the late 19th century was championed by Henri Bergson.

The idea of evolution by natural selection proposed by Charles Darwin put an end to these metaphysical theologies. In a letter to Joseph Dalton Hooker on 1 February 1871,[70] Darwin discussed the suggestion that the original spark of life may have begun in a "warm little pond, with all sorts of ammonia and phosphoric salts, light, heat, electricity, &c., present, that a proteine compound was chemically formed ready to undergo still more complex changes." He went on to explain that "at the present day such matter would be instantly devoured or absorbed, which would not have been the case before living creatures were formed." He had written to Hooker in 1863 stating that, "It is mere rubbish, thinking at present of the origin of life; one might as well think of the origin of matter." In *On the Origin of Species*, he had referred to life having been "created", by which he "really meant 'appeared' by some wholly unknown process", but had soon regretted using the Old Testament term "creation".[71]

### 1.2.4 "Primordial soup" hypothesis

Main article: primordial soup
Further information: Miller–Urey experiment

No new notable research or theory on the subject appeared until 1924, when Alexander Oparin reasoned that atmospheric oxygen prevents the synthesis of certain organic compounds that are necessary building blocks for the evolution of life. In his book *The Origin of Life*,[72][73] Oparin proposed that the "spontaneous generation of life" that had been attacked by Louis Pasteur did in fact occur once, but was now impossible because the conditions found on the early Earth had changed, and preexisting organisms would immediately consume any spontaneously generated organism. Oparin argued that a "primeval soup" of organic molecules could be created in an oxygenless atmosphere through the action of sunlight. These would combine in ever more complex ways until they formed coacervate droplets. These droplets would "grow" by fusion with other droplets,

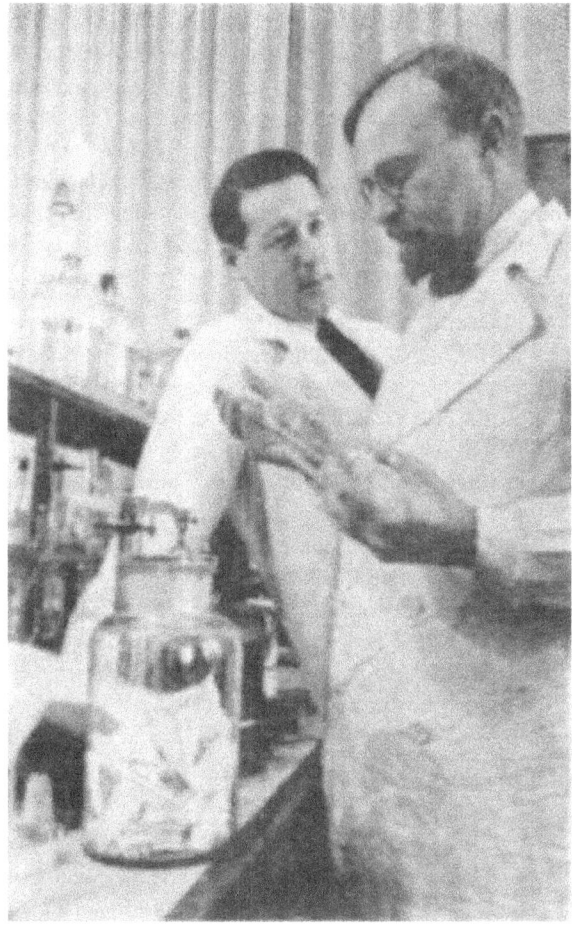

*Alexander Oparin (right) at the laboratory*

About this time, Haldane suggested that the Earth's prebiotic oceans (quite different from their modern counterparts) would have formed a "hot dilute soup" in which organic compounds could have formed. Bernal called this idea *biopoiesis* or *biopoesis*, the process of living matter evolving from self-replicating but non-living molecules,[62][75] and proposed that biopoiesis passes through a number of intermediate stages.

One of the most important pieces of experimental support for the "soup" theory came in 1952. Stanley L. Miller and Harold C. Urey performed an experiment that demonstrated how organic molecules could have spontaneously formed from inorganic precursors under conditions like those posited by the Oparin-Haldane hypothesis. The now-famous Miller–Urey experiment used a highly reducing mixture of gases - methane, ammonia, and hydrogen, as well as water vapor - to form basic organic monomers such as amino acids.[76] The mixture of gases was cycled through an apparatus that delivered electrical sparks to the mixture. After one week, it was found that about 10% to 15% of the carbon in the system was now in the form of a racemic mixture of organic compounds, including amino acids, which are the building blocks of proteins. This provided direct experimental support for the second point of the "soup" theory, and it is around the remaining two points of the theory that much of the debate now centers.

Bernal shows that based upon this and subsequent work there is no difficulty in principle in forming most of the molecules we recognize as the basic molecules of life from their inorganic precursors. The underlying hypothesis held by Oparin, Haldane, Bernal, Miller and Urey, for instance, was that multiple conditions on the primeval Earth favored chemical reactions that synthesized the same set of complex organic compounds from such simple precursors. A 2011 reanalysis of the saved vials containing the original extracts that resulted from the Miller and Urey experiments, using current and more advanced analytical equipment and technology, has uncovered more biochemicals than originally discovered in the 1950s. One of the more important findings was 23 amino acids, far more than the five originally found.[77] However, Bernal said that "it is not enough to explain the formation of such molecules, what is necessary, is a physical-chemical explanation of the origins of these molecules that suggests the presence of suitable sources and sinks for free energy."[78]

### 1.2.5 Proteinoid microspheres

Main article: Proteinoid

In trying to uncover the intermediate stages of abiogenesis mentioned by Bernal, Sidney W. Fox in the 1950s and

and "reproduce" through fission into daughter droplets, and so have a primitive metabolism in which factors that promote "cell integrity" survive, and those that do not become extinct. Many modern theories of the origin of life still take Oparin's ideas as a starting point.

Robert Shapiro has summarized the "primordial soup" theory of Oparin and J. B. S. Haldane in its "mature form" as follows:[74]

1. The early Earth had a chemically reducing atmosphere.

2. This atmosphere, exposed to energy in various forms, produced simple organic compounds ("monomers").

3. These compounds accumulated in a "soup" that may have concentrated at various locations (shorelines, oceanic vents etc.).

4. By further transformation, more complex organic polymers - and ultimately life - developed in the soup.

1960s studied the spontaneous formation of peptide structures under conditions that might plausibly have existed early in Earth's history. He demonstrated that amino acids could spontaneously form small chains called peptides. In one of his experiments, he allowed amino acids to dry out as if puddled in a warm, dry spot in prebiotic conditions. He found that, as they dried, the amino acids formed long, often cross-linked, thread-like, submicroscopic polypeptide molecules now named "proteinoid microspheres."[79]

In another experiment using a similar method to set suitable conditions for life to form, Fox collected volcanic material from a cinder cone in Hawaii. He discovered that the temperature was over 100 °C (212 °F) just 4 inches (100 mm) beneath the surface of the cinder cone, and suggested that this might have been the environment in which life was created—molecules could have formed and then been washed through the loose volcanic ash and into the sea. He placed lumps of lava over amino acids derived from methane, ammonia and water, sterilized all materials, and baked the lava over the amino acids for a few hours in a glass oven. A brown, sticky substance formed over the surface and when the lava was drenched in sterilized water a thick, brown liquid leached out. It turned out that the amino acids had combined to form proteinoids, and the proteinoids had combined to form small globules that Fox called "microspheres." His proteinoids were not cells, although they formed clumps and chains reminiscent of cyanobacteria, but they contained no functional nucleic acids or any encoded information. Based upon such experiments, Colin S. Pittendrigh stated in December 1967 that "laboratories will be creating a living cell within ten years," a remark that reflected the typical contemporary levels of innocence of the complexity of cell structures.[80]

## 1.3 Current models

There is still no "standard model" of the origin of life. Most currently accepted models draw at least some elements from the framework laid out by Alexander Oparin (in 1924) and J. B. S. Haldane (in 1925), who postulated the molecular or chemical evolution theory of life.[81] According to them, the first molecules constituting the earliest cells "were synthesized under natural conditions by a slow process of molecular evolution, and these molecules then organized into the first molecular system with properties with biological order."[81] Oparin and Haldane suggested that the atmosphere of the early Earth may have been chemically reducing in nature, composed primarily of methane ($CH_4$), ammonia ($NH_3$), water ($H_2O$), hydrogen sulfide ($H_2S$), carbon dioxide ($CO_2$) or carbon monoxide (CO), and phosphate ($PO_4^{3-}$), with molecular oxygen ($O_2$) and ozone ($O_3$) either rare or absent. According to later models, the atmosphere in the late Hadean period consisted largely of nitrogen ($N_2$) and carbon dioxide, with smaller amounts of carbon monoxide, hydrogen ($H_2$), and sulfur compounds;[82] while it did lack molecular oxygen and ozone,[83] it wasn't as chemically reducing as Oparin and Haldane supposed. In the atmosphere proposed by Oparin and Haldane, electrical activity can produce certain basic small molecules (monomers) of life, such as amino acids. This was demonstrated in the Miller–Urey experiment reported in 1953.

Bernal coined the term *biopoiesis* in 1949 to refer to the origin of life.[84] In 1967, he suggested that it occurred in three "stages": 1) the origin of biological monomers; 2) the origin of biological polymers; and 3) the evolution from molecules to cells. He suggested that evolution commenced between stage 1 and 2. The first stage is now fairly well understood, and the discovery of alkaline vents and the similarity with the "proton pump" found as the basis of biological life has begun to provide evidence about the second stage. Bernal considered the third, the discovery of methods by which biological reactions were incorporated behind cell walls, to be the most difficult. Modern work on the self organising capacities by which cell membranes self-assemble, and the work on micropores in various substrates is seen as a halfway house towards the development of independent free-living cells, and research into this is an ongoing effort.[85][86]

The chemical processes that took place on the early Earth are called *chemical evolution*. Both Manfred Eigen and Sol Spiegelman demonstrated that evolution, including replication, variation, and natural selection, can occur in populations of molecules as well as in organisms.[43] Spiegelman took advantage of natural selection to synthesize the Spiegelman Monster, which had a genome with just 218 nucleotide bases, having deconstructively evolved from a 4500 base bacterial RNA. Eigen built on Spiegelman's work and produced a similar system further degraded to just 48 or 54 nucleotides, which was the minimum required for the binding of the replication enzyme.[87]

Chemical evolution was followed by the initiation of biological evolution, which led to the first cells.[43] No one has yet synthesized a "protocell" using basic components with the necessary properties of life (the so-called "bottom-up-approach"). Without such a proof-of-principle, explanations have tended to focus on chemosynthesis.[88] However, some researchers are working in this field, notably Steen Rasmussen and Jack W. Szostak. Others have argued that a "top-down approach" is more feasible. One such approach, successfully attempted by Craig Venter and others at J. Craig Venter Institute, involves engineering existing prokaryotic cells with progressively fewer genes, attempting to discern at which point the most minimal requirements for life were reached.[89][90][91]

## 1.4 Chemical origin of organic molecules

The elements, except for hydrogen, ultimately derive from stellar nucleosynthesis. Complex molecules, including organic molecules, form naturally both in space and on planets.[19] There are two possible sources of organic molecules on the early Earth:

1. Terrestrial origins – organic molecule synthesis driven by impact shocks or by other energy sources (such as UV light, redox coupling, or electrical discharges) (e.g., Miller's experiments)

2. Extraterrestrial origins – formation of organic molecules in interstellar dust clouds, which rain down on planets.[92][93] (See pseudo-panspermia)

Estimates of the production of organics from these sources suggest that the Late Heavy Bombardment before 3.5 Ga within the early atmosphere made available quantities of organics comparable to those produced by terrestrial sources.[94][95]

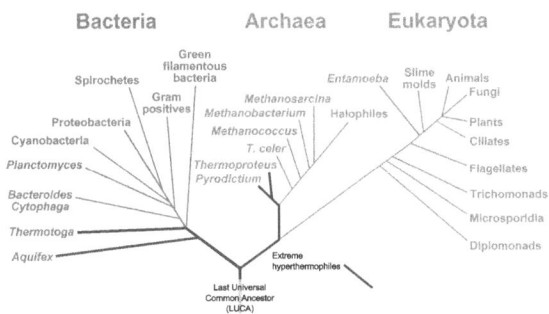

*A cladogram demonstrating extreme hyperthermophiles at the base of the phylogenetic tree of life.*

It has been estimated that the Late Heavy Bombardment may also have effectively sterilised the Earth's surface to a depth of tens of metres. If life evolved deeper than this, it would have also been shielded from the early high levels of ultraviolet radiation from the T Tauri stage of the Sun's evolution. Simulations of geothermically heated oceanic crust yield far more organics than those found in the Miller-Urey experiments (see below). In the deep hydrothermal vents, Everett Shock has found "there is an enormous thermodynamic drive to form organic compounds, as seawater and hydrothermal fluids, which are far from equilibrium, mix and move towards a more stable state."[96] Shock has found that the available energy is maximised at around 100 – 150 degrees Celsius, precisely the temperatures at which the hyperthermophilic bacteria and thermoacidophilic archaea have been found, at the base of the phylogenetic tree of life closest to the Last Universal Common Ancestor (LUCA).[97]

### 1.4.1 Chemical synthesis

While features of self-organization and self-replication are often considered the hallmark of living systems, there are many instances of abiotic molecules exhibiting such characteristics under proper conditions. Stan Palasek showed that self-assembly of ribonucleic acid (RNA) molecules can occur spontaneously due to physical factors in hydrothermal vents.[98] Virus self-assembly within host cells has implications for the study of the origin of life,[99] as it lends further credence to the hypothesis that life could have started as self-assembling organic molecules.[100][101]

Multiple sources of energy were available for chemical reactions on the early Earth. For example, heat (such as from geothermal processes) is a standard energy source for chemistry. Other examples include sunlight and electrical discharges (lightning), among others.[43] Unfavorable reactions can also be driven by highly favorable ones, as in the case of iron-sulfur chemistry. For example, this was probably important for carbon fixation (the conversion of carbon from its inorganic form to an organic one).[note 1] Carbon fixation via iron-sulfur chemistry is highly favorable, and occurs at neutral pH and 100 °C (212 °F). Iron-sulfur surfaces, which are abundant near hydrothermal vents, are also capable of producing small amounts of amino acids and other biological metabolites.[43]

Formamide produces all four ribonucleotides and other biological molecules when warmed in the presence of various terrestrial minerals. Formamide is ubiquitous in the Universe, produced by the reaction of water and hydrogen cyanide (HCN). It has several advantages as a biotic precursor, including the ability to easily become concentrated through the evaporation of water.[102][103] Although HCN is poisonous, it only affects aerobic organisms (eukaryotes and aerobic bacteria), which did not yet exist. It can play roles in other chemical processes as well, such as the synthesis of the amino acid glycine.[43]

In 1961, it was shown that the nucleic acid purine base adenine can be formed by heating aqueous ammonium cyanide solutions.[104] Other pathways for synthesizing bases from inorganic materials were also reported.[105] Leslie E. Orgel and colleagues have shown that freezing temperatures are advantageous for the synthesis of purines, due to the concentrating effect for key precursors such as hydrogen cyanide.[106] Research by Stanley L. Miller and colleagues suggested that while adenine and guanine require freezing conditions for synthesis, cytosine and uracil may require boiling temperatures.[107] Research by the Miller group notes the formation of seven different amino acids and 11 types of nucleobases in ice when ammonia and cyanide were left in a freezer from 1972 to 1997.[108][109] Other work demonstrated the formation of s-triazines (alternative nucleobases), pyrimidines (including cytosine and uracil), and adenine from urea solutions subjected to freeze-thaw cycles under a reductive atmosphere (with spark discharges as an energy source).[110] The explanation given for the unusual speed of these reactions at such a low temperature is eutectic freezing. As an ice crystal forms, it stays pure: only molecules of water join the growing crystal, while impurities like salt or cyanide are excluded. These impurities become crowded in microscopic pockets of liquid within the ice, and this crowding causes the molecules to collide more often. Mechanistic exploration using quantum chemical methods provide a more detailed understanding of some of the chemical processes involved in chemical evolution, and a partial answer to the fundamental question of molecular biogenesis.[111]

At the time of the Miller–Urey experiment, scientific consensus was that the early Earth had a reducing atmosphere with compounds relatively rich in hydrogen and poor in oxygen (e.g., $CH_4$ and $NH_3$ as opposed to $CO_2$ and nitrogen dioxide ($NO_2$)). However, current scientific consensus describes the primitive atmosphere as either weakly reducing or neutral[112][113] (see also Oxygen Catastrophe). Such an atmosphere would diminish both the amount and variety of amino acids that could be produced, although studies that include iron and carbonate minerals (thought present in early oceans) in the experimental conditions have again produced a diverse array of amino acids.[112] Other scientific research has focused on two other potential reducing environments: outer space and deep-sea thermal vents.[114][115][116]

The spontaneous formation of complex polymers from abiotically generated monomers under the conditions posited by the "soup" theory is not at all a straightforward process. Besides the necessary basic organic monomers, compounds that would have prohibited the formation of polymers were also formed in high concentration during the Miller–Urey and Joan Oró experiments.[117] The Miller–Urey experiment, for example, produces many substances that would react with the amino acids or terminate their coupling into peptide chains.[118]

A research project completed in March 2015 by John D. Sutherland and others found that a network of reactions beginning with hydrogen cyanide and hydrogen sulfide, in streams of water irradiated by UV light, could produce the chemical components of proteins and lipids, as well as those of RNA,[119][120] while not producing a wide range of other compounds.[121] The researchers used the term "cyanosulfidic" to describe this network of reactions.[120]

### 1.4.2 Autocatalysis

Main article: Autocatalysis

Autocatalysts are substances that catalyze the production of themselves and therefore are "molecular replicators." The simplest self-replicating chemical systems are autocatalytic, and typically contain three components: a product molecule and two precursor molecules. The product molecule joins together the precursor molecules, which in turn produce more product molecules from more precursor molecules. The product molecule catalyzes the reaction by providing a complementary template that binds to the precursors, thus bringing them together. Such systems have been demonstrated both in biological macromolecules and in small organic molecules.[122][123] Systems that do not proceed by template mechanisms, such as the self-reproduction of micelles and vesicles, have also been observed.[123]

It has been proposed that life initially arose as autocatalytic chemical networks.[124] British ethologist Richard Dawkins wrote about autocatalysis as a potential explanation for the origin of life in his 2004 book *The Ancestor's Tale*.[125] In his book, Dawkins cites experiments performed by Julius Rebek, Jr. and his colleagues in which they combined amino adenosine and pentafluorophenyl esters with the autocatalyst amino adenosine triacid ester (AATE). One product was a variant of AATE, which catalysed the synthesis of themselves. This experiment demonstrated the possibility that autocatalysts could exhibit competition within a population of entities with heredity, which could be interpreted as a rudimentary form of natural selection.[126][127]

In the early 1970s, Manfred Eigen and Peter Schuster examined the transient stages between the molecular chaos and a self-replicating hypercycle in a prebiotic soup.[128] In a hypercycle, the information storing system (possibly RNA) produces an enzyme, which catalyzes the formation of another information system, in sequence until the product of the last aids in the formation of the first information system. Mathematically treated, hypercycles could create quasispecies, which through natural selection entered into a form of Darwinian evolution. A boost to hypercycle theory was the discovery of ribozymes capable of catalyzing their own chemical reactions. The hypercycle theory requires the existence of complex biochemicals, such as nucleotides, which do not form under the conditions proposed by the Miller–Urey experiment.

It has been shown that early error prone translation machinery can be stable against an error catastrophe of the type that had been envisaged as problematical known as

"Orgel's paradox" caused by catalytic activities that would be disruptive.[129][130][131]

### 1.4.3 Information theory

A theory that speaks to the origin of life on Earth and other rocky planets posits life as an information system in which information content grows because of selection. Life must start with minimum possible information, or minimum possible departure from thermodynamic equilibrium, and it requires thermodynamically free energy accessible by means of its information content. The most benign circumstances, minimum entropy variations with abundant free energy, suggest the pore space in the first few kilometers of the surface. Free energy is derived from the condensed products of the chemical reactions taking place in the cooling nebula.[132]

### 1.4.4 Homochirality

Main article: Homochirality

Homochirality refers to the geometric property of some materials that are composed of chiral units. Chiral refers to nonsuperimposable 3D forms that are mirror images of one another, as are left and right hands. Living organisms use molecules that have the same chirality ("handedness"): with almost no exceptions,[133] amino acids are left-handed while nucleotides and sugars are right-handed. Chiral molecules can be synthesized, but in the absence of a chiral source or a chiral catalyst, they are formed in a 50/50 mixture of both enantiomers (called a racemic mixture). Known mechanisms for the production of non-racemic mixtures from racemic starting materials include: asymmetric physical laws, such as the electroweak interaction; asymmetric environments, such as those caused by circularly polarized light, quartz crystals, or the Earth's rotation; and statistical fluctuations during racemic synthesis.[134]

Once established, chirality would be selected for.[135] A small bias (enantiomeric excess) in the population can be amplified into a large one by asymmetric autocatalysis, such as in the Soai reaction.[136] In asymmetric autocatalysis, the catalyst is a chiral molecule, which means that a chiral molecule is catalysing its own production. An initial enantiomeric excess, such as can be produced by polarized light, then allows the more abundant enantiomer to outcompete the other.[137]

Clark has suggested that homochirality may have started in outer space, as the studies of the amino acids on the Murchison meteorite showed that L-alanine is more than twice as frequent as its D form, and L-glutamic acid was more than three times prevalent than its D counterpart. Various chiral crystal surfaces can also act as sites for possible concentration and assembly of chiral monomer units into macromolecules.[138] Compounds found on meteorites suggest that the chirality of life derives from abiogenic synthesis, since amino acids from meteorites show a left-handed bias, whereas sugars show a predominantly right-handed bias, the same as found in living organisms.[139]

## 1.5 Self-enclosement, reproduction, duplication and the RNA world

### 1.5.1 Protocells

Main article: Protocell

A protocell is a self-organized, self-ordered, spherical col-

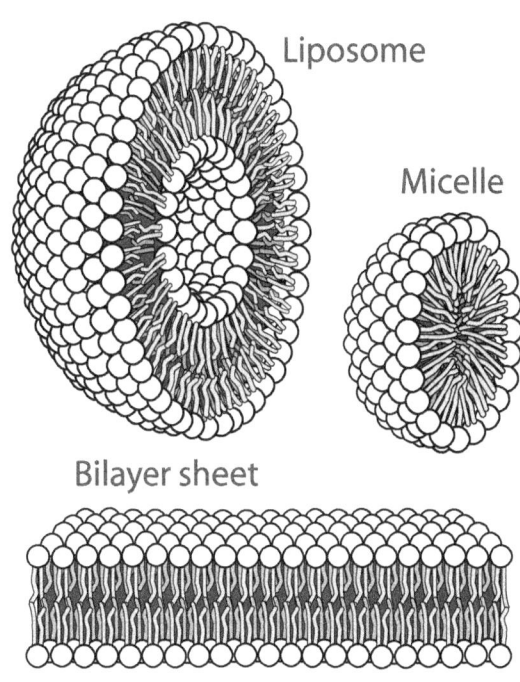

*The three main structures phospholipids form spontaneously in solution: the liposome (a closed bilayer), the micelle and the bilayer.*

lection of lipids proposed as a stepping-stone to the origin of life.[140] A central question in evolution is how simple protocells first arose and differed in reproductive contribution to the following generation driving the evolution of life. Although a functional protocell has not yet been achieved in a laboratory setting, there are scientists who think the goal is well within reach.[141][142][143]

Self-assembled vesicles are essential components of primitive cells.[140] The second law of thermodynamics requires that the Universe move in a direction in which entropy increases, yet life is distinguished by its great degree of organization. Therefore, a boundary is needed to separate life processes from non-living matter.[144] Researchers Irene A. Chen and Jack W. Szostak amongst others, suggest that simple physicochemical properties of elementary protocells can give rise to essential cellular behaviors, including primitive forms of differential reproduction competition and energy storage. Such cooperative interactions between the membrane and its encapsulated contents could greatly simplify the transition from simple replicating molecules to true cells.[142] Furthermore, competition for membrane molecules would favor stabilized membranes, suggesting a selective advantage for the evolution of cross-linked fatty acids and even the phospholipids of today.[142] Such microencapsulation would allow for metabolism within the membrane, the exchange of small molecules but the prevention of passage of large substances across it.[145] The main advantages of encapsulation include the increased solubility of the contained cargo within the capsule and the storage of energy in the form of a electrochemical gradient.

A 2012 study led by Armen Y. Mulkidjanian of Germany's University of Osnabrück, suggests that inland pools of condensed and cooled geothermal vapour have the ideal characteristics for the origin of life.[146] Scientists confirmed in 2002 that by adding a montmorillonite clay to a solution of fatty acid micelles (lipid spheres), the clay sped up the rate of vesicles formation 100-fold.[143]

Another protocell model is the Jeewanu. First synthesized in 1963 from simple minerals and basic organics while exposed to sunlight, it is still reported to have some metabolic capabilities, the presence of semipermeable membrane, amino acids, phospholipids, carbohydrates and RNA-like molecules.[147][148] However, the nature and properties of the Jeewanu remains to be clarified.

Electrostatic interactions induced by short, positively charged, hydrophobic peptides containing 7 amino acids in length or fewer, can attach RNA to a vesicle membrane, the basic cell membrane.[149]

### 1.5.2 RNA world

Main article: RNA world

The RNA world hypothesis describes an early Earth with self-replicating and catalytic RNA but no DNA or proteins.[151] It is generally accepted that current life on Earth descends from an RNA world,[15][152] although RNA-based life may not have been the first life to exist.[16][17] This conclusion is drawn from many independent lines of evidence, such as the observations that RNA is central to the translation process and that small RNAs can catalyze all of the chemical groups and information transfers required for

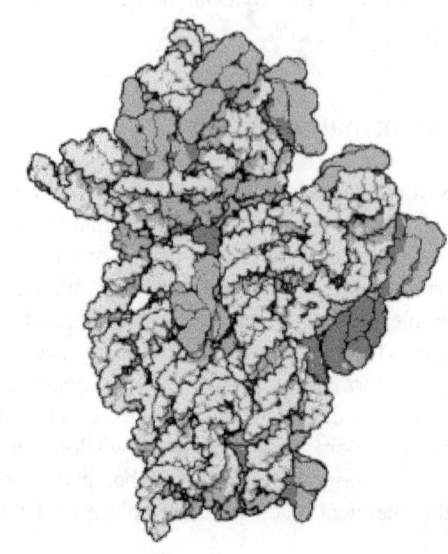

*Molecular structure of the ribosome 30S subunit from* Thermus thermophilus.*[150] Proteins are shown in blue and the single RNA chain in orange.*

life.[17][153] The structure of the ribosome has been called the "smoking gun," as it showed that the ribosome is a ribozyme, with a central core of RNA and no amino acid side chains within 18 angstroms of the active site where peptide bond formation is catalyzed.[16] The concept of the RNA world was first proposed in 1962 by Alexander Rich,[154] and the term was coined by Walter Gilbert in 1986.[17][155]

Possible precursors for the evolution of protein synthesis include a mechanism to synthesize short peptide cofactors or form a mechanism for the duplication of RNA. It is likely that the ancestral ribosome was composed entirely of RNA, although some roles have since been taken over by proteins. Major remaining questions on this topic include identifying the selective force for the evolution of the ribosome and determining how the genetic code arose.[156]

Eugene Koonin said, "Despite considerable experimental and theoretical effort, no compelling scenarios currently exist for the origin of replication and translation, the key processes that together comprise the core of biological systems and the apparent pre-requisite of biological evolution. The RNA World concept might offer the best chance for the resolution of this conundrum but so far cannot adequately account for the emergence of an efficient RNA replicase or the translation system. The MWO [Ed.: "many worlds in one"] version of the cosmological model of eternal inflation could suggest a way out of this conundrum because, in an infinite multiverse with a finite number of distinct macroscopic histories (each repeated an infinite number of times), emergence of even highly complex systems by chance is not just possible but inevitable."[157]

**Viral origins and the RNA World**

Recent evidence for a "virus first" hypothesis, which may support theories of the RNA world have been suggested in new research.[158] One of the difficulties for the study viral origins and evolution is their high rate of mutation; this is particularly the case in RNA retroviruses like HIV.[159] A 2015 study compared protein fold structures across different branches of the tree of life, where researchers can reconstruct the evolutionary histories of the folds and of the organisms whose genomes code for those folds. They argue that protein folds are better markers of ancient events as their three-dimensional structures can be maintained even as the sequences that code for those begin to change.[158] Thus, the viral protein repertoire retain traces of ancient evolutionary history that can be recovered using advanced bioinformatics approaches. Those researchers have concluded that, "the prolonged pressure of genome and particle size reduction eventually reduced virocells into modern viruses (identified by the complete loss of cellular makeup), meanwhile other coexisting cellular lineages diversified into modern cells.[160] The data suggest that viruses originated from ancient cells that co-existed with the ancestors of modern cells.[158] These ancient cells likely contained segmented RNA genomes.[158][161]

### 1.5.3 RNA synthesis and replication

The RNA world hypothesis has spurred scientists to determine if RNA molecules could have spontaneously formed able to catalyze their own replication.[162][163][164] Evidence suggests that the chemical conditions, including the presence of boron, molybdenum and oxygen needed for the initial production of RNA molecules, may have been better on the planet Mars than on the planet Earth.[162][163] If so, life-suitable molecules originating on Mars, may have later migrated to Earth via meteor ejections.[162][163]

A number of hypotheses of formation of RNA have been put forward. As of 1994, there are difficulties in the explanation of the abiotic synthesis of the nucleotides cytosine and uracil.[165] Subsequent research has shown possible routes of synthesis; for example, formamide produces all four ribonucleotides and other biological molecules when warmed in the presence of various terrestrial minerals.[102][103] Early cell membranes could have formed spontaneously from proteinoids, which are protein-like molecules produced when amino acid solutions are heated while in the correct concentration of aqueous solution. These are seen to form micro-spheres which are observed to behave similarly to membrane-enclosed compartments. Other possible means of producing more complicated organic molecules include chemical reactions that take place on clay substrates or on the surface of the mineral pyrite.

Factors supportive of an important role for RNA in early life include its ability to act both to store information and to catalyze chemical reactions (as a ribozyme); its many important roles as an intermediate in the expression of and maintenance of the genetic information (in the form of DNA) in modern organisms; and the ease of chemical synthesis of at least the components of the RNA molecule under the conditions that approximated the early Earth. Relatively short RNA molecules have been artificially produced in labs, which are capable of replication.[166] Such replicase RNA, which functions as both code and catalyst provides its own template upon which copying can occur. Jack W. Szostak has shown that certain catalytic RNAs can join smaller RNA sequences together, creating the potential for self-replication. If these conditions were present, Darwinian natural selection would favour the proliferation of such autocatalytic sets, to which further functionalities could be added.[167] Such autocatalytic systems of RNA capable of self-sustained replication have been identified.[168] The RNA replication systems, which include two ribozymes that catalyze each other's synthesis, showed a doubling time of the product of about one hour, and were subject to natural selection under the conditions that existed in the experiment.[169] In evolutionary competition experiments, this led to the emergence of new systems which replicated more efficiently.[16] This was the first demonstration of evolutionary adaptation occurring in a molecular genetic system.[169]

Depending on the specific definition used, life can be considered to have emerged when RNA chains began to express the basic conditions necessary for natural selection to operate as conceived by Darwin: heritability, variation of type, and differential reproductive output. The fitness of an RNA replicator (its per capita rate of increase) would likely be a function of its adaptive capacities that are intrinsic (in the sense that they were determined by the nucleotide sequence) and the availability of its resources.[170][171] The three primary adaptive capacities may have been (1) the capacity to replicate with moderate fidelity, giving rise to both heritability while allowing variation of type, (2) the capacity to avoid decay, and (3) the capacity to acquire and process resources.[170][171] These capacities would have been determined initially by the folded configurations of the RNA replicators that, in turn, would be encoded in their individual nucleotide sequences. Relative reproductive success, competition, between different replicators would have depended on the relative values of their adaptive capacities.

### 1.5.4 Pre-RNA world

It is possible that a different type of nucleic acid, such as PNA, TNA or GNA, was the first to emerge as a self-reproducing molecule, only later replaced by RNA.[172][173] Larralde *et al.*, say that "the generally accepted prebiotic synthesis of ribose, the formose reaction, yields numerous sugars without any selectivity."[174] and they conclude that their "results suggest that the backbone of the first genetic material could not have contained ribose or other sugars because of their instability." The ester linkage of ribose and phosphoric acid in RNA is known to be prone to hydrolysis.[175]

Pyrimidine ribonucleosides and their respective nucleotides have been prebiotically synthesised by a sequence of reactions which by-pass the free sugars, and are assembled in a stepwise fashion by using nitrogenous or oxygenous chemistries. Sutherland has demonstrated high yielding routes to cytidine and uridine ribonucleotides built from small 2 and 3 carbon fragments such as glycolaldehyde, glyceraldehyde or glyceraldehyde-3-phosphate, cyanamide and cyanoacetylene. One of the steps in this sequence allows the isolation of enantiopure ribose aminooxazoline if the enantiomeric excess of glyceraldehyde is 60% or greater.[176] This can be viewed as a prebiotic purification step, where the said compound spontaneously crystallised out from a mixture of the other pentose aminooxazolines. Ribose aminooxazoline can then react with cyanoacetylene in a mild and highly efficient manner to give the alpha cytidine ribonucleotide. Photoanomerization with UV light allows for inversion about the 1' anomeric centre to give the correct beta stereochemistry.[177] In 2009 they showed that the same simple building blocks allow access, via phosphate controlled nucleobase elaboration, to 2',3'-cyclic pyrimidine nucleotides directly, which are known to be able to polymerise into RNA. This paper also highlights the possibility for the photo-sanitization of the pyrimidine-2',3'-cyclic phosphates.[178]

## 1.6 Origin of biological metabolism

Research suggests that metabolism-like reactions could have occurred naturally in early oceans, before the first organisms evolved.[18][179] The findings suggests that metabolism predates the origin of life and evolved through the chemical conditions that prevailed in the world's earliest oceans. Reconstructions in laboratories show that some of these reactions can produce RNA, and some others resemble two essential reaction cascades of metabolism: glycolysis and the pentose phosphate pathway, that provide essential precursors for nucleic acids, amino acids and lipids.[179] Following are some observed discoveries and related hypotheses.

### 1.6.1 Iron–sulfur world

Main article: Iron–sulfur world theory

In the 1980s, Günter Wächtershäuser, encouraged and supported by Karl R. Popper,[180][181][182] postulated in his iron–sulfur world, a theory of the evolution of pre-biotic chemical pathways as the starting point in the evolution of life. It presents a consistent system of tracing today's biochemistry back to ancestral reactions that provide alternative pathways to the synthesis of organic building blocks from simple gaseous compounds.

In contrast to the classical Miller experiments, which depend on external sources of energy (such as simulated lightning or ultraviolet irradiation), "Wächtershäuser systems" come with a built-in source of energy, sulfides of iron (iron pyrite) and other minerals . The energy released from redox reactions of these metal sulfides is available for the synthesis of organic molecules. It is therefore hypothesized that such systems may be able to evolve into autocatalytic sets of self-replicating, metabolically active entities that predate the life forms known today.[18][179] Experiments with such sulfides in an aqueous environment at 100 °C produced a relatively small yield of dipeptides (0.4% to 12.4%) and a smaller yield of tripeptides (0.10%) although under the same conditions, dipeptides were quickly broken down.[183]

Several models reject the idea of the self-replication of a "naked-gene" but postulate the emergence of a primitive metabolism which could provide a safe environment for the later emergence of RNA replication. The centrality of the Krebs cycle (citric acid cycle) to energy production in aerobic organisms, and in drawing in carbon dioxide and hydrogen ions in biosynthesis of complex organic chemicals, suggests that it was one of the first parts of the metabolism to evolve.[184] Somewhat in agreement with these notions, geochemist Michael Russell has proposed that "the purpose of life is to hydrogenate carbon dioxide" (as part of a "metabolism-first," rather than a "genetics-first," scenario).[185][186] Physicist Jeremy England of MIT has proposed that thermodynamically, life was bound to eventually arrive, as based on established physics, he mathematically indicates "...that when a group of atoms is driven by an external source of energy (like the sun or chemical fuel) and surrounded by a heat bath (like the ocean or atmosphere), it will often gradually restructure itself in order to dissipate increasingly more energy. This could mean that under certain conditions, matter inexorably acquires the key physical attribute associated with life."[187][188]

One of the earliest incarnations of this idea was put forward in 1924 with Oparin's notion of primitive self-replicating vesicles which predated the discovery of the structure of DNA. Variants in the 1980s and 1990s include Wächtershäuser's iron–sulfur world theory and models introduced by Christian de Duve based on the chemistry of thioesters. More abstract and theoretical arguments for the plausibility of the emergence of metabolism without the presence of genes include a mathematical model introduced by Freeman Dyson in the early 1980s and Stuart Kauffman's notion of collectively autocatalytic sets, discussed later in that decade.

Orgel summarized his analysis of the proposal by stating, "There is at present no reason to expect that multistep cycles such as the reductive citric acid cycle will self-organize on the surface of FeS/FeS$_2$ or some other mineral."[189] It is possible that another type of metabolic pathway was used at the beginning of life. For example, instead of the reductive citric acid cycle, the "open" acetyl-CoA pathway (another one of the five recognised ways of carbon dioxide fixation in nature today) would be compatible with the idea of self-organisation on a metal sulfide surface. The key enzyme of this pathway, carbon monoxide dehydrogenase/acetyl-CoA synthase harbours mixed nickel-iron-sulfur clusters in its reaction centers and catalyses the formation of acetyl-CoA (which may be regarded as a modern form of acetyl-thiol) in a single step. There are increasing concerns, however, that prebiotic thiolated (i.e.Thioacetic acid) and Thioester compounds are thermodynamically and kinetically unfavourable to accumulate in presumed prebiotic conditions (i.e Hydrothermal vents).[190]

### 1.6.2 Zn-world hypothesis

The Zn-world (zinc world) theory of Armen Y. Mulkidjanian[191] is an extension of Wächtershäuser's pyrite hypothesis. Wächtershäuser based his theory of the initial chemical processes leading to informational molecules (i.e., RNA, peptides) on a regular mesh of electric charges at the surface of pyrite that may have made the primeval polymerization thermodynamically more favourable by attracting reactants and arranging them appropriately relative to each other.[192] The Zn-world theory specifies and differentiates further.[191][193] Hydrothermal fluids rich in H$_2$S interacting with cold primordial ocean (or Darwin's "warm little pond") water leads to the precipitation of metal sulfide particles. Oceanic vent systems and other hydrothermal systems have a zonal structure reflected in ancient volcanogenic massive sulfide deposits (VMS) of hydrothermal origin. They reach many kilometers in diameter and date back to the Archean Eon. Most abundant are pyrite (FeS$_2$), chalcopyrite (CuFeS$_2$), and sphalerite (ZnS), with additions of galena (PbS) and alabandite (MnS). ZnS and MnS have a unique ability to store radiation energy, e.g., provided by UV light. Since during the relevant time window of the origins of replicating molecules the primordial atmospheric pressure was high enough (>100 bar, about 100 atmospheres) to precipitate near the Earth's surface and UV irradiation was 10 to 100 times more intense than now, the unique photosynthetic properties mediated by ZnS provided just the right energy conditions to energize the synthesis of informational and metabolic molecules and the selection of photostable nucleobases.

The Zn-world theory has been further filled out with experimental and theoretical evidence for the ionic constitution of the interior of the first proto-cells before archaea, bacteria and proto-eukaryotes evolved. Archibald Macallum noted the resemblance of organism fluids such as blood, and lymph to seawater;[194] however, the inorganic composition of all cells differ from that of modern seawater, which led Mulkidjanian and colleagues to reconstruct the "hatcheries" of the first cells combining geochemical analysis with phylogenomic scrutiny of the inorganic ion requirements of universal components of modern cells. The authors conclude that ubiquitous, and by inference primordial, proteins and functional systems show affinity to and functional requirement for K$^+$, Zn$^{2+}$, Mn$^{2+}$, and phosphate. Geochemical reconstruction shows that the ionic composition conducive to the origin of cells could not have existed in what we today call marine settings but is compatible with emissions of vapor-dominated zones of what we today call inland geothermal systems. Under the oxygen depleted, CO$_2$-dominated primordial atmosphere, the chemistry of water condensates and exhalations near geothermal fields would resemble the internal milieu of modern cells. Therefore, the precellular stages of evolution may have taken place in shallow "Darwin ponds" lined with porous silicate minerals mixed with metal sulfides and enriched in K$^+$, Zn$^{2+}$, and phosphorus compounds.[195][196]

### 1.6.3 Deep sea vent hypothesis

The deep sea vent, or alkaline hydrothermal vent, theory for the origin of life on Earth posits that life may have begun at submarine hydrothermal vents,[197] William Martin and Michael Russell have suggested "that life evolved in structured iron monosulphide precipitates in a seepage site hydrothermal mound at a redox, pH and temperature gradient between sulphide-rich hydrothermal fluid and iron(II)-containing waters of the Hadean ocean floor. The naturally arising, three-dimensional compartmentation observed within fossilized seepage-site metal sulphide precipitates indicates that these inorganic compartments were the precursors of cell walls and membranes found in free-living prokaryotes. The known capability of FeS and NiS to catalyse the synthesis of the acetyl-methylsulphide

## 1.6. ORIGIN OF BIOLOGICAL METABOLISM

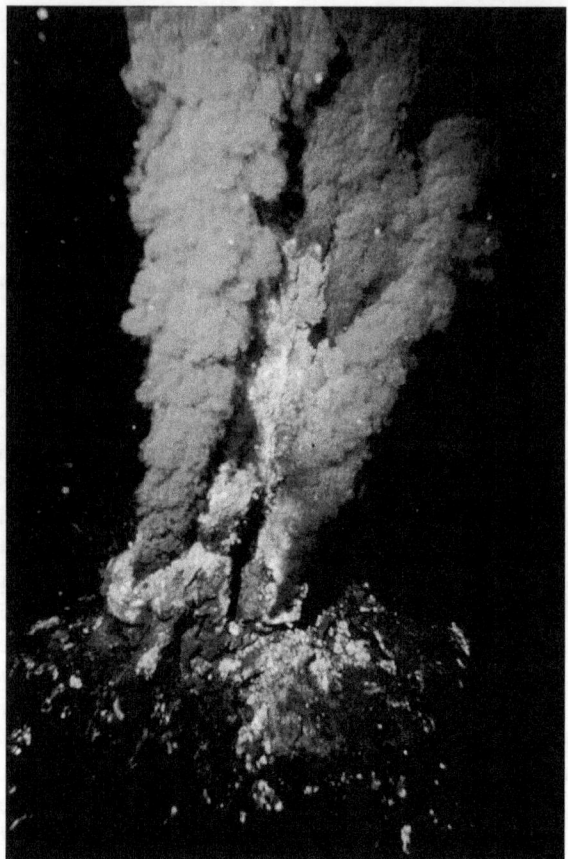

*Deep-sea hydrothermal vent or 'black smoker'*

1. Diffusion force caused by concentration gradient—all particles including ions tend to diffuse from higher concentration to lower.

2. Electrostatic force caused by electrical potential gradient—cations like protons H⁺ tend to diffuse down the electrical potential, anions in the opposite direction.

These two gradients taken together can be expressed as an electrochemical gradient, providing energy for abiogenic synthesis. The proton motive force can be described as the measure of the potential energy stored as a combination of proton and voltage gradients across a membrane (differences in proton concentration and electrical potential).

*White smokers emitting liquid carbon dioxide ($CO_2$) at the Champagne vent, Marianas Trench Marine National Monument*

from carbon monoxide and methylsulphide, constituents of hydrothermal fluid, indicates that pre-biotic syntheses occurred at the inner surfaces of these metal-sulphide-walled compartments,..."[198] These form where hydrogen-rich fluids emerge from below the sea floor, as a result of serpentinization of ultra-mafic olivine with seawater and a pH interface with carbon dioxide-rich ocean water. The vents form a sustained chemical energy source derived from redox reactions, in which electron donors, such as molecular hydrogen, react with electron acceptors, such as carbon dioxide (see Iron–sulfur world theory). These are highly exothermic reactions.[note 2]

Michael Russell demonstrated that alkaline vents created an abiogenic proton motive force (PMF) chemiosmotic gradient,[198] in which conditions are ideal for an abiogenic hatchery for life. Their microscopic compartments "provide a natural means of concentrating organic molecules," composed of iron-sulfur minerals such as mackinawite, endowed these mineral cells with the catalytic properties envisaged by Wächtershäuser.[184] This movement of ions across the membrane depends on a combination of two factors:

Jack W. Szostak suggested that geothermal activity provides greater opportunities for the origination of life in open lakes where there is a buildup of minerals. In 2010, based on spectral analysis of sea and hot mineral water, Ignat Ignatov and Oleg Mosin demonstrated that life may have predominantly originated in hot mineral water. The hot mineral water that contains bicarbonate and calcium ions has the most optimal range.[199] This case is similar to the origin of life in hydrothermal vents, but with bicarbonate and calcium ions in hot water. This water has a pH of 9–11 and is possible to have the reactions in seawater. According to Melvin Calvin, certain reactions of condensation-dehydration of amino acids and nucleotides in individual blocks of peptides and nucleic acids can take place in the primary hydrosphere with pH 9-11 at a later evolutionary stage.[200] Some of these compounds like hydrocyanic acid (HCN) have been proven in the experiments of Miller. This is the environment in which the stromatolites have been created. David Ward of Montana State University described the formation of stromatolites in hot mineral water at the Yellowstone National Park. Stromatolites survive in hot mineral water and in proximity to areas with volcanic activity.[201] Processes have evolved in the sea near geysers of hot mineral water. In 2011, Tadashi Sugawara from the University of Tokyo created a protocell in hot water.[202]

Experimental research and computer modeling suggest that the surfaces of mineral particles inside hydrothermal vents have catalytic properties similar to those of enzymes and are able to create simple organic molecules, such as methanol ($CH_3OH$) and formic, acetic and pyruvic acid out of the dissolved $CO_2$ in the water.[203][204]

### 1.6.4 Thermosynthesis

Today's bioenergetic process of fermentation is carried out by either the aforementioned citric acid cycle or the Acetyl-CoA pathway, both of which have been connected to the primordial Iron–sulfur world. In a different approach, the thermosynthesis hypothesis considers the bioenergetic process of chemiosmosis, which plays an essential role in cellular respiration and photosynthesis, more basal than fermentation: the ATP synthase enzyme, which sustains chemiosmosis, is proposed as the currently extant enzyme most closely related to the first metabolic process.[205][206]

First, life needed an energy source to bring about the condensation reaction that yielded the peptide bonds of proteins and the phosphodiester bonds of RNA. In a generalization and thermal variation of the binding change mechanism of today's ATP synthase, the "first protein" would have bound substrates (peptides, phosphate, nucleosides, RNA 'monomers') and condensed them to a reaction product that remained bound until after a temperature change it was released by thermal unfolding.

The energy source under the thermosynthesis hypothesis was thermal cycling, the result of suspension of protocells in a convection current, as is plausible in a volcanic hot spring; the convection accounts for the self-organization and dissipative structure required in any origin of life model. The still ubiquitous role of thermal cycling in germination and cell division is considered a relic of primordial thermosynthesis.

By phosphorylating cell membrane lipids, this "first protein" gave a selective advantage to the lipid protocell that contained the protein. This protein also synthesized a library of many proteins, of which only a minute fraction had thermosynthesis capabilities. As proposed by Dyson,[13] it propagated functionally: it made daughters with similar capabilities, but it did not copy itself. Functioning daughters consisted of different amino acid sequences.

Whereas the Iron–sulfur world identifies a circular pathway as the most simple, the thermosynthesis hypothesis does not even invoke a pathway: ATP synthase's binding change mechanism resembles a physical adsorption process that yields free energy,[207] rather than a regular enzyme's mechanism, which decreases the free energy. It has been claimed that the emergence of cyclic systems of protein catalysts is implausible.[208]

## 1.7 Other models of abiogenesis

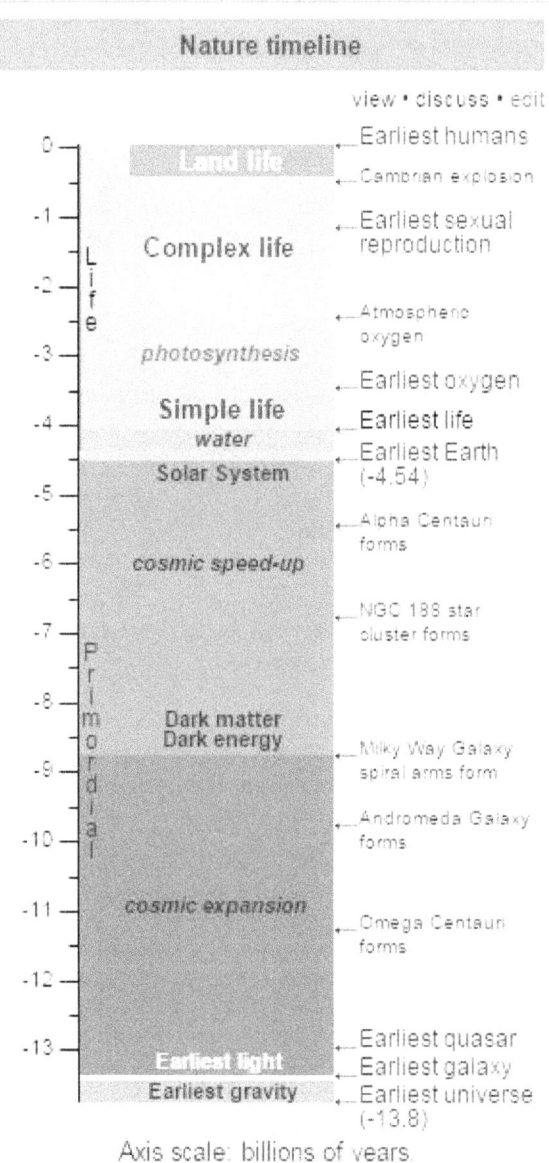

### 1.7.1 Clay hypothesis

Montmorillonite, an abundant clay, is a catalyst for the polymerization of RNA and for the formation of membranes from lipids.[209] A model for the origin of life using clay was forwarded by Alexander Graham Cairns-Smith in 1985 and explored as a plausible mechanism by several scientists.[210] The clay hypothesis postulates that complex organic molecules arose gradually on a pre-existing, non-organic replication surfaces of silicate crystals in solution.

At the Rensselaer Polytechnic Institute, James P. Ferris' studies have also confirmed that clay minerals of montmorillonite catalyze the formation of RNA in aqueous solution, by joining nucleotides to form longer chains.[211]

In 2007, Bart Kahr from the University of Washington and colleagues reported their experiments that tested the idea that crystals can act as a source of transferable information, using crystals of potassium hydrogen phthalate. "Mother" crystals with imperfections were cleaved and used as seeds to grow "daughter" crystals from solution. They then examined the distribution of imperfections in the new crystals and found that the imperfections in the mother crystals were reproduced in the daughters, but the daughter crystals also had many additional imperfections. For gene-like behavior to be observed, the quantity of inheritance of these imperfections should have exceeded that of the mutations in the successive generations, but it did not. Thus Kahr concluded that the crystals "were not faithful enough to store and transfer information from one generation to the next."[212]

### 1.7.2 Gold's "deep-hot biosphere" model

In the 1970s, Thomas Gold proposed the theory that life first developed not on the surface of the Earth, but several kilometers below the surface. It is claimed that discovery of microbial life below the surface of another body in our Solar System would lend significant credence to this theory. Thomas Gold also asserted that a trickle of food from a deep, unreachable, source is needed for survival because life arising in a puddle of organic material is likely to consume all of its food and become extinct. Gold's theory is that the flow of such food is due to out-gassing of primordial methane from the Earth's mantle; more conventional explanations of the food supply of deep microbes (away from sedimentary carbon compounds) is that the organisms subsist on hydrogen released by an interaction between water and (reduced) iron compounds in rocks.

### 1.7.3 Panspermia

Main article: Panspermia

Panspermia is the hypothesis that life exists throughout the Universe, distributed by meteoroids, asteroids, comets,[213][214] planetoids,[215] and, also, by spacecraft in the form of unintended contamination by microorganisms.[216][217]

Panspermia hypothesis does not attempt to explain how life first originated, but merely shifts it to another planet or a comet. The advantage of an extraterrestrial origin of primitive life is that life is not required to have formed on each planet it occurs on, but rather in a single location, and then spread about the galaxy to other star systems via cometary and/or meteorite impact. Evidence to support the hypothesis is scant, but it finds support in studies of Martian meteorites found in Antarctica and in studies of extremophile microbes' survival in outer space tests.[218][219][220][221] (See also: List of microorganisms tested in outer space.)

### 1.7.4 Extraterrestrial organic molecules

See also: List of interstellar and circumstellar molecules and Panspermia § Pseudo-panspermia

An organic compound is any member of a large class

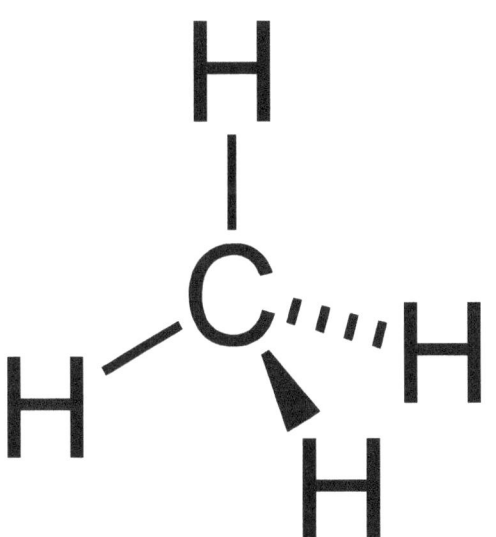

*Methane is one of the simplest organic compounds*

of gaseous, liquid, or solid chemicals whose molecules contain carbon. Carbon is the fourth most abundant element in the Universe by mass after hydrogen, helium, and oxygen.[222] Carbon is abundant in the Sun, stars, comets, and in the atmospheres of most planets.[223] Organic compounds are relatively common in space, formed by "factories of complex molecular synthesis" which occur in molecular clouds and circumstellar envelopes, and chemically evolve after reactions are initiated mostly by ionizing radiation.[19][224][225][226] Based on computer model studies, the complex organic molecules necessary for life may have formed on dust grains in the protoplanetary disk surrounding the Sun before the formation of the Earth.[40] According to the computer studies, this same process may also occur around other stars that acquire planets.[40]

Observations suggest that the majority of organic compounds introduced on Earth by interstellar dust particles are considered principal agents in the formation of complex molecules, thanks to their peculiar surface-catalytic activities.[227][228] Studies reported in 2008, based on $^{12}C/^{13}C$ isotopic ratios of organic compounds found in the Murchison meteorite, suggested that the RNA component uracil and related molecules, including xanthine, were formed extraterrestrially.[229][230] On 8 August 2011, a report based on NASA studies of meteorites found on Earth was published suggesting DNA components (adenine, guanine and related organic molecules) were made in outer space.[227][231][232] Scientists also found that the cosmic dust permeating the Universe contains complex organics ("amorphous organic solids with a mixed aromatic–aliphatic structure") that could be created naturally, and rapidly, by stars.[233][234][235] Sun Kwok of The University of Hong Kong suggested that these compounds may have been related to the development of life on Earth said that "If this is the case, life on Earth may have had an easier time getting started as these organics can serve as basic ingredients for life."[233]

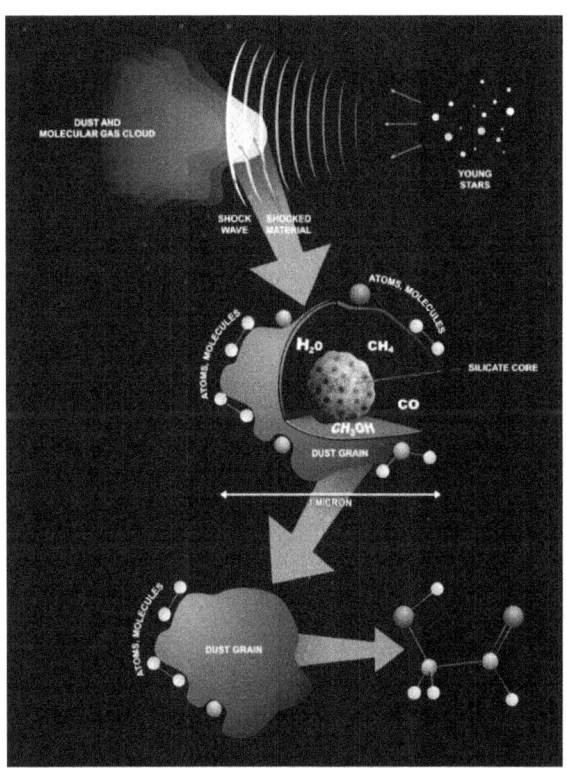

*Formation of glycolaldehyde in stardust*

## 1.7. OTHER MODELS OF ABIOGENESIS

Glycolaldehyde, the first example of an interstellar sugar molecule, was detected in the star-forming region near the center of our galaxy. It was discovered in 2000 by Jes Jørgensen and Jan M. Hollis.[236] In 2012, Jørgensen's team reported the detection of glycolaldehyde in a distant star system. The molecule was found around the protostellar binary IRAS 16293-2422 400 light years from Earth.[237][238][239] Glycolaldehyde is needed to form RNA, which is similar in function to DNA. These findings suggest that complex organic molecules may form in stellar systems prior to the formation of planets, eventually arriving on young planets early in their formation.[240] Because sugars are associated with both metabolism and the genetic code, two of the most basic aspects of life, it is thought the discovery of extraterrestrial sugar increases the likelihood that life may exist elsewhere in our galaxy.[236]

NASA announced in 2009 that scientists had identified another fundamental chemical building block of life in a comet for the first time, glycine, an amino acid, which was detected in material ejected from comet Wild 2 in 2004 and grabbed by NASA's *Stardust* probe. Glycine has been detected in meteorites before. Carl Pilcher, who leads the NASA Astrobiology Institute commented that "The discovery of glycine in a comet supports the idea that the fundamental building blocks of life are prevalent in space, and strengthens the argument that life in the Universe may be common rather than rare."[241] Comets are encrusted with outer layers of dark material, thought to be a tar-like substance composed of complex organic material formed from simple carbon compounds after reactions initiated mostly by ionizing radiation. It is possible that a rain of material from comets could have brought significant quantities of such complex organic molecules to Earth.[242][243][244] Amino acids which were formed extraterrestrially may also have arrived on Earth via comets.[43] It is estimated that during the Late Heavy Bombardment, meteorites may have delivered up to five million tons of organic prebiotic elements to Earth per year.[43]

Polycyclic aromatic hydrocarbons (PAH) are the most common and abundant of the known polyatomic molecules in the observable universe, and are considered a likely constituent of the primordial sea.[245][246][247] In 2010, PAHs, along with fullerenes (or "buckyballs"), have been detected in nebulae.[248][249]

In March 2015, NASA scientists reported that, for the first time, complex DNA and RNA organic compounds of life, including uracil, cytosine and thymine, have been formed in the laboratory under outer space conditions, using starting chemicals, such as pyrimidine, found in meteorites. Pyrimidine, like PAHs, the most carbon-rich chemical found in the Universe, may have been formed in red giant stars or in interstellar dust and gas clouds.[250]

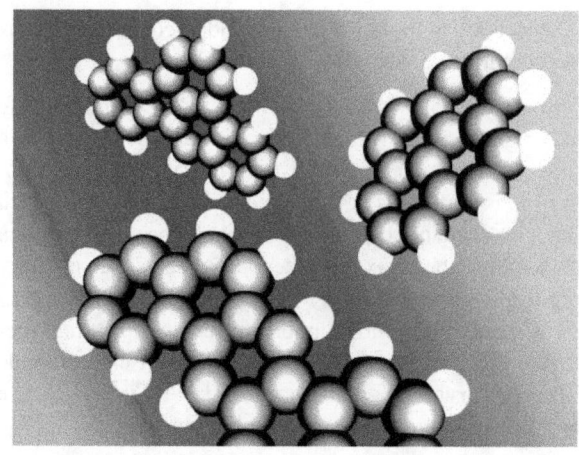

*An illustration of typical polycyclic aromatic hydrocarbons. Clockwise from top left: benz(e)acephenanthrylene, pyrene and dibenz(ah)anthracene.*

### 1.7.5 Lipid world

Main article: Gard model

The lipid world theory postulates that the first self-replicating object was lipid-like.[251][252] It is known that phospholipids form lipid bilayers in water while under agitation—the same structure as in cell membranes. These molecules were not present on early Earth, but other amphiphilic long-chain molecules also form membranes. Furthermore, these bodies may expand (by insertion of additional lipids), and under excessive expansion may undergo spontaneous splitting which preserves the same size and composition of lipids in the two progenies. The main idea in this theory is that the molecular composition of the lipid bodies is the preliminary way for information storage, and evolution led to the appearance of polymer entities such as RNA or DNA that may store information favorably. Studies on vesicles from potentially prebiotic amphiphiles have so far been limited to systems containing one or two types of amphiphiles. This in contrast to the output of simulated prebiotic chemical reactions, which typically produce very heterogeneous mixtures of compounds.[140] Within the hypothesis of a lipid bilayer membrane composed of a mixture of various distinct amphiphilic compounds there is the opportunity of a huge number of theoretically possible combinations in the arrangements of these amphiphiles in the membrane. Among all these potential combinations, a specific local arrangement of the membrane would have favored the constitution of an hypercycle,[253][254] actually a positive feedback composed of two mutual catalysts represented by a membrane site and a specific compound trapped in the vesicle. Such site/compound pairs are transmissible to the daughter vesicles leading to the emergence of distinct lineages of vesicles which would have allowed Darwinian natural selection.[255]

### 1.7.6 Polyphosphates

A problem in most scenarios of abiogenesis is that the thermodynamic equilibrium of amino acid versus peptides is in the direction of separate amino acids. What has been missing is some force that drives polymerization. The resolution of this problem may well be in the properties of polyphosphates.[256][257] Polyphosphates are formed by polymerization of ordinary monophosphate ions $PO_4^{-3}$. Several mechanisms for such polymerization have been suggested. Polyphosphates cause polymerization of amino acids into peptides. They are also logical precursors in the synthesis of such key biochemical compounds as adenosine triphosphate (ATP). A key issue seems to be that calcium reacts with soluble phosphate to form insoluble calcium phosphate (apatite), so some plausible mechanism must be found to keep calcium ions from causing precipitation of phosphate. There has been much work on this topic over the years, but an interesting new idea is that meteorites may have introduced reactive phosphorus species on the early Earth.[258]

### 1.7.7 PAH world hypothesis

Main article: PAH world hypothesis

Polycyclic aromatic hydrocarbons (PAH) are known to be abundant in the Universe,[245][246][247] including in the interstellar medium, in comets, and in meteorites, and are some of the most complex molecules so far found in space.[223]

Other sources of complex molecules have been postulated, including extraterrestrial stellar or interstellar origin. For example, from spectral analyses, organic molecules are known to be present in comets and meteorites. In 2004, a team detected traces of PAHs in a nebula.[259] In 2010, another team also detected PAHs, along with fullerenes, in nebulae.[248] The use of PAHs has also been proposed as a precursor to the RNA world in the PAH world hypothesis. The Spitzer Space Telescope has detected a star, HH 46-IR, which is forming by a process similar to that by which the Sun formed. In the disk of material surrounding the star, there is a very large range of molecules, including cyanide compounds, hydrocarbons, and carbon monoxide. In September 2012, NASA scientists reported that PAHs, subjected to interstellar medium conditions, are transformed, through hydrogenation, oxygenation and hydroxylation, to more complex organics—"a step along the path toward amino acids and nucleotides, the raw materials of proteins and DNA, respectively."[260][261] Further, as a result of these transformations, the PAHs lose their spectroscopic signature which could be one of the reasons "for the lack of PAH detection in interstellar ice grains, particularly the outer regions of cold, dense clouds or the upper molecular layers of protoplanetary disks."[260][261]

NASA maintains a database for tracking PAHs in the Universe.[223][262] More than 20% of the carbon in the Universe may be associated with PAHs,[223][223] possible starting materials for the formation of life. PAHs seem to have been formed shortly after the Big Bang, are widespread throughout the Universe,[245][246][247] and are associated with new stars and exoplanets.[223]

### 1.7.8 Radioactive beach hypothesis

Zachary Adam claims that tidal processes that occurred during a time when the Moon was much closer may have concentrated grains of uranium and other radioactive elements at the high-water mark on primordial beaches, where they may have been responsible for generating life's building blocks.[263] According to computer models reported in *Astrobiology*,[264] a deposit of such radioactive materials could show the same self-sustaining nuclear reaction as that found in the Oklo uranium ore seam in Gabon. Such radioactive beach sand might have provided sufficient energy to generate organic molecules, such as amino acids and sugars from acetonitrile in water. Radioactive monazite material also has released soluble phosphate into the regions between sand-grains, making it biologically "accessible." Thus amino acids, sugars, and soluble phosphates might have been produced simultaneously, according to Adam. Radioactive actinides, left behind in some concentration by the reaction, might have formed part of organometallic complexes. These complexes could have been important early catalysts to living processes.

John Parnell has suggested that such a process could provide part of the "crucible of life" in the early stages of any early wet rocky planet, so long as the planet is large enough to have generated a system of plate tectonics which brings radioactive minerals to the surface. As the early Earth is thought to have had many smaller plates, it might have provided a suitable environment for such processes.[265]

### 1.7.9 Thermodynamic dissipation

Karo Michaelian from the National Autonomous University of Mexico (UNAM) points out that any model for the origin of life must take into account the fact that life is an irreversible thermodynamic process and, like all irreversible processes, its origin and persistence as a "self-organized" system is due to its dissipation an imposed generalized chemical potential, i.e., the production of entropy. That is, entropy production is not incidental to the process of life, but rather the fundamental reason for its existence. Present day life augments the entropy production of Earth in its solar environment by dissipating ultraviolet and visible photons into heat through organic pigments in water. This heat then catalyzes a host of secondary dissipative processes such as the water cycle, ocean and wind currents, hurricanes, etc.[266][267] Michaelian argues that if the thermodynamic function of life today is to produce entropy through photon dissipation, then this probably was its function at its very beginnings.[268] It turns out that both RNA and DNA when in water solution are very strong absorbers and extremely rapid dissipaters of UV light within the 230–290 nm wavelength region, which is a part of the Sun's spectrum that could have penetrated the prebiotic atmosphere.[269] The amount of ultraviolet (UV-C) light reaching the Earth's surface within this spectral range in the Archean could have been on the order of 4 W/m$^2$,[270] or some 31 orders of magnitude greater than it is today at 260 nm where RNA and DNA absorb most strongly.[269] In fact, not only RNA and DNA, but many fundamental molecules of life (those common to all three domains of life, archea, bacteria, and eucaryote) are also pigments that absorb in the UV-C, and many of these also have a chemical affinity to RNA and DNA.[271][272] Nucleic acids may thus have acted as acceptor molecules to the UV-C photon excited antenna pigment donor molecules by providing an ultrafast channel for dissipation. Michaelian has shown that there would have existed a non-linear, non-equilibrium thermodynamic imperative to the abiogenic UV-C photochemical synthesis [178] and proliferation of these pigments over the entire Earth surface if they augmented the solar photon dissipation rate.[273]

A simple mechanism to explain enzyme-less replication of RNA and DNA can be given within the same dissipative thermodynamic framework by assuming that life arose when the temperature of the primitive seas had cooled to somewhat below the denaturing temperature of RNA or DNA. The ratio of $^{18}O/^{16}O$ found in cherts of the Barberton greenstone belt of South Africa indicates that the Earth's surface temperature was around 80 °C at 3.8 Ga,[274][275] falling to 70±15 °C about 3.5 to 3.2 Ga,[276] suggestively close to RNA or DNA denaturing (uncoiling and separation) temperatures. During the night, the surface water temperature would drop below the denaturing temperature and single strand RNA/DNA could act as extension template for the formation of double strand RNA/DNA. During the daylight hours, RNA and DNA would absorb UV-C light and convert this directly into heat at the ocean surface, thereby raising the local temperature enough to allow for denaturing of RNA and DNA. Direct experimental evidence for the denaturing of DNA through UV-C light dissipation has now been obtained.[277]

The copying process would have been repeated with each diurnal cycle.[268] Such an ultraviolet and temperature assisted RNA/DNA reproduction (UVTAR) bears similarity to polymerase chain reaction (PCR), a routine laboratory procedure employed to multiply DNA segments. Since denaturation would be most probable in the late afternoon when the Archean sea surface temperature would be highest, and since late afternoon submarine sunlight is somewhat circularly polarized, the homochirality of the organic molecules of life can also be explained within the proposed thermodynamic framework.[268]

The fact that the aromatic amino acids have been shown to have chemical affinity to their codons, or anti-codons, and that they also absorb strongly in the UV-C, suggests that they might have originally acted as antenna pigments to increase dissipation and to provide more local heat for UVTAR replication of RNA and DNA as the sea surface temperature cooled. The accumulation of information, e.g., coding for the aromatic amino acids, in RNA or DNA would thus be related to reproductive success under this mechanism. Michaelian suggests that the traditional origin of life research, that expects to describe the emergence of life without overwhelming reference to entropy production through dissipation, is erroneous and that imposed environmental potentials, such as the solar photon flux, and the dissipation of this flux, must be considered to understand the emergence, proliferation, and evolution of life.

### 1.7.10 Multiple genesis

Different forms of life with variable origin processes may have appeared quasi-simultaneously in the early history of Earth.[278] The other forms may be extinct (having left distinctive fossils through their different biochemistry—e.g., hypothetical types of biochemistry). It has been proposed that:

> The first organisms were self-replicating iron-rich clays which fixed carbon dioxide into oxalic and other dicarboxylic acids. This system of replicating clays and their metabolic phenotype then evolved into the sulfide rich region of the hotspring acquiring the ability to fix nitrogen. Finally phosphate was incorporated into the evolving system which allowed the synthesis of nucleotides and phospholipids. If biosynthesis recapitulates biopoiesis, then the synthesis of amino acids preceded the synthesis of the purine and pyrimidine bases. Furthermore the polymerization of the amino acid thioesters into polypeptides preceded the directed polymerization of amino acid esters by polynucleotides.[279]

## 1.7.11 Fluctuating hydrothermal pools on volcanic islands

Bruce Damer and David Deamer have come to the conclusion that cell membranes cannot be formed in salty seawater, and must therefore have originated in freshwater. Before the continents formed, the only dry land on Earth would be volcanic islands, where rainwater would form ponds where lipids could form the first stages towards cell membranes. These predecessors of true cells are assumed to have behaved more like a superorganism rather than individual structures, where the porous membranes would house molecules which would leak out and enter other protocells. Only when true cells had evolved would they gradually adapt to saltier environments and enter the ocean.[280]

## 1.8 See also

- Anthropic principle
- Artificial cell
- Astrochemistry
- Biological immortality
- Common descent
- Emergence
- Entropy and life
- GADV protein world
- Mediocrity principle
- Mycoplasma laboratorium
- Nexus for Exoplanet System Science
- Planetary habitability
- Rare Earth hypothesis
- Shadow biosphere
- Stromatolite

## 1.9 Notes

[1] The reactions are:

$$FeS + H_2S \rightarrow FeS_2 + 2H^+ + 2e^-$$
$$FeS + H_2S + CO_2 \rightarrow FeS_2 + HCOOH$$

[2] The reactions are:
**Reaction 1**: *Fayalite + water → magnetite + aqueous silica + hydrogen*

$$3Fe_2SiO_4 + 2H_2O \rightarrow 2Fe_3O_4 + 3SiO_2 + 2H_2$$

**Reaction 2**: *Forsterite + aqueous silica → serpentine*

$$3Mg_2SiO_4 + SiO_2 + 4H_2O \rightarrow 2Mg_3Si_2O_5(OH)_4$$

**Reaction 3**: *Forsterite + water → serpentine + brucite*

$$2Mg_2SiO_4 + 3H_2O \rightarrow Mg_3Si_2O_5(OH)_4 + Mg(OH)_2$$

Reaction 3 describes the hydration of olivine with water only to yield serpentine and $Mg(OH)_2$ (brucite). Serpentine is stable at high pH in the presence of brucite like calcium silicate hydrate, (C-S-H) phases formed along with portlandite ($Ca(OH)_2$) in hardened Portland cement paste after the hydration of belite ($Ca_2SiO_4$), the artificial calcium equivalent of forsterite. Analogy of reaction 3 with belite hydration in ordinary Portland cement: *Belite + water → C-S-H phase + portlandite*

$$2\,Ca_2SiO_4 + 4\,H_2O \rightarrow 3\,CaO \cdot 2\,SiO_2 \cdot 3\,H_2O + Ca(OH)_2$$

## 1.10 References

[1] Pronunciation: "/ˌeɪbʌɪə(ʊ)ˈdʒɛnɪsɪs/". Pearsall, Judy; Hanks, Patrick, eds. (1998). "abiogenesis". *The New Oxford Dictionary of English* (1st ed.). Oxford, UK: Oxford University Press. p. 3. ISBN 0-19-861263-X.

[2] OED On-line (2003)

[3] "Abiogenesis". *Dictionary.com Unabridged*. Random House.

[4] "Abiogenesis". *Merriam-Webster Dictionary*.

[5] Bernal 1960, p. 30

[6] Scharf, Caleb; et al. (18 December 2015). "A Strategy for Origins of Life Research". *Astrobiology (journal)* **15** (12): 1031–1042. doi:10.1089/ast.2015.1113. Retrieved 20 December 2015.

[7] Oparin 1953, p. vi

[8] Warmflash, David; Warmflash, Benjamin (November 2005). "Did Life Come from Another World?". *Scientific American* (Stuttgart: Georg von Holtzbrinck Publishing Group) **293** (5): 64–71. doi:10.1038/scientificamerican1105-64. ISSN 0036-8733.

[9] Yarus 2010, p. 47

[10] Peretó, Juli (2005). "Controversies on the origin of life" (PDF). *International Microbiology* (Barcelona: Spanish Society for Microbiology) **8** (1): 23–31. ISSN 1139-6709. PMID 15906258. Retrieved 2015-06-01.

## 1.10. REFERENCES

[11] Elizabeth A. Bell. "Potentially biogenic carbon preserved in a 4.1 billion-year-old zircon".

[12] Voet & Voet 2004, p. 29

[13] Dyson 1999

[14] Davies, Paul (1998) "The Fifth Miracle, Search for the origin and meaning of life" 9Penguin)

[15]
- Copley, Shelley D.; Smith, Eric; Morowitz, Harold J. (December 2007). "The origin of the RNA world: Co-evolution of genes and metabolism" (PDF). *Bioorganic Chemistry* (Amsterdam, the Netherlands: Elsevier) **35** (6): 430–443. doi:10.1016/j.bioorg.2007.08.001. ISSN 0045-2068. PMID 17897696. Retrieved 2015-06-08. The proposal that life on Earth arose from an RNA world is widely accepted.
- Orgel, Leslie E. (April 2003). "Some consequences of the RNA world hypothesis". *Origins of Life and Evolution of the Biosphere* (Kluwer Academic Publishers) **33** (2): 211–218. doi:10.1023/A:1024616317965. ISSN 0169-6149. PMID 12967268. It now seems very likely that our familiar DNA/RNA/protein world was preceded by an RNA world...
- Robertson & Joyce 2012: "There is now strong evidence indicating that an RNA World did indeed exist before DNA- and protein-based life."
- Neveu, Kim & Benner 2013: "[The RNA world's existence] has broad support within the community today."

[16] Robertson, Michael P.; Joyce, Gerald F. (May 2012). "The origins of the RNA world". *Cold Spring Harbor Perspectives in Biology* (Cold Spring Harbor, NY: Cold Spring Harbor Laboratory Press) **4** (5): a003608. doi:10.1101/cshperspect.a003608. ISSN 1943-0264. PMC 3331698. PMID 20739415.

[17] Cech, Thomas R. (July 2012). "The RNA Worlds in Context". *Cold Spring Harbor Perspectives in Biology* (Cold Spring Harbor, NY: Cold Spring Harbor Laboratory Press) **4** (7): a006742. doi:10.1101/cshperspect.a006742. ISSN 1943-0264. PMC 3385955. PMID 21441585.

[18] Keller, Markus A.; Turchyn, Alexandra V.; Ralser, Markus (25 March 2014). "Non-enzymatic glycolysis and pentose phosphate pathway-like reactions in a plausible Archean ocean". *Molecular Systems Biology* (Heidelberg, Germany: EMBO Press on behalf of the European Molecular Biology Organization) **10** (725). doi:10.1002/msb.20145228. ISSN 1744-4292. PMC 4023395. PMID 24771084.

[19] Ehrenfreund, Pascale; Cami, Jan (December 2010). "Cosmic carbon chemistry: from the interstellar medium to the early Earth.". *Cold Spring Harbor Perspectives in Biology* (Cold Spring Harbor, NY: Cold Spring Harbor Laboratory Press) **2** (12): a002097. doi:10.1101/cshperspect.a002097. ISSN 1943-0264. PMC 2982172. PMID 20554702.

[20] Perkins, Sid (8 April 2015). "Organic molecules found circling nearby star". *Science* (News) (Washington, D.C.: American Association for the Advancement of Science). ISSN 1095-9203. Retrieved 2015-06-02.

[21] King, Anthony (14 April 2015). "Chemicals formed on meteorites may have started life on Earth". *Chemistry World* (News) (London: Royal Society of Chemistry). ISSN 1473-7604. Retrieved 2015-04-17.

[22] Saladino, Raffaele; Carota, Eleonora; Botta, Giorgia; et al. (13 April 2015). "Meteorite-catalyzed syntheses of nucleosides and of other prebiotic compounds from formamide under proton irradiation". *Proc. Natl. Acad. Sci. U.S.A.* (Washington, D.C.: National Academy of Sciences) **112** (21): E2746–E2755. doi:10.1073/pnas.1422225112. ISSN 1091-6490. PMID 25870268.

[23] Rampelotto, Pabulo Henrique (26 April 2010). *Panspermia: A Promising Field Of Research* (PDF). Astrobiology Science Conference 2010. Houston, TX: Lunar and Planetary Institute. p. 5224. Bibcode:2010LPICo1538.5224R. Retrieved 2014-12-03. Conference held at League City, TX

[24] Loeb, Abraham (October 2014). "The habitable epoch of the early Universe". *International Journal of Astrobiology* (Cambridge, UK: Cambridge University Press) **13** (4): 337–339. arXiv:1312.0613. Bibcode:2014IJAsB..13..337L. doi:10.1017/S1473550414000196. ISSN 1473-5504.

- Loeb, Abraham (3 June 2014). "The Habitable Epoch of the Early Universe". arXiv:1312.0613v3 [astro-ph.CO].

[25] Dreifus, Claudia (2 December 2014). "Much-Discussed Views That Go Way Back". *The New York Times* (New York: The New York Times Company). p. D2. ISSN 0362-4331. Retrieved 2014-12-03.

[26] Graham, Robert W. (February 1990). "Extraterrestrial Life in the Universe" (PDF) (NASA Technical Memorandum 102363). Lewis Research Center, Cleveland, Ohio: NASA. Retrieved 2015-06-02.

[27] Altermann 2009, p. xvii

[28] "Age of the Earth". United States Geological Survey. 9 July 2007. Retrieved 2006-01-10.

[29] Dalrymple 2001, pp. 205–221

[30] Manhesa, Gérard; Allègre, Claude J.; Dupréa, Bernard; Hamelin, Bruno (May 1980). "Lead isotope study of basic-ultrabasic layered complexes: Speculations about the age of the earth and primitive mantle characteristics". *Earth and Planetary Science Letters* (Amsterdam, the Netherlands: Elsevier) **47** (3): 370–382. Bibcode:1980E&PSL..47..370M. doi:10.1016/0012-821X(80)90024-2. ISSN 0012-821X.

[31] Schopf, J. William; Kudryavtsev, Anatoliy B.; Czaja, Andrew D.; Tripathi, Abhishek B. (5 October 2007).

"Evidence of Archean life: Stromatolites and microfossils". *Precambrian Research* (Amsterdam, the Netherlands: Elsevier) **158** (3–4): 141–155. doi:10.1016/j.precamres.2007.04.009. ISSN 0301-9268.

[32] Schopf, J. William (29 June 2006). "Fossil evidence of Archaean life". *Philosophical Transactions of the Royal Society B* (London: Royal Society) **361** (1470): 869–885. doi:10.1098/rstb.2006.1834. ISSN 0962-8436. PMC 1578735. PMID 16754604.

[33] Raven & Johnson 2002, p. 68

[34] Borenstein, Seth (13 November 2013). "Oldest fossil found: Meet your microbial mom". *Excite* (Yonkers, NY: Mindspark Interactive Network). Associated Press. Retrieved 2015-06-02.

[35] Pearlman, Jonathan (13 November 2013). "'Oldest signs of life on Earth found'". *The Daily Telegraph* (London: Telegraph Media Group). Retrieved 2014-12-15.

[36] Noffke, Nora; Christian, Daniel; Wacey, David; Hazen, Robert M. (16 November 2013). "Microbially Induced Sedimentary Structures Recording an Ancient Ecosystem in the ca. 3.48 Billion-Year-Old Dresser Formation, Pilbara, Western Australia". *Astrobiology* (New Rochelle, NY: Mary Ann Liebert, Inc.) **13** (12): 1103–1124. Bibcode:2013AsBio..13.1103N. doi:10.1089/ast.2013.1030. ISSN 1531-1074. PMC 3870916. PMID 24205812.

[37] Ohtomo, Yoko; Kakegawa, Takeshi; Ishida, Akizumi; et al. (January 2014). "Evidence for biogenic graphite in early Archaean Isua metasedimentary rocks". *Nature Geoscience* (London: Nature Publishing Group) **7** (1): 25–28. Bibcode:2014NatGe...7...25O. doi:10.1038/ngeo2025. ISSN 1752-0894.

[38] Borenstein, Seth (19 October 2015). "Hints of life on what was thought to be desolate early Earth". *Excite* (Yonkers, NY: Mindspark Interactive Network). Associated Press. Retrieved 2015-10-20.

[39] Bell, Elizabeth A.; Boehnike, Patrick; Harrison, T. Mark; et al. (19 October 2015). "Potentially biogenic carbon preserved in a 4.1 billion-year-old zircon" (PDF). *Proc. Natl. Acad. Sci. U.S.A.* (Washington, D.C.: National Academy of Sciences) **112**: 201517557. doi:10.1073/pnas.1517557112. ISSN 1091-6490. PMC 4664351. PMID 26483481. Retrieved 2015-10-20. Early edition, published online before print.

[40] Moskowitz, Clara (29 March 2012). "Life's Building Blocks May Have Formed in Dust Around Young Sun". *Space.com* (Salt Lake City, UT: Purch). Retrieved 2012-03-30.

[41] Fesenkov 1959, p. 9

[42] Kasting, James F. (12 February 1993). "Earth's Early Atmosphere" (PDF). *Science* (Washington, D.C.: American Association for the Advancement of Science) **259** (5097): 922. doi:10.1126/science.11536547. ISSN 0036-8075. PMID 11536547. Retrieved 2015-07-28.

[43] Follmann, Hartmut; Brownson, Carol (November 2009). "Darwin's warm little pond revisited: from molecules to the origin of life". *Naturwissenschaften* (Berlin: Springer-Verlag) **96** (11): 1265–1292. Bibcode:2009NW.....96.1265F. doi:10.1007/s00114-009-0602-1. ISSN 0028-1042. PMID 19760276.

[44] Morse, John W.; MacKenzie, Fred T. (1998). "Hadean Ocean Carbonate Geochemistry". *Aquatic Geochemistry* (Kluwer Academic Publishers) **4** (3–4): 301–319. doi:10.1023/A:1009632230875. ISSN 1380-6165.

[45] Wilde, Simon A.; Valley, John W.; Peck, William H.; Graham, Colin M. (11 January 2001). "Evidence from detrital zircons for the existence of continental crust and oceans on the Earth 4.4 Gyr ago" (PDF). *Nature* (London: Nature Publishing Group) **409** (6817): 175–178. doi:10.1038/35051550. ISSN 0028-0836. PMID 11196637. Retrieved 2015-06-03.

[46] Rosing, Minik T.; Bird, Dennis K.; Sleep, Norman H.; et al. (22 March 2006). "The rise of continents—An essay on the geologic consequences of photosynthesis" (PDF). *Palaeogeography, Palaeoclimatology, Palaeoecology* (Amsterdam, the Netherlands: Elsevier) **232** (2–4): 99–113. doi:10.1016/j.palaeo.2006.01.007. ISSN 0031-0182. Retrieved 2015-06-08.

[47] Sleep, Norman H.; Zahnle, Kevin J.; Kasting, James F.; et al. (9 November 1989). "Annihilation of ecosystems by large asteroid impacts on early Earth". *Nature* (London: Nature Publishing Group) **342** (6246): 139–142. Bibcode:1989Natur.342..139S. doi:10.1038/342139a0. ISSN 0028-0836. PMID 11536616.

[48] Davies 1999

[49] O'Donoghue, James (21 August 2011). "Oldest reliable fossils show early life was a beach". *New Scientist* (London: Reed Business Information). ISSN 0262-4079. Retrieved 2014-10-13.

- Wacey, David; Kilburn, Matt R.; Saunders, Martin; et al. (October 2011). "Microfossils of sulphur-metabolizing cells in 3.4-billion-year-old rocks of Western Australia". *Nature Geoscience* (London: Nature Publishing Group) **4** (10): 698–702. Bibcode:2011NatGe...4..698W. doi:10.1038/ngeo1238. ISSN 1752-0894.

[50] Wolpert, Stuart (19 October 2015). "Life on Earth likely started at least 4.1 billion years ago — much earlier than scientists had thought". ULCA. Retrieved 20 October 2015.

[51] Gomes, Rodney; Levison, Hal F.; Tsiganis, Kleomenis; Morbidelli, Alessandro (26 May 2005). "Origin of the cataclysmic Late Heavy Bombardment period of the terrestrial planets". *Nature* (London: Nature Publishing Group) **435** (7041): 466–469. Bibcode:2005Natur.435..466G.

## 1.10. REFERENCES

doi:10.1038/nature03676. ISSN 0028-0836. PMID 15917802.

[52] Davies 2007, pp. 61–73

[53] Maher, Kevin A.; Stevenson, David J. (18 February 1988). "Impact frustration of the origin of life". *Nature* (London: Nature Publishing Group) **331** (6157): 612–614. Bibcode:1988Natur.331..612M. doi:10.1038/331612a0. ISSN 0028-0836. PMID 11536595.

[54] Wade, Nicholas (25 July 2016). "Meet Luca, the Ancestor of All Living Things". *New York Times*. Retrieved 25 July 2016.

[55] Sheldon 2005

[56] Vartanian 1973, pp. 307–312

[57] Lennox 2001, pp. 229–258

[58] Balme, D. M. (1962). "Development of Biology in Aristotle and Theophrastus: Theory of Spontaneous Generation". *Phronesis* (Leiden, the Netherlands: Brill Publishers) **7** (1–2): 91–104. doi:10.1163/156852862X00052. ISSN 0031-8868.

[59] Ross 1652

[60] Dobell 1960

[61] Bondeson 1999

[62] Bernal 1967

[63] "Biogenesis". *Hmolpedia*. Ancaster, Ontario, Canada: WikiFoundry, Inc. Retrieved 2014-05-19.

[64] Huxley 1968

[65] Bastian 1871

[66] Bastian 1871, p. xi–xii

[67] Oparin 1953, p. 196

[68] Tyndall 1905, IV, XII (1876), XIII (1878)

[69] Bernal 1967, p. 139

[70] Priscu, John C. "Origin and Evolution of Life on a Frozen Earth". Arlington County, VA: National Science Foundation. Retrieved 2014-03-01.

[71] Darwin 1887, p. 18: "It is often said that all the conditions for the first production of a living organism are now present, which could ever have been present. But if (and oh! what a big if!) we could conceive in some warm little pond, with all sorts of ammonia and phosphoric salts, light, heat, electricity, &c., present, that a proteine compound was chemically formed ready to undergo still more complex changes, at the present day such matter would be instantly devoured or absorbed, which would not have been the case before living creatures were formed." — Charles Darwin, 1 February 1871

[72] Bernal 1967, *The Origin of Life* (A. I. Oparin, 1924), pp. 199–234

[73] Oparin 1953

[74] Shapiro 1987, p. 110

[75] Bryson 2004, pp. 300–302

[76] Miller, Stanley L. (15 May 1953). "A Production of Amino Acids Under Possible Primitive Earth Conditions". *Science* (Washington, D.C.: American Association for the Advancement of Science) **117** (3046): 528–529. Bibcode:1953Sci...117..528M. doi:10.1126/science.117.3046.528. ISSN 0036-8075. PMID 13056598.

[77] Parker, Eric T.; Cleaves, Henderson J.; Dworkin, Jason P.; et al. (5 April 2011). "Primordial synthesis of amines and amino acids in a 1958 Miller $H_2S$-rich spark discharge experiment" (PDF). *Proc. Natl. Acad. Sci. U.S.A.* (Washington, D.C.: National Academy of Sciences) **108** (14): 5526–5531. Bibcode:2011PNAS..108.5526P. doi:10.1073/pnas.1019191108. ISSN 0027-8424. PMC 3078417. PMID 21422282. Retrieved 2015-06-08.

[78] Bernal 1967, p. 143

[79] Walsh, J. Bruce (1995). "Part 4: Experimental studies of the origins of life". *Origins of life* (Lecture notes). Tucson, AZ: University Of Arizona. Archived from the original on 2008-01-13. Retrieved 2015-06-08.

[80] Woodward 1969, p. 287

[81] Bahadur, Krishna (1973). "Photochemical Formation of Self–sustaining Coacervates" (PDF). *Proceedings of the Indian National Science Academy* (New Delhi: Indian National Science Academy) **39B** (4): 455–467. ISSN 0370-0046.

- Bahadur, Krishna (1975). "Photochemical Formation of Self-Sustaining Coacervates". *Zentralblatt für Bakteriologie, Parasitenkunde, Infektionskrankheiten und Hygiene* (Jena, Germany: Gustav Fischer Verlag) **130** (3): 211–218. doi:10.1016/S0044-4057(75)80076-1. OCLC 641018092. PMID 1242552.

[82] Kasting 1993, p. 922

[83] Kasting 1993, p. 920

[84] Bernal 1951

[85] Bernal, John Desmond (September 1949). "The Physical Basis of Life". *Proceedings of the Physical Society. Section A* (Bristol, UK: Physical Society) **62** (9): 537–558. Bibcode:1949PPSA...62..537B. doi:10.1088/0370-1298/62/9/301. ISSN 0370-1298.

[86] Kauffman 1995

[87] Oehlenschläger, Frank; Eigen, Manfred (December 1997). "30 Years Later – a New Approach to Sol Spiegelman's and Leslie Orgel's in vitro EVOLUTIONARY STUDIES Dedicated to Leslie Orgel on the occasion of his 70th birthday". *Origins of Life and Evolution of Biospheres* (Kluwer Academic Publishers) **27** (5-6): 437–457. doi:10.1023/A:1006501326129. ISSN 0169-6149. PMID 9394469.

[88] McCollom, Thomas; Mayhew, Lisa; Scott, Jim (7 October 2014). "NASA awards CU-Boulder-led team $7 million to study origins, evolution of life in universe" (Press release). Boulder, CO: University of Colorado Boulder. Retrieved 2015-06-08.

[89] Gibson, Daniel G.; Glass, John I.; Lartigue, Carole; et al. (2 July 2010). "Creation of a Bacterial Cell Controlled by a Chemically Synthesized Genome". *Science* (Washington, D.C.: American Association for the Advancement of Science) **329** (5987): 52–56. Bibcode:2010Sci...329...52G. doi:10.1126/science.1190719. ISSN 0036-8075. PMID 20488990.

[90] Swaby, Rachel (20 May 2010). "Scientists Create First Self-Replicating Synthetic Life". *Wired* (New York: Condé Nast). Retrieved 2015-06-08.

[91] Coughlan, Andy (2016) "Smallest ever genome comes to life: Humans built it but we don't know what a third of its genes actually do" (New Scientist 2nd April 2016 No 3067)p.6

[92] Gawlowicz, Susan (6 November 2011). "Carbon-based organic 'carriers' in interstellar dust clouds? Newly discovered diffuse interstellar bands". *Science Daily* (Rockville, MD: ScienceDaily, LLC). Retrieved 2015-06-08. Post is reprinted from materials provided by the Rochester Institute of Technology.

- Geballe, Thomas R.; Najarro, Francisco; Figer, Donald F.; et al. (10 November 2011). "Infrared diffuse interstellar bands in the Galactic Centre region". *Nature* (London: Nature Publishing Group) **479** (7372): 200–202. arXiv:1111.0613. Bibcode:2011Natur.479..200G. doi:10.1038/nature10527. ISSN 0028-0836. PMID 22048316.

[93] Klyce 2001

[94] Chyba, Christopher; Sagan, Carl (9 January 1992). "Endogenous production, exogenous delivery and impact-shock synthesis of organic molecules: an inventory for the origins of life". *Nature* (London: Nature Publishing Group) **355** (6356): 125–132. Bibcode:1992Natur.355..125C. doi:10.1038/355125a0. ISSN 0028-0836. PMID 11538392.

[95] Furukawa, Yoshihiro; Sekine, Toshimori; Oba, Masahiro; et al. (January 2009). "Biomolecule formation by oceanic impacts on early Earth". *Nature Geoscience* (London: Nature Publishing Group) **2** (1): 62–66. Bibcode:2009NatGe...2...62F. doi:10.1038/NGEO383. ISSN 1752-0894.

[96] Davies 1999, p. 155

[97] Bock & Goode 1996

[98] Palasek, Stan (23 May 2013). "Primordial RNA Replication and Applications in PCR Technology". arXiv:1305.5581v1 [q-bio.BM].

[99] Koonin, Eugene V.; Senkevich, Tatiana G.; Dolja, Valerian V. (19 September 2006). "The ancient Virus World and evolution of cells". *Biology Direct* (London: BioMed Central) **1**: 29. doi:10.1186/1745-6150-1-29. ISSN 1745-6150. PMC 1594570. PMID 16984643.

[100] Vlassov, Alexander V.; Kazakov, Sergei A.; Johnston, Brian H.; et al. (August 2005). "The RNA World on Ice: A New Scenario for the Emergence of RNA Information". *Journal of Molecular Evolution* (Berlin: Springer-Verlag) **61** (2): 264–273. doi:10.1007/s00239-004-0362-7. ISSN 0022-2844. PMID 16044244.

[101] Nussinov, Mark D.; Otroshchenko, Vladimir A.; Santoli, Salvatore (1997). "The emergence of the non-cellular phase of life on the fine-grained clayish particles of the early Earth's regolith". *BioSystems* (Amsterdam, the Netherlands: Elsevier) **42** (2–3): 111–118. doi:10.1016/S0303-2647(96)01699-1. ISSN 0303-2647. PMID 9184757.

[102] Saladino, Raffaele; Crestini, Claudia; Pino, Samanta; et al. (March 2012). "Formamide and the origin of life.". *Physics of Life Reviews* (Amsterdam, the Netherlands: Elsevier) **9** (1): 84–104. Bibcode:2012PhLRv...9...84S. doi:10.1016/j.plrev.2011.12.002. ISSN 1571-0645. PMID 22196896.

[103] Saladino, Raffaele; Botta, Giorgia; Pino, Samanta; et al. (July 2012). "From the one-carbon amide formamide to RNA all the steps are prebiotically possible". *Biochimie* (Amsterdam, the Netherlands: Elsevier) **94** (7): 1451–1456. doi:10.1016/j.biochi.2012.02.018. ISSN 0300-9084. PMID 22738728.

[104] Oró, Joan (16 September 1961). "Mechanism of Synthesis of Adenine from Hydrogen Cyanide under Possible Primitive Earth Conditions". *Nature* (London: Nature Publishing Group) **191** (4794): 1193–1194. Bibcode:1961Natur.191.1193O. doi:10.1038/1911193a0. ISSN 0028-0836. PMID 13731264.

[105] Basile, Brenda; Lazcano, Antonio; Oró, Joan (1984). "Prebiotic syntheses of purines and pyrimidines". *Advances in Space Research* (Amsterdam, the Netherlands: Elsevier) **4** (12): 125–131. Bibcode:1984AdSpR...4..125B. doi:10.1016/0273-1177(84)90554-4. ISSN 0273-1177. PMID 11537766.

[106] Orgel, Leslie E. (August 2004). "Prebiotic Adenine Revisited: Eutectics and Photochemistry". *Origins of Life*

## 1.10. REFERENCES

and *Evolution of Biospheres* (Kluwer Academic Publishers) **34** (4): 361–369. Bibcode:2004OLEB...34..361O. doi:10.1023/B:ORIG.0000029882.52156.c2. ISSN 0169-6149. PMID 15279171.

[107] Robertson, Michael P.; Miller, Stanley L. (29 June 1995). "An efficient prebiotic synthesis of cytosine and uracil". *Nature* (London: Nature Publishing Group) **375** (6534): 772–774. Bibcode:1995Natur.375..772R. doi:10.1038/375772a0. ISSN 0028-0836. PMID 7596408.

[108] Fox, Douglas (February 2008). "Did Life Evolve in Ice?". *Discover* (Waukesha, WI: Kalmbach Publishing). ISSN 0274-7529. Retrieved 2008-07-03.

[109] Levy, Matthew; Miller, Stanley L.; Brinton, Karen; Bada, Jeffrey L. (June 2000). "Prebiotic Synthesis of Adenine and Amino Acids Under Europa-like Conditions". *Icarus* (Amsterdam, the Netherlands: Elsevier) **145** (2): 609–613. Bibcode:2000Icar..145..609L. doi:10.1006/icar.2000.6365. ISSN 0019-1035. PMID 11543508.

[110] Menor-Salván, César; Ruiz-Bermejo, Marta; Guzmán, Marcelo I.; Osuna-Esteban, Susana; Veintemillas-Verdaguer, Sabino (20 April 2009). "Synthesis of Pyrimidines and Triazines in Ice: Implications for the Prebiotic Chemistry of Nucleobases". *Chemistry: A European Journal* (Weinheim, Germany: Wiley-VCH on behalf of ChemPubSoc Europe) **15** (17): 4411–4418. doi:10.1002/chem.200802656. ISSN 0947-6539. PMID 19288488.

[111] Roy, Debjani; Najafian, Katayoun; von Ragué Schleyer, Paul (30 October 2007). "Chemical evolution: The mechanism of the formation of adenine under prebiotic conditions". *Proc. Natl. Acad. Sci. U.S.A.* (Washington, D.C.: National Academy of Sciences) **104** (44): 17272–17277. Bibcode:2007PNAS..10417272R. doi:10.1073/pnas.0708434104. ISSN 0027-8424. PMC 2077245. PMID 17951429.

[112] Cleaves, H. James; Chalmers, John H.; Lazcano, Antonio; et al. (April 2008). "A Reassessment of Prebiotic Organic Synthesis in Neutral Planetary Atmospheres". *Origins of Life and Evolution of Biospheres* (Dordrecht, the Netherlands: Springer) **38** (2): 105–115. Bibcode:2008OLEB...38..105C. doi:10.1007/s11084-007-9120-3. ISSN 0169-6149. PMID 18204914.

[113] Chyba, Christopher F. (13 May 2005). "Rethinking Earth's Early Atmosphere". *Science* (Washington, D.C.: American Association for the Advancement of Science) **308** (5724): 962–963. doi:10.1126/science.1113157. ISSN 0036-8075. PMID 15890865.

[114] Barton et al. 2007, pp. 93–95

[115] Bada & Lazcano 2009, pp. 56–57

[116] Bada, Jeffrey L.; Lazcano, Antonio (2 May 2003). "Prebiotic Soup--Revisiting the Miller Experiment" (PDF). *Science* (Washington, D.C.: American Association for the Advancement of Science) **300** (5620): 745–746. doi:10.1126/science.1085145. ISSN 0036-8075. PMID 12730584. Retrieved 2015-06-13.

[117] Oró, Joan; Kimball, Aubrey P. (February 1962). "Synthesis of purines under possible primitive earth conditions: II. Purine intermediates from hydrogen cyanide". *Archives of Biochemistry and Biophysics* (Amsterdam, the Netherlands: Elsevier) **96** (2): 293–313. doi:10.1016/0003-9861(62)90412-5. ISSN 0003-9861. PMID 14482339.

[118] Ahuja, Mukesh, ed. (2006). "Origin of Life". *Life Science* **1**. Delhi: Isha Books. p. 11. ISBN 81-8205-386-2. OCLC 297208106.

[119] Service, Robert F. (16 March 2015). "Researchers may have solved origin-of-life conundrum". *Science* (News) (Washington, D.C.: American Association for the Advancement of Science). ISSN 1095-9203. Retrieved 2015-07-26.

[120] Patel, Bhavesh H.; Percivalle, Claudia; Ritson, Dougal J.; Duffy, Colm D.; Sutherland, John D. (April 2015). "Common origins of RNA, protein and lipid precursors in a cyanosulfidic protometabolism". *Nature Chemistry* (London: Nature Publishing Group) **7** (4): 301–307. Bibcode:2015NatCh...7..301P. doi:10.1038/nchem.2202. ISSN 1755-4330. PMC 4568310. PMID 25803468. Retrieved 2015-07-22.

[121] Patel et al. 2015, p. 302

[122] Paul, Natasha; Joyce, Gerald F. (December 2004). "Minimal self-replicating systems". *Current Opinion in Chemical Biology* (Amsterdam, the Netherlands: Elsevier) **8** (6): 634–639. doi:10.1016/j.cbpa.2004.09.005. ISSN 1367-5931. PMID 15556408.

[123] Bissette, Andrew J.; Fletcher, Stephen P. (2 December 2013). "Mechanisms of Autocatalysis". *Angewandte Chemie International Edition* (Weinheim, Germany: Wiley-VCH on behalf of the German Chemical Society) **52** (49): 12800–12826. doi:10.1002/anie.201303822. ISSN 1433-7851. PMID 24127341.

[124] Kauffman 1993, chpt. 7

[125] Dawkins 2004

[126] Tjivikua, T.; Ballester, Pablo; Rebek, Julius, Jr. (January 1990). "Self-replicating system". *Journal of the American Chemical Society* (Washington, D.C.: American Chemical Society) **112** (3): 1249–1250. doi:10.1021/ja00159a057. ISSN 0002-7863.

[127] Browne, Malcolm W. (30 October 1990). "Chemists Make Molecule With Hint of Life". *The New York Times* (New York: The New York Times Company). ISSN 0362-4331. Retrieved 2015-07-14.

[128] Eigen & Schuster 1979

[129] Hoffmann, Geoffrey W. (25 June 1974). "On the origin of the genetic code and the stability of the translation apparatus". *Journal of Molecular Biology* (Amsterdam, the Netherlands: Elsevier) **86** (2): 349–362. doi:10.1016/0022-2836(74)90024-2. ISSN 0022-2836. PMID 4414916.

[130] Orgel, Leslie E. (April 1963). "The Maintenance of the Accuracy of Protein Synthesis and its Relevance to Ageing". *Proc. Natl. Acad. Sci. U.S.A.* (Washington, D.C.: National Academy of Sciences) **49** (4): 517–521. Bibcode:1963PNAS...49..517O. doi:10.1073/pnas.49.4.517. ISSN 0027-8424. PMC 299893. PMID 13940312.

[131] Hoffmann, Geoffrey W. (October 1975). "The Stochastic Theory of the Origin of the Genetic Code". *Annual Review of Physical Chemistry* (Palo Alto, CA: Annal Reviews) **26**: 123–144. Bibcode:1975ARPC...26..123H. doi:10.1146/annurev.pc.26.100175.001011. ISSN 0066-426X.

[132] Colgate, S. A.; Rasmussen, S.; Solem, J. C.; Lackner, K. (2003). "An astrophysical basis for a universal origin of life". *Advances in Complex Systems* **6** (4): 487–505. doi:10.1142/s0219525903001079.

[133] Chaichian, Rojas & Tureanu 2014, pp. 353–364

[134] Plasson, Raphaël; Kondepudi, Dilip K.; Bersini, Hugues; et al. (August 2007). "Emergence of homochirality in far-from-equilibrium systems: Mechanisms and role in prebiotic chemistry". *Chirality* (Hoboken, NJ: John Wiley & Sons) **19** (8): 589–600. doi:10.1002/chir.20440. ISSN 0899-0042. PMID 17559107. "Special Issue: Proceedings from the Eighteenth International Symposium on Chirality (ISCD-18), Busan, Korea, 2006"

[135] Clark, Stuart (July–August 1999). "Polarized Starlight and the Handedness of Life". *American Scientist* (Research Triangle Park, NC: Sigma Xi) **87** (4): 336. Bibcode:1999AmSci..87..336C. doi:10.1511/1999.4.336. ISSN 0003-0996.

[136] Shibata, Takanori; Morioka, Hiroshi; Hayase, Tadakatsu; et al. (17 January 1996). "Highly Enantioselective Catalytic Asymmetric Automultiplication of Chiral Pyrimidyl Alcohol". *Journal of the American Chemical Society* (Washington, D.C.: American Chemical Society) **118** (2): 471–472. doi:10.1021/ja953066g. ISSN 0002-7863.

[137] Soai, Kenso; Sato, Itaru; Shibata, Takanori (2001). "Asymmetric autocatalysis and the origin of chiral homogeneity in organic compounds". *The Chemical Record* (Hoboken, NJ: John Wiley & Sons on behalf of The Japan Chemical Journal Forum) **1** (4): 321–332. doi:10.1002/tcr.1017. ISSN 1528-0691. PMID 11893072.

[138] Hazen 2005

[139] Mullen, Leslie (5 September 2005). "Building Life from Star-Stuff". *Astrobiology Magazine* (New York: NASA). Retrieved 2015-06-15.

[140] Chen, Irene A.; Walde, Peter (July 2010). "From Self-Assembled Vesicles to Protocells" (PDF). *Cold Spring Harbor Perspectives in Biology* (Cold Spring Harbor, NY: Cold Spring Harbor Laboratory Press) **2** (7): a002170. doi:10.1101/cshperspect.a002170. ISSN 1943-0264. PMC 2890201. PMID 20519344. Retrieved 2015-06-15.

[141] "Exploring Life's Origins: Protocells". *Exploring Life's Origins: A Virtual Exhibit*. Arlington County, VA: National Science Foundation. Retrieved 2014-03-18.

[142] Chen, Irene A. (8 December 2006). "The Emergence of Cells During the Origin of Life". *Science* (Washington, D.C.: American Association for the Advancement of Science) **314** (5805): 1558–1559. doi:10.1126/science.1137541. ISSN 0036-8075. PMID 17158315. Retrieved 2015-06-15.

[143] Zimmer, Carl (26 June 2004). "What Came Before DNA?". *Discover* (Waukesha, WI: Kalmbach Publishing). ISSN 0274-7529.

[144] Shapiro, Robert (June 2007). "A Simpler Origin for Life". *Scientific American* (Stuttgart: Georg von Holtzbrinck Publishing Group) **296** (6): 46–53. doi:10.1038/scientificamerican0607-46. ISSN 0036-8733. PMID 17663224. Retrieved 2015-06-15.

[145] Chang 2007

[146] Switek, Brian (13 February 2012). "Debate bubbles over the origin of life". *Nature* (London: Nature Publishing Group). doi:10.1038/nature.2012.10024. ISSN 0028-0836.

[147] Grote, Mathias (September 2011). "*Jeewanu*, or the 'particles of life'" (PDF). *Journal of Biosciences* (Bangalore, India: Indian Academy of Sciences; Springer) **36** (4): 563–570. doi:10.1007/s12038-011-9087-0. ISSN 0250-5991. PMID 21857103. Retrieved 2015-06-15.

[148] Gupta, V. K.; Rai, R. K. (August 2013). "Histochemical localisation of RNA-like material in photochemically formed self-sustaining, abiogenic supramolecular assemblies 'Jeewanu'". *International Research Journal of Science & Engineering* (Amravati, India) **1** (1): 1–4. ISSN 2322-0015. Retrieved 2015-06-15.

[149] Welter, Kira (10 August 2015). "Peptide glue may have held first protocell components together". *Chemistry World* (News) (London: Royal Society of Chemistry). ISSN 1473-7604. Retrieved 2015-08-29.

- Kamat, Neha P.; Tobé, Sylvia; Hill, Ian T.; Szostak, Jack W. (29 July 2015). "Electrostatic Localization of RNA to Protocell Membranes by Cationic Hydrophobic Peptides". *Angewandte Chemie International Edition* (Weinheim, Germany: Wiley-VCH on behalf of the German Chemical Society). doi:10.1002/anie.201505742. ISSN 1433-7851. "Early View (Online Version of Record published before inclusion in an issue)"

## 1.10. REFERENCES

[150] Wimberly, Brian T.; Brodersen, Ditlev E.; Clemons, William M., Jr.; et al. (21 September 2000). "Structure of the 30S ribosomal subunit". *Nature* (London: Nature Publishing Group) **407** (6802): 327–339. doi:10.1038/35030006. ISSN 0028-0836. PMID 11014182.

[151] Zimmer, Carl (25 September 2014). "A Tiny Emissary From the Ancient Past". *The New York Times* (New York: The New York Times Company). ISSN 0362-4331. Retrieved 2014-09-26.

[152] Wade, Nicholas (4 May 2015). "Making Sense of the Chemistry That Led to Life on Earth". *The New York Times* (New York: The New York Times Company). ISSN 0362-4331. Retrieved 2015-05-10.

[153] Yarus, Michael (April 2011). "Getting Past the RNA World: The Initial Darwinian Ancestor". *Cold Spring Harbor Perspectives in Biology* (Cold Spring Harbor, NY: Cold Spring Harbor Laboratory Press) **3** (4): a003590. doi:10.1101/cshperspect.a003590. ISSN 1943-0264. PMC 3062219. PMID 20719875.

[154] Neveu, Marc; Kim, Hyo-Joong; Benner, Steven A. (22 April 2013). "The 'Strong' RNA World Hypothesis: Fifty Years Old". *Astrobiology* (New Rochelle, NY: Mary Ann Liebert, Inc.) **13** (4): 391–403. Bibcode:2013AsBio..13..391N. doi:10.1089/ast.2012.0868. ISSN 1531-1074. PMID 23551238.

[155] Gilbert, Walter (20 February 1986). "Origin of life: The RNA world". *Nature* (London: Nature Publishing Group) **319** (6055): 618. Bibcode:1986Natur.319..618G. doi:10.1038/319618a0. ISSN 0028-0836.

[156] Noller, Harry F. (April 2012). "Evolution of protein synthesis from an RNA world.". *Cold Spring Harbor Perspectives in Biology* (Cold Spring Harbor, NY: Cold Spring Harbor Laboratory Press) **4** (4): a003681. doi:10.1101/cshperspect.a003681. ISSN 1943-0264. PMC 3312679. PMID 20610545.

[157] Koonin, Eugene V. (31 May 2007). "The cosmological model of eternal inflation and the transition from chance to biological evolution in the history of life". *Biology Direct* (London: BioMed Central) **2**: 15. doi:10.1186/1745-6150-2-15. ISSN 1745-6150. PMC 1892545. PMID 17540027.

[158] Yates, Diana (25 September 2015). "Study adds to evidence that viruses are alive" (Press release). Champaign, IL: University of Illinois at Urbana–Champaign. Retrieved 2015-10-20.

[159] Katzourakis, Aris (2013)"Paleovirology: inferring viral evolution from host genome sequence data" (Philosophical Transactions of the Royal Society Published 12 August 2013.DOI: 10.1098/rstb.2012.0493)

[160] Arshan, Nasir; Caetano-Anollés, Gustavo (25 September 2015). "A phylogenomic data-driven exploration of viral origins and evolution". *Science Advances* (Washington, D.C.: American Association for the Advancement of Science) **1** (8): e1500527. doi:10.1126/sciadv.1500527. ISSN 2375-2548.

[161] Nasir, Arshan; Naeem, Aisha; Jawad Khan, Muhammad; et al. (December 2011). "Annotation of Protein Domains Reveals Remarkable Conservation in the Functional Make up of Proteomes Across Superkingdoms". *Genes* (Basel, Switzerland: MDPI) **2** (4): 869–911. doi:10.3390/genes2040869. ISSN 2073-4425. PMC 3927607. PMID 24710297.

[162] Zimmer, Carl (12 September 2013). "A Far-Flung Possibility for the Origin of Life". *The New York Times* (New York: The New York Times Company). ISSN 0362-4331. Retrieved 2015-06-15.

[163] Webb, Richard (29 August 2013). "Primordial broth of life was a dry Martian cup-a-soup". *New Scientist* (London: Reed Business Information). ISSN 0262-4079. Retrieved 2015-06-16.

[164] Wentao Ma; Chunwu Yu; Wentao Zhang; et al. (November 2007). "Nucleotide synthetase ribozymes may have emerged first in the RNA world". *RNA* (Cold Spring Harbor, NY: Cold Spring Harbor Laboratory Press on behalf of the RNA Society) **13** (11): 2012–2019. doi:10.1261/rna.658507. ISSN 1355-8382. PMC 2040096. PMID 17878321.

[165] Orgel, Leslie E. (October 1994). "The origin of life on Earth". *Scientific American* (Stuttgart: Georg von Holtzbrinck Publishing Group) **271** (4): 76–83. doi:10.1038/scientificamerican1094-76. ISSN 0036-8733. PMID 7524147.

[166] Johnston, Wendy K.; Unrau, Peter J.; Lawrence, Michael S.; et al. (18 May 2001). "RNA-Catalyzed RNA Polymerization: Accurate and General RNA-Templated Primer Extension". *Science* (Washington, D.C.: American Association for the Advancement of Science) **292** (5520): 1319–1325. Bibcode:2001Sci...292.1319J. doi:10.1126/science.1060786. ISSN 0036-8075. PMID 11358999.

[167] Szostak, Jack W. (5 February 2015). "The Origins of Function in Biological Nucleic Acids, Proteins, and Membranes". Chevy Chase (CDP), MD: Howard Hughes Medical Institute. Retrieved 2015-06-16.

[168] Lincoln, Tracey A.; Joyce, Gerald F. (27 February 2009). "Self-Sustained Replication of an RNA Enzyme". *Science* (Washington, D.C.: American Association for the Advancement of Science) **323** (5918): 1229–1232. Bibcode:2009Sci...323.1229L. doi:10.1126/science.1167856. ISSN 0036-8075. PMC 2652413. PMID 19131595.

[169] Joyce, Gerald F. (2009). "Evolution in an RNA world" (PDF). *Cold Spring Harbor Perspectives in Biology* (Cold Spring Harbor, NY: Cold Spring Harbor Laboratory Press) **74** (Evolution: The Molecular Landscape): 17–23.

doi:10.1101/sqb.2009.74.004. ISSN 1943-0264. PMC 2891321. PMID 19667013. Retrieved 2015-06-16.

[170] Bernstein, Harris; Byerly, Henry C.; Hopf, Frederick A.; et al. (June 1983). "The Darwinian Dynamic". *The Quarterly Review of Biology* (Chicago, IL: University of Chicago Press) **58** (2): 185–207. doi:10.1086/413216. ISSN 0033-5770. JSTOR 2828805.

[171] Michod 1999

[172] Orgel, Leslie E. (17 November 2000). "A Simpler Nucleic Acid". *Science* (Washington, D.C.: American Association for the Advancement of Science) **290** (5495): 1306–1307. doi:10.1126/science.290.5495.1306. ISSN 0036-8075. PMID 11185405.

[173] Nelson, Kevin E.; Levy, Matthew; Miller, Stanley L. (11 April 2000). "Peptide nucleic acids rather than RNA may have been the first genetic molecule". *Proc. Natl. Acad. Sci. U.S.A.* (Washington, D.C.: National Academy of Sciences) **97** (8): 3868–3871. Bibcode:2000PNAS...97.3868N. doi:10.1073/pnas.97.8.3868. ISSN 0027-8424. PMC 18108. PMID 10760258.

[174] Larralde, Rosa; Robertson, Michael P.; Miller, Stanley L. (29 August 1995). "Rates of Decomposition of Ribose and Other Sugars: Implications for Chemical Evolution" (PDF). *Proc. Natl. Acad. Sci. U.S.A.* (Washington, D.C.: National Academy of Sciences) **92** (18): 8158–8160. Bibcode:1995PNAS...92.8158L. doi:10.1073/pnas.92.18.8158. ISSN 0027-8424. PMC 41115. PMID 7667262.

[175] Lindahl, Tomas (22 April 1993). "Instability and decay of the primary structure of DNA". *Nature* (London: Nature Publishing Group) **362** (6422): 709–715. Bibcode:1993Natur.362..709L. doi:10.1038/362709a0. ISSN 0028-0836. PMID 8469282.

[176] Anastasi, Carole; Crowe, Michael A.; Powner, Matthew W.; Sutherland, John D. (18 September 2006). "Direct Assembly of Nucleoside Precursors from Two- and Three-Carbon Units". *Angewandte Chemie International Edition* (Weinheim, Germany: Wiley-VCH on behalf of the German Chemical Society) **45** (37): 6176–6179. doi:10.1002/anie.200601267. ISSN 1433-7851. PMID 16917794.

[177] Powner, Matthew W.; Sutherland, John D. (13 October 2008). "Potentially Prebiotic Synthesis of Pyrimidine β-D-Ribonucleotides by Photoanomerization/Hydrolysis of α-D-Cytidine-2′-Phosphate". *ChemBioChem* (Weinheim, Germany: Wiley-VCH) **9** (15): 2386–2387. doi:10.1002/cbic.200800391. ISSN 1439-4227. PMID 18798212.

[178] Powner, Matthew W.; Gerland, Béatrice; Sutherland, John D. (14 May 2009). "Synthesis of activated pyrimidine ribonucleotides in prebiotically plausible conditions". *Nature* (London: Nature Publishing Group) **459** (7244): 239–242. Bibcode:2009Natur.459..239P. doi:10.1038/nature08013. ISSN 0028-0836. PMID 19444213.

[179] Senthilingam, Meera (25 April 2014). "Metabolism May Have Started in Early Oceans Before the Origin of Life" (Press release). Wellcome Trust. EurekAlert!. Retrieved 2015-06-16.

[180] Yue-Ching Ho, Eugene (July–September 1990). "Evolutionary Epistemology and Sir Karl Popper's Latest Intellectual Interest: A First-Hand Report". *Intellectus* (Hong Kong: Hong Kong Institute of Economic Science) **15**: 1–3. OCLC 26878740. Retrieved 2012-08-13.

[181] Wade, Nicholas (22 April 1997). "Amateur Shakes Up Ideas on Recipe for Life". *The New York Times* (New York: The New York Times Company). ISSN 0362-4331. Retrieved 2015-06-16.

[182] Popper, Karl R. (29 March 1990). "Pyrite and the origin of life". *Nature* (London: Nature Publishing Group) **344** (6265): 387. Bibcode:1990Natur.344..387P. doi:10.1038/344387a0. ISSN 0028-0836.

[183] Huber, Claudia; Wächtershäuser, Günter (31 July 1998). "Peptides by Activation of Amino Acids with CO on (Ni,Fe)S Surfaces: Implications for the Origin of Life". *Science* (Washington, D.C.: American Association for the Advancement of Science) **281** (5377): 670–672. Bibcode:1998Sci...281..670H. doi:10.1126/science.281.5377.670. ISSN 0036-8075. PMID 9685253.

[184] Lane 2009

[185] Musser, George (23 September 2011). "How Life Arose on Earth, and How a Singularity Might Bring It Down". *Observations* (Blog). Scientific American. ISSN 0036-8733. Retrieved 2015-06-17.

[186] Carroll, Sean (10 March 2010). "Free Energy and the Meaning of Life". *Cosmic Variance* (Blog). Discover. ISSN 0274-7529. Retrieved 2015-06-17.

[187] Wolchover, Natalie (22 January 2014). "A New Physics Theory of Life". *Quanta Magazine* (New York: Simons Foundation). Retrieved 2015-06-17.

[188] England, Jeremy L. (28 September 2013). "Statistical physics of self-replication" (PDF). *Journal of Chemical Physics* (College Park, MD: American Institute of Physics) **139**: 121923. arXiv:1209.1179. Bibcode:2013JChPh.139l1923E. doi:10.1063/1.4818538. ISSN 0021-9606. Retrieved 2015-06-18.

[189] Orgel, Leslie E. (7 November 2000). "Self-organizing biochemical cycles". *Proc. Natl. Acad. Sci. U.S.A.* (Washington, D.C.: National Academy of Sciences) **97** (23): 12503–12507. Bibcode:2000PNAS...9712503O. doi:10.1073/pnas.220406697. ISSN 0027-8424. PMC 18793. PMID 11058157.

## 1.10. REFERENCES

[190] Chandru, Kuhan; Gilbert, Alexis; Butch, Christopher; Aono, Masashi; Cleaves, Henderson James II (21 July 2016). "The Abiotic Chemistry of Thiolated Acetate Derivatives and the Origin of Life". *Scientific Reports* **6** (29883). doi:10.1038/srep29883.

[191] Mulkidjanian, Armen Y. (24 August 2009). "On the origin of life in the zinc world: 1. Photosynthesizing, porous edifices built of hydrothermally precipitated zinc sulfide as cradles of life on Earth". *Biology Direct* (London: BioMed Central) **4**: 26. doi:10.1186/1745-6150-4-26. ISSN 1745-6150.

[192] Wächtershäuser, Günter (December 1988). "Before Enzymes and Templates: Theory of Surface Metabolism" (PDF). *Microbiological Reviews* (Washington, D.C.: American Society for Microbiology) **52** (4): 452–484. ISSN 0146-0749. PMC 373159. PMID 3070320.

[193] Mulkidjanian, Armen Y.; Galperin, Michael Y. (24 August 2009). "On the origin of life in the zinc world. 2. Validation of the hypothesis on the photosynthesizing zinc sulfide edifices as cradles of life on Earth". *Biology Direct* (London: BioMed Central) **4**: 27. doi:10.1186/1745-6150-4-27. ISSN 1745-6150.

[194] Macallum, A. B. (1 April 1926). "The Paleochemistry of the body fluids and tissues". *Physiological Reviews* (Bethesda, MD: American Physiological Society) **6** (2): 316–357. ISSN 0031-9333. Retrieved 2015-06-18.

[195] Mulkidjanian, Armen Y.; Bychkov, Andrew Yu.; Dibrova, Daria V.; et al. (3 April 2012). "Origin of first cells at terrestrial, anoxic geothermal fields". *Proc. Natl. Acad. Sci. U.S.A.* (Washington, D.C.: National Academy of Sciences) **109** (14): E821–E830. Bibcode:2012PNAS..109E.821M. doi:10.1073/pnas.1117774109. ISSN 1091-6490. PMC 3325685. PMID 22331915.

[196] For a deeper integrative version of this hypothesis, see in particular Lankenau 2011, pp. 225–286, interconnecting the "Two RNA worlds" concept and other detailed aspects; and Davidovich, Chen; Belousoff, Matthew; Bashan, Anat; Yonath, Ada (September 2009). "The evolving ribosome: from non-coded peptide bond formation to sophisticated translation machinery". *Research in Microbiology* (Amsterdam, the Netherlands: Elsevier) **160** (7): 487–492. doi:10.1016/j.resmic.2009.07.004. ISSN 1769-7123. PMID 19619641.

[197] Schirber, Michael (24 June 2014). "Hydrothermal Vents Could Explain Chemical Precursors to Life". *NASA Astrobiology: Life in the Universe*. NASA. Retrieved 2015-06-19.

[198] Martin, William; Russell, Michael J. (29 January 2003). "On the origins of cells: a hypothesis for the evolutionary transitions from abiotic geochemistry to chemoautotrophic prokaryotes, and from prokaryotes to nucleated cells". *Philosophical Transactions of the Royal Society B* (London: Royal Society) **358** (1429): 59–83; discussion 83–85. doi:10.1098/rstb.2002.1183. ISSN 0962-8436. PMC 1693102. PMID 12594918.

[199] Ignatov, Ignat; Mosin, Oleg V. (2013). "Possible Processes for Origin of Life and Living Matter with modeling of Physiological Processes of Bacterium *Bacillus Subtilis* in Heavy Water as Model System". *Journal of Natural Sciences Research* (New York: International Institute for Science, Technology and Education) **3** (9): 65–76. ISSN 2225-0921.

[200] Calvin 1969

[201] Schirber, Michael (1 March 2010). "First Fossil-Makers in Hot Water". *Astrobiology Magazine* (New York: NASA). Retrieved 2015-06-19.

[202] Kurihara, Kensuke; Tamura, Mieko; Shohda, Kohichiroh; et al. (October 2011). "Self-Reproduction of supramolecular giant vesicles combined with the amplification of encapsulated DNA". *Nature Chemistry* (London: Nature Publishing Group) **3** (10): 775–781. Bibcode:2011NatCh...3..775K. doi:10.1038/nchem.1127. ISSN 1755-4330. PMID 21941249.

[203] Usher, Oli (27 April 2015). "Chemistry of seabed's hot vents could explain emergence of life" (Press release). University College London. Retrieved 2015-06-19.

[204] Roldan, Alberto; Hollingsworth, Nathan; Roffey, Anna; Islam, Husn-Ubayda; et al. (May 2015). "Bio-inspired CO2 conversion by iron sulfide catalysts under sustainable conditions" (PDF). *Chemical Communications* (London: Royal Society of Chemistry) **51** (35): 7501–7504. doi:10.1039/C5CC02078F. ISSN 1359-7345. PMID 25835242. Retrieved 2015-06-19.

[205] Muller, Anthonie W. J. (7 August 1985). "Thermosynthesis by biomembranes: Energy gain from cyclic temperature changes". *Journal of Theoretical Biology* (Amsterdam, the Netherlands: Elsevier) **115** (3): 429–453. doi:10.1016/S0022-5193(85)80202-2. ISSN 0022-5193. PMID 3162066.

[206] Muller, Anthonie W. J. (1995). "Were the first organisms heat engines? A new model for biogenesis and the early evolution of biological energy conversion". *Progress in Biophysics and Molecular Biology* (Oxford, UK; New York: Pergamon Press) **63** (2): 193–231. doi:10.1016/0079-6107(95)00004-7. ISSN 0079-6107. PMID 7542789.

[207] Muller, Anthonie W. J.; Schulze-Makuch, Dirk (1 April 2006). "Sorption heat engines: Simple inanimate negative entropy generators". *Physica A: Statistical Mechanics and its Applications* (Utrecht, the Netherlands: Elsevier) **362** (2): 369–381. arXiv:physics/0507173. Bibcode:2006PhyA..362..369M. doi:10.1016/j.physa.2005.12.003. ISSN 0378-4371.

[208] Orgel 1987, pp. 9–16

[209] Perry, Caroline (7 February 2011). "Clay-armored bubbles may have formed first protocells" (Press release). Cambridge, MA: Harvard University. EurekAlert!. Retrieved 2015-06-20.

[210] Dawkins 1996, pp. 148–161

[211] Wenhua Huang; Ferris, James P. (12 July 2006). "One-Step, Regioselective Synthesis of up to 50-mers of RNA Oligomers by Montmorillonite Catalysis". *Journal of the American Chemical Society* (Washington, D.C.: American Chemical Society) **128** (27): 8914–8919. doi:10.1021/ja061782k. ISSN 0002-7863. PMID 16819887.

[212] Moore, Caroline (16 July 2007). "Crystals as genes?". *Highlights in Chemical Science* (London: Royal Society of Chemistry). ISSN 2041-5818. Retrieved 2015-06-21.

- Bullard, Theresa; Freudenthal, John; Avagyan, Serine; et al. (2007). "Test of Cairns-Smith's 'crystals-as-genes' hypothesis". *Faraday Discussions* **136**: 231–245. Bibcode:2007FaDi..136..231B. doi:10.1039/b616612c. ISSN 1359-6640.

[213] Wickramasinghe, Chandra (2011). "Bacterial morphologies supporting cometary panspermia: a reappraisal". *International Journal of Astrobiology* **10** (1): 25–30. Bibcode:2011IJAsB..10...25W. doi:10.1017/S1473550410000157.

[214] Napier, William (October 2011). "Exchange of Biomaterial Between Planetary Systems" (PDF) **16**: 6616–6642.

[215] Rampelotto, P. H. (2010). Panspermia: A promising field of research. In: Astrobiology Science Conference. Abs 5224.

[216] Forward planetary contamination like *Tersicoccus phoenicis*, that has shown resistance to methods usually used in spacecraft assembly clean rooms: Madhusoodanan, Jyoti (May 19, 2014). "Microbial stowaways to Mars identified". *Nature*. doi:10.1038/nature.2014.15249. Retrieved May 23, 2014.

[217] Webster, Guy (November 6, 2013). "Rare New Microbe Found in Two Distant Clean Rooms". *NASA.gov*. Retrieved November 6, 2013.

[218] Clark, Stuart (25 September 2002). "Tough Earth bug may be from Mars". *New Scientist* (London: Reed Business Information). ISSN 0262-4079. Retrieved 2015-06-21.

[219] Horneck, Gerda; Klaus, David M.; Mancinelli, Rocco L. (March 2010). "Space Microbiology". *Microbiology and Molecular Biology Reviews* (Washington, D.C.: American Society for Microbiology) **74** (1): 121–156. doi:10.1128/MMBR.00016-09. ISSN 1092-2172. PMC 2832349. PMID 20197502.

[220] Rabbow, Elke; Horneck, Gerda; Rettberg, Petra; et al. (December 2009). "EXPOSE, an Astrobiological Exposure Facility on the International Space Station – from Proposal to Flight". *Origins of Life and Evolution of Biospheres* (Dordrecht, the Netherlands: Springer) **39** (6): 581–598. Bibcode:2009OLEB...39..581R. doi:10.1007/s11084-009-9173-6. ISSN 0169-6149. PMID 19629743.

[221] Onofri, Silvano; de la Torre, Rosa; de Vera, Jean-Pierre; et al. (May 2012). "Survival of Rock-Colonizing Organisms After 1.5 Years in Outer Space". *Astrobiology* (New Rochelle, NY: Mary Ann Liebert, Inc.) **12** (5): 508–516. Bibcode:2012AsBio..12..508O. doi:10.1089/ast.2011.0736. ISSN 1531-1074. PMID 22680696.

[222] "biological abundance of elements". *Encyclopedia of Science*. Dundee, Scotland: David Darling Enterprises. Retrieved 2008-10-09.

[223] Hoover, Rachel (21 February 2014). "Need to Track Organic Nano-Particles Across the Universe? NASA's Got an App for That". *Ames Research Center*. Mountain View, CA: NASA. Retrieved 2015-06-22.

[224] Chang, Kenneth (18 August 2009). "From a Distant Comet, a Clue to Life". *The New York Times* (New York: The New York Times Company). p. A18. ISSN 0362-4331. Retrieved 2015-06-22.

[225] Goncharuk, Vladislav V.; Zui, O. V. (February 2015). "Water and carbon dioxide as the main precursors of organic matter on Earth and in space". *Journal of Water Chemistry and Technology* (Dordrecht, the Netherlands: Springer on behalf of Allerton Press) **37** (1): 2–3. doi:10.3103/S1063455X15010026. ISSN 1063-455X.

[226] Abou Mrad, Ninette; Vinogradoff, Vassilissa; Duvernay, Fabrice; et al. (2015). "Laboratory experimental simulations: Chemical evolution of the organic matter from interstellar and cometary ice analogs" (PDF). *Bulletin de la Société Royale des Sciences de Liège* (Liège, Belgium: Société royale des sciences de Liège) **84**: 21–32. Bibcode:2015BSRSL..84...21A. ISSN 0037-9565. Retrieved 2015-04-06.

[227] Gallori, Enzo (June 2011). "Astrochemistry and the origin of genetic material". *Rendiconti Lincei* (Milan, Italy: Springer) **22** (2): 113–118. doi:10.1007/s12210-011-0118-4. ISSN 2037-4631. "Paper presented at the Symposium 'Astrochemistry: molecules in space and time' (Rome, 4–5 November 2010), sponsored by Fondazione 'Guido Donegani', Accademia Nazionale dei Lincei."

[228] Martins, Zita (February 2011). "Organic Chemistry of Carbonaceous Meteorites". *Elements* (Chantilly, VA: Mineralogical Society of America et al.) **7** (1): 35–40. doi:10.2113/gselements.7.1.35. ISSN 1811-5209.

[229] Martins, Zita; Botta, Oliver; Fogel, Marilyn L.; et al. (15 June 2008). "Extraterrestrial nucleobases in the Murchison meteorite". *Earth and Planetary Science Letters* (Amsterdam, the Netherlands: Elsevier) **270** (1–2): 130–136. arXiv:0806.2286. Bibcode:2008E&PSL.270..130M. doi:10.1016/j.epsl.2008.03.026. ISSN 0012-821X.

[230] "We may all be space aliens: study". *ABC News* (Sydney: Australian Broadcasting Corporation). AFP. 14 June 2008. Retrieved 2015-06-22.

## 1.10. REFERENCES

[231] Callahan, Michael P.; Smith, Karen E.; Cleaves, H. James, II; et al. (23 August 2011). "Carbonaceous meteorites contain a wide range of extraterrestrial nucleobases". *Proc. Natl. Acad. Sci. U.S.A.* (Washington, D.C.: National Academy of Sciences) **108** (34): 13995–13998. Bibcode:2011PNAS..10813995C. doi:10.1073/pnas.1106493108. ISSN 0027-8424. PMC 3161613. PMID 21836052.

[232] Steigerwald, John (8 August 2011). "NASA Researchers: DNA Building Blocks Can Be Made in Space". *Goddard Space Flight Center*. Greenbelt, MD: NASA. Retrieved 2015-06-23.

[233] Chow, Denise (26 October 2011). "Discovery: Cosmic Dust Contains Organic Matter from Stars". *Space.com* (Ogden, UT: Purch). Retrieved 2015-06-23.

[234] "Astronomers Discover Complex Organic Matter Exists Throughout the Universe". Rockville, MD: ScienceDaily, LLC. 26 October 2011. Retrieved 2015-06-23. Post is reprinted from materials provided by The University of Hong Kong.

[235] Sun Kwok; Yong Zhang (3 November 2011). "Mixed aromatic–aliphatic organic nanoparticles as carriers of unidentified infrared emission features". *Nature* (London: Nature Publishing Group) **479** (7371): 80–83. Bibcode:2011Natur.479...80K. doi:10.1038/nature10542. ISSN 0028-0836. PMID 22031328.

[236] Clemence, Lara; Cohen, Jarrett (7 February 2005). "Space Sugar's a Sweet Find". *Goddard Space Flight Center*. Greenbelt, MD: NASA. Retrieved 2015-06-23.

[237] Than, Ker (30 August 2012). "Sugar Found In Space: A Sign of Life?". *National Geographic News* (Washington, D.C.: National Geographic Society). Retrieved 2015-06-23.

[238] "Sweet! Astronomers spot sugar molecule near star". *Excite* (Yonkers, NY: Mindspark Interactive Network). Associated Press. 29 August 2012. Retrieved 2015-06-23.

[239] "Building blocks of life found around young star". *News & Events*. Leiden, the Netherlands: Leiden University. 30 September 2012. Retrieved 2013-12-11.

[240] Jørgensen, Jes K.; Favre, Cécile; Bisschop, Suzanne E.; et al. (20 September 2012). "Detection of the simplest sugar, glycolaldehyde, in a solar-type protostar with ALMA" (PDF). *The Astrophysical Journal Letters* (Bristol, England: IOP Publishing for the American Astronomical Society) **757** (1): L4. arXiv:1208.5498. Bibcode:2012ApJ...757L...4J. doi:10.1088/2041-8205/757/1/L4. ISSN 2041-8213. L4. Retrieved 2015-06-23.

[241] "'Life chemical' detected in comet". *BBC News* (London: BBC). 18 August 2009. Retrieved 2015-06-23.

[242] Thompson, William Reid; Murray, B. G.; Khare, Bishun Narain; Sagan, Carl (30 December 1987). "Coloration and darkening of methane clathrate and other ices by charged particle irradiation: Applications to the outer solar system". *Journal of Geophysical Research* (Washington, D.C.: American Geophysical Union) **92** (A13): 14933–14947. Bibcode:1987JGR....9214933T. doi:10.1029/JA092iA13p14933. ISSN 0148-0227. PMID 11542127.

[243] Stark, Anne M. (5 June 2013). "Life on Earth shockingly comes from out of this world". Livermore, CA: Lawrence Livermore National Laboratory. Retrieved 2015-06-23.

[244] Goldman, Nir; Tamblyn, Isaac (20 June 2013). "Prebiotic Chemistry within a Simple Impacting Icy Mixture". *Journal of Physical Chemistry A* (Washington, D.C.: American Chemical Society) **117** (24): 5124–5131. doi:10.1021/jp402976n. ISSN 1089-5639. PMID 23639050.

[245] Carey, Bjorn (18 October 2005). "Life's Building Blocks 'Abundant in Space'". *Space.com* (Watsonville, CA: Imaginova). Retrieved 2015-06-23.

[246] Hudgins, Douglas M.; Bauschlicher, Charles W., Jr.; Allamandola, Louis J. (10 October 2005). "Variations in the Peak Position of the 6.2 µm Interstellar Emission Feature: A Tracer of N in the Interstellar Polycyclic Aromatic Hydrocarbon Population" (PDF). *The Astrophysical Journal* (Bristol, England: IOP Publishing for the American Astronomical Society) **632** (1): 316–332. Bibcode:2005ApJ...632..316H. doi:10.1086/432495. ISSN 0004-637X.

[247] Des Marais, David J.; Allamandola, Louis J.; Sandford, Scott; et al. (2009). "Cosmic Distribution of Chemical Complexity". *Ames Research Center*. Mountain View, CA: NASA. Retrieved 2015-06-24. See the Ames Research Center 2009 annual team report to the NASA Astrobiology Institute here .

[248] García-Hernández, Domingo. A.; Manchado, Arturo; García-Lario, Pedro; et al. (20 November 2010). "Formation of Fullerenes in H-Containing Planetary Nebulae". *The Astrophysical Journal Letters* (Bristol, England: IOP Publishing for the American Astronomical Society) **724** (1): L39–L43. arXiv:1009.4357. Bibcode:2010ApJ...724L..39G. doi:10.1088/2041-8205/724/1/L39. ISSN 2041-8213.

[249] Atkinson, Nancy (27 October 2010). "Buckyballs Could Be Plentiful in the Universe". *Universe Today* (Courtenay, British Columbia: Fraser Cain). Retrieved 2015-06-24.

[250] Marlaire, Ruth, ed. (3 March 2015). "NASA Ames Reproduces the Building Blocks of Life in Laboratory". *Ames Research Center*. Moffett Field, CA: NASA. Retrieved 2015-03-05.

[251] Lancet, Doron (30 December 2014). "Systems Prebiology-Studies of the origin of Life". *The Lancet Lab*. Rehovot,

Israel: Department of Molecular Genetics; Weizmann Institute of Science. Retrieved 2015-06-26.

[252] Segré, Daniel; Ben-Eli, Dafna; Deamer, David W.; Lancet, Doron (February 2001). "The Lipid World" (PDF). *Origins of Life and Evolution of the Biosphere* (Kluwer Academic Publishers) **31** (1–2): 119–145. doi:10.1023/A:1006746807104. ISSN 0169-6149. PMID 11296516. Retrieved 2008-09-11.

[253] Eigen, Manfred; Schuster, Peter (November 1977). "The Hypercycle. A Principle of Natural Self-Organization. Part A: Emergence of the Hypercycle" (PDF). *Naturwissenschaften* (Berlin: Springer-Verlag) **64** (11): 541–565. Bibcode:1977NW.....64..541E. doi:10.1007/bf00450633. ISSN 0028-1042. PMID 593400. Retrieved 2015-06-13.

- Eigen, Manfred; Schuster, Peter (1978). "The Hypercycle. A Principle of Natural Self-Organization. Part B: The Abstract Hypercycle" (PDF). *Naturwissenschaften* (Berlin: Springer-Verlag) **65**: 7–41. Bibcode:1978NW.....65....7E. doi:10.1007/bf00420631. ISSN 0028-1042. Retrieved 2015-06-13.

- Eigen, Manfred; Schuster, Peter (July 1978). "The Hypercycle. A Principle of Natural Self-Organization. Part C: The Realistic Hypercycle" (PDF). *Naturwissenschaften* (Berlin: Springer-Verlag) **65** (7): 341–369. Bibcode:1978NW.....65..341E. doi:10.1007/bf00439699. ISSN 0028-1042. Retrieved 2015-06-13.

[254] Markovitch, Omer; Lancet, Doron (Summer 2012). "Excess Mutual Catalysis Is Required for Effective Evolvability" (PDF). *Artificial Life* (Cambridge, MA: MIT Press) **18** (3): 243–266. doi:10.1162/artl_a_00064. ISSN 1064-5462. PMID 22662913. Retrieved 2015-06-26.

[255] Tessera, Marc (2011). "Origin of Evolution *versus* Origin of Life: A Shift of Paradigm". *International Journal of Molecular Sciences* (Basel, Switzerland: MDPI) **12** (6): 3445–3458. doi:10.3390/ijms12063445. ISSN 1422-0067. PMC 3131571. PMID 21747687. Special Issue: "Origin of Life 2011"

[256] Brown, Michael R. W.; Kornberg, Arthur (16 November 2004). "Inorganic polyphosphate in the origin and survival of species". *Proc. Natl. Acad. Sci. U.S.A.* (Washington, D.C.: National Academy of Sciences) **101** (46): 16085–16087. Bibcode:2004PNAS..10116085B. doi:10.1073/pnas.0406909101. ISSN 0027-8424. PMC 528972. PMID 15520374.

[257] Clark, David P. (3 August 1999). "The Origin of Life". *Microbiology 425: Biochemistry and Physiology of Microorganism* (Lecture). Carbondale, IL: College of Science; Southern Illinois University Carbondale. Archived from the original on 2000-10-02. Retrieved 2015-06-26.

[258] Pasek, Matthew A. (22 January 2008). "Rethinking early Earth phosphorus geochemistry". *Proc. Natl. Acad. Sci. U.S.A.* (Washington, D.C.: National Academy of Sciences) **105** (3): 853–858. Bibcode:2008PNAS..105..853P. doi:10.1073/pnas.0708205105. ISSN 0027-8424. PMC 2242691. PMID 18195373.

[259] Witt, Adolf N.; Vijh, Uma P.; Gordon, Karl D. (2003). "Discovery of Blue Fluorescence by Polycyclic Aromatic Hydrocarbon Molecules in the Red Rectangle". *Bulletin of the American Astronomical Society* (Washington, D.C.: American Astronomical Society) **35**: 1381. Bibcode:2003AAS...20311017W. Archived from the original on 2003-12-19. Retrieved 2015-06-26. American Astronomical Society Meeting 203, #110.17, January 2004.

[260] "NASA Cooks Up Icy Organics to Mimic Life's Origins". *Space.com*. Ogden, UT: Purch. 20 September 2012. Retrieved 2015-06-26.

[261] Gudipati, Murthy S.; Rui Yang (1 September 2012). "In-situ Probing of Radiation-induced Processing of Organics in Astrophysical Ice Analogs—Novel Laser Desorption Laser Ionization Time-of-flight Mass Spectroscopic Studies". *The Astrophysical Journal Letters* (Bristol, England: IOP Publishing for the American Astronomical Society) **756** (1): L24. Bibcode:2012ApJ...756L..24G. doi:10.1088/2041-8205/756/1/L24. ISSN 2041-8213. L24.

[262] "NASA Ames PAH IR Spectroscopic Database". NASA. Retrieved 2015-06-17.

[263] Dartnell, Lewis (12 January 2008). "Did life begin on a radioactive beach?". *New Scientist* (London: Reed Business Information) (2638): 8. ISSN 0262-4079. Retrieved 2015-06-26.

[264] Adam, Zachary (2007). "Actinides and Life's Origins". *Astrobiology* (New Rochelle, NY: Mary Ann Liebert, Inc.) **7** (6): 852–872. Bibcode:2007AsBio...7..852A. doi:10.1089/ast.2006.0066. ISSN 1531-1074. PMID 18163867.

[265] Parnell, John (December 2004). "Mineral Radioactivity in Sands as a Mechanism for Fixation of Organic Carbon on the Early Earth". *Origins of Life and Evolution of Biospheres* (Kluwer Academic Publishers) **34** (6): 533–547. Bibcode:2004OLEB...34..533P. doi:10.1023/B:ORIG.0000043132.23966.a1. ISSN 0169-6149. PMID 15570707.

[266] Michaelian, Karo (30 June 2009). "Thermodynamic Function of Life". arXiv:0907.0040 [physics.gen-ph].

[267] Michaelian, Karo (25 January 2011). "Biological catalysis of the hydrological cycle: life's thermodynamic function". *Hydrology and Earth System Sciences Discussions* (Göttingen, Germany: Copernicus Publications on behalf of the European Geosciences Union) **8**: 1093–1123. Bibcode:2011HESSD...8.1093M. doi:10.5194/hessd-8-1093-2011. ISSN 1812-2116.

[268] Michaelian, Karo (11 March 2011). "Thermodynamic Dissipation Theory for the Origin of Life" (PDF). *Earth System Dynamics* (Göttingen, Germany: Copernicus Publications on behalf of the European Geosciences Union) **2**: 37–51. arXiv:0907.0042. Bibcode:2011ESD.....2...37M. doi:10.5194/esd-2-37-2011. ISSN 2190-4987. Retrieved 2015-06-28.

[269] Cnossen, Ingrid; Sanz-Forcada, Jorge; Favata, Fabio; et al. (February 2007). "Habitat of early life: Solar X-ray and UV radiation at Earth's surface 4–3.5 billion years ago". *Journal of Geophysical Research* (Washington, D.C.: American Geophysical Union) **112** (E2): E02008. arXiv:astro-ph/0702529. Bibcode:2007JGRE..112.2008C. doi:10.1029/2006JE002784. ISSN 0148-0227.

[270] Sagan, Carl (April 1973). "Ultraviolet Selection Pressure on the Earliest Organisms". *Journal of Theoretical Biology* (Amsterdam, the Netherlands: Elsevier) **39** (1): 195–200. doi:10.1016/0022-5193(73)90216-6. ISSN 0022-5193. PMID 4741712.

[271] Michaelian, Karo; Simeonov, Aleksander (19 August 2015). "Fundamental molecules of life are pigments which arose and co-evolved as a response to the thermodynamic imperative of dissipating the prevailing solar spectrum". *Biogeosciences* **12**: 4913–4937. doi:10.5194/bg-12-4913-2015.

[272] Michaelian, Karo; Simeonov, Aleksandar (16 May 2014). "Fundamental Molecules of Life are Pigments which Arose and Evolved to Dissipate the Solar Spectrum". arXiv:1405.4059 [physics.bio-ph].

[273] Michaelian, Karo (2013). "A non-linear irreversible thermodynamic perspective on organic pigment proliferation and biological evolution" (PDF). *Journal of Physics: Conference Series* (Bristol, England: IOP Publishing) **475** (conference 1): 012010. arXiv:1307.5924. Bibcode:2013JPhCS.475a2010M. doi:10.1088/1742-6596/475/1/012010. ISSN 1742-6596. "4th National Meeting in Chaos, Complex System and Time Series 29 November to 2 December 2011, Xalapa, Veracruz, Mexico"

[274] Knauth 1992, pp. 123–152

[275] Knauth, L. Paul; Lowe, Donald R. (May 2003). "High Archean climatic temperature inferred from oxygen isotope geochemistry of cherts in the 3.5 Ga Swaziland group, South Africa". *Geological Society of America Bulletin* (Boulder, CO: Geological Society of America) **115**: 566–580. Bibcode:2003GSAB..115..566K. doi:10.1130/0016-7606(2003)115<0566:hactif>2.0.co;2. ISSN 0016-7606.

[276] Lowe, Donald R.; Tice, Michael M. (June 2004). "Geologic evidence for Archean atmospheric and climatic evolution: Fluctuating levels of $CO_2$, $CH_4$, and $O_2$ with an overriding tectonic control". *Geology* (Boulder, CO: Geological Society of America) **32** (6): 493–496. Bibcode:2004Geo....32..493L. doi:10.1130/G20342.1. ISSN 0091-7613.

[277] Michaelian, Karo; Santillán Padilla, Norberto (24 November 2014). "DNA Denaturing through UV-C Photon Dissipation: A Possible Route to Archean Non-enzymatic Replication" (PDF). *bioRxiv* (Cold Spring Harbor, NY: Cold Spring Harbor Laboratory). doi:10.1101/009126. Retrieved 2015-06-29.

[278] Davies, Paul (December 2007). "Are Aliens Among Us?" (PDF). *Scientific American* (Stuttgart: Georg von Holtzbrinck Publishing Group) **297** (6): 62–69. doi:10.1038/scientificamerican1207-62. ISSN 0036-8733. Retrieved 2015-07-16. ...if life does emerge readily under terrestrial conditions, then perhaps it formed many times on our home planet. To pursue this possibility, deserts, lakes and other extreme or isolated environments have been searched for evidence of "alien" life-forms—organisms that would differ fundamentally from known organisms because they arose independently.

[279] Hartman, Hyman (October 1998). "Photosynthesis and the Origin of Life". *Origins of Life and Evolution of Biospheres* (Kluwer Academic Publishers) **28** (4–6): 515–521. Bibcode:1998OLEB...28..515H. doi:10.1023/A:1006548904157. ISSN 0169-6149. PMID 11536891.

[280] Damer, Bruce; Deamer, David (13 March 2015). "Coupled Phases and Combinatorial Selection in Fluctuating Hydrothermal Pools: A Scenario to Guide Experimental Approaches to the Origin of Cellular Life". *Life* (Basel, Switzerland: MDPI) **5** (1): 872–887. doi:10.3390/life5010872. ISSN 2075-1729. PMC 4390883. PMID 25780958.

## 1.11 Bibliography

- Altermann, Wladyslaw (2009). "From Fossils to Astrobiology – A Roadmap to Fata Morgana?" (PDF). In Seckbach, Joseph; Walsh, Maud. *From Fossils to Astrobiology: Records of Life on Earth and the Search for Extraterrestrial Biosignatures*. Cellular Origin, Life in Extreme Habitats and Astrobiology **12**. Dordrecht, the Netherlands; London: Springer Science+Business Media. ISBN 978-1-4020-8836-0. LCCN 2008933212. Retrieved 2015-06-05.

- Bada, Jeffrey L.; Lazcano, Antonio (2009). "The Origin of Life". In Ruse, Michael; Travis, Joseph. *Evolution: The First Four Billion Years*. Foreword by Edward O. Wilson. Cambridge, MA: Belknap Press of Harvard University Press. ISBN 978-0-674-03175-3. LCCN 2008030270. OCLC 225874308.

- Barton, Nicholas H.; Briggs, Derek E. G.; Eisen, Jonathan A.; et al. (2007). *Evolution*. Cold Spring Harbor, NY: Cold Spring Harbor Laboratory Press.

ISBN 978-0-87969-684-9. LCCN 2007010767. OCLC 86090399.

- Bastian, H. Charlton (1871). *The Modes of Origin of Lowest Organisms*. London; New York: Macmillan and Company. LCCN 11004276. OCLC 42959303. Retrieved 2015-06-06.

- Bernal, J. D. (1951). *The Physical Basis of Life*. London: Routledge & Kegan Paul. LCCN 51005794.

- Bernal, J. D. (1960). "The Problem of Stages in Biopoesis". In Florkin, M. *Aspects of the Origin of Life*. International Series of Monographs on Pure and Applied Biology. Oxford, UK; New York: Pergamon Press. ISBN 978-1-4831-3587-8. LCCN 60013823.

- Bernal, J. D. (1967) [Reprinted work by A. I. Oparin originally published 1924; Moscow: The Moscow Worker]. *The Origin of Life*. The Weidenfeld and Nicolson Natural History. Translation of Oparin by Ann Synge. London: Weidenfeld & Nicolson. LCCN 67098482.

- Bock, Gregory R.; Goode, Jamie A., eds. (1996). *Evolution of Hydrothermal Ecosystems on Earth (and Mars?)*. Ciba Foundation Symposium **202**. Chichester, UK; New York: John Wiley & Sons. ISBN 0-471-96509-X. LCCN 96031351.

- Bondeson, Jan (1999). *The Feejee Mermaid and Other Essays in Natural and Unnatural History*. Ithaca, NY: Cornell University Press. ISBN 0-8014-3609-5. LCCN 98038295.

- Bryson, Bill (2004). *A Short History of Nearly Everything*. London: Black Swan. ISBN 978-0-552-99704-1. OCLC 55589795.

- Calvin, Melvin (1969). *Chemical Evolution: Molecular Evolution Towards the Origin of Living Systems on the Earth and Elsewhere*. Oxford, UK: Clarendon Press. ISBN 0-19-855342-0. LCCN 70415289. OCLC 25220.

- Chaichian, Masud; Rojas, Hugo Perez; Tureanu, Anca (2014). "Physics and Life". *Basic Concepts in Physics: From the Cosmos to Quarks*. Undergraduate Lecture Notes in Physics. Berlin; Heidelberg: Springer Berlin Heidelberg. doi:10.1007/978-3-642-19598-3_12. ISBN 978-3-642-19597-6. ISSN 2192-4791. LCCN 2013950482. OCLC 900189038.

- Chang, Thomas Ming Swi (2007). *Artificial Cells: Biotechnology, Nanomedicine, Regenerative Medicine, Blood Substitutes, Bioencapsulation, and Cell/Stem Cell Therapy*. Regenerative Medicine, Artificial Cells and Nanomedicine **1**. Hackensack, NJ: World Scientific. ISBN 978-981-270-576-1. LCCN 2007013738. OCLC 173522612.

- Clancy, Paul; Brack, André; Horneck, Gerda (2005). *Looking for Life, Searching the Solar System*. Cambridge, UK: Cambridge University Press. ISBN 978-0-521-82450-7. LCCN 2006271630. OCLC 57574490.

- Dalrymple, G. Brent (2001). "The age of the Earth in the twentieth century: a problem (mostly) solved". In Lewis, C. L. E.; Knell, S. J. *The Age of the Earth: from 4004 BC to AD 2002*. Geological Society Special Publication **190**. London: Geological Society of London. Bibcode:2001GSLSP.190..205D. doi:10.1144/gsl.sp.2001.190.01.14. ISBN 1-86239-093-2. ISSN 0305-8719. LCCN 2003464816. OCLC 48570033.

- Darwin, Charles (1887). Darwin, Francis, ed. *The Life and Letters of Charles Darwin, Including an Autobiographical Chapter* **3** (3rd ed.). London: John Murray. OCLC 834491774.

- Davies, Geoffrey F. (2007). "Chapter 2.3 Dynamics of the Hadean and Archaean Mantle". In van Kranendonk, Martin J.; Smithies, R. Hugh; Bennett, Vickie C. *Earth's Oldest Rocks*. Developments in Precambrian Geology **15**. Amsterdam, the Netherlands; Boston: Elsevier. doi:10.1016/S0166-2635(07)15023-4. ISBN 978-0-444-52810-0. LCCN 2009525003.

- Davies, Paul (1999). *The Fifth Miracle: The Search for the Origin of Life*. London: Penguin Books. ISBN 0-14-028226-2.

- Dawkins, Richard (1996). *The Blind Watchmaker* (Reissue with a new introduction ed.). New York: W. W. Norton & Company. ISBN 0-393-31570-3. LCCN 96229669. OCLC 35648431.

- Dawkins, Richard (2004). *The Ancestor's Tale: A Pilgrimage to the Dawn of Evolution*. Boston, MA: Houghton Mifflin. ISBN 0-618-00583-8. LCCN 2004059864. OCLC 56617123.

- Dobell, Clifford (1960) [Originally published 1932; New York: Harcourt, Brace & Company]. *Antony van Leeuwenhoek and His 'Little Animals'*. New York: Dover Publications. LCCN 60002548.

- Dyson, Freeman (1999). *Origins of Life* (Revised ed.). Cambridge, UK; New York: Cambridge University Press. ISBN 0-521-62668-4. LCCN 99021079.

## 1.11. BIBLIOGRAPHY

- Eigen, M.; Schuster, P. (1979). *The Hypercycle: A Principle of Natural Self-Organization*. Berlin; New York: Springer-Verlag. ISBN 0-387-09293-5. LCCN 79001315. OCLC 4665354.

- Fesenkov, V. G. (1959). "Some Considerations about the Primaeval State of the Earth". In Oparin, A. I.; et al. *The Origin of Life on the Earth*. I.U.B. Symposium Series **1**. Edited for the International Union of Biochemistry by Frank Clark and R. L. M. Synge (English-French-German ed.). London; New York: Pergamon Press. ISBN 978-1-4832-2240-0. LCCN 59012060. Retrieved 2015-06-03. International Symposium on the Origin of Life on the Earth (held at Moscow, 19–24 August 1957)

- Hazen, Robert M. (2005). *Genesis: The Scientific Quest for Life's Origin*. Washington, D.C.: Joseph Henry Press. ISBN 0-309-09432-1. LCCN 2005012839. OCLC 60321860.

- Huxley, Thomas Henry (1968) [Originally published 1897]. "VIII Biogenesis and Abiogenesis [1870]". *Discourses, Biological and Geological*. Collected Essays **VIII** (Reprint ed.). New York: Greenwood Press. LCCN 70029958. Retrieved 2014-05-19.

- Kauffman, Stuart (1993). *The Origins of Order: Self-Organization and Selection in Evolution*. New York: Oxford University Press. ISBN 978-0-19-507951-7. LCCN 91011148. OCLC 23253930.

- Kauffman, Stuart (1995). *At Home in the Universe: The Search for Laws of Self-Organization and Complexity*. New York: Oxford University Press. ISBN 0-19-509599-5. LCCN 94025268.

- Klyce, Brig (22 January 2001). Kingsley, Stuart A.; Bhathal, Ragbir, eds. *Panspermia Asks New Questions*. The Search for Extraterrestrial Intelligence (SETI) in the Optical Spectrum III. Bellingham, WA: SPIE. doi:10.1117/12.435366. ISBN 0-8194-3951-7. LCCN 2001279159. Retrieved 2015-06-09. Proceedings of the SPIE held at San Jose, CA, 22–24 January 2001

- Knauth, L. Paul (1992). "Origin and diagenesis of cherts: An isotopic perspective". In Clauer, Norbert; Chaudhuri, Sambhu. *Isotopic Signatures and Sedimentary Records*. Lecture Notes in Earth Sciences **43**. Berlin; New York: Springer-Verlag. doi:10.1007/BFb0009863. ISBN 3-540-55828-4. ISSN 0930-0317. LCCN 92025372. OCLC 26262469.

- Lane, Nick (2009). *Life Ascending: The 10 Great Inventions of Evolution* (1st American ed.). New York: W. W. Norton & Company. ISBN 978-0-393-06596-1. LCCN 2009005046. OCLC 286488326.

- Lankenau, Dirk-Henner (2011). "Two RNA Worlds: Toward the Origin of Replication, Genes, Recombination and Repair". In Egel, Richard; Lankenau, Dirk-Henner; Mulkidjanian,, Armen Y. *Origins of Life: The Primal Self-Organization*. Heidelberg: Springer. doi:10.1007/978-3-642-21625-1. ISBN 978-3-642-21624-4. LCCN 2011935879. OCLC 733245537.

- Lennox, James G. (2001). *Aristotle's Philosophy of Biology: Studies in the Origins of Life Science*. Cambridge Studies in Philosophy and Biology. Cambridge, UK; New York: Cambridge University Press. ISBN 0-521-65976-0. LCCN 00026070.

- McKinney, Michael L. (1997). "How do rare species avoid extinction? A paleontological view". In Kunin, William E.; Gaston, Kevin J. *The Biology of Rarity: Causes and consequences of rare—common differences* (1st ed.). London; New York: Chapman & Hall. ISBN 0-412-63380-9. LCCN 96071014. OCLC 36442106.

- Michod, Richard E. (1999). "Darwinian Dynamics: Evolutionary Transitions in Fitness and Individuality". Princeton, NJ: Princeton University Press. ISBN 0-691-02699-8. LCCN 98004166. OCLC 38948118.

- Miller, G. Tyler; Spoolman, Scott E. (2012). *Environmental Science* (14th ed.). Belmont, CA: Brooks/Cole. ISBN 978-1-111-98893-7. LCCN 2011934330. OCLC 741539226.

- Oparin, A. I. (1953) [Originally published 1938; New York: The Macmillan Company]. *The Origin of Life*. Translation and new introduction by Sergius Morgulis (2nd ed.). Mineola, NY: Dover Publications. ISBN 0-486-49522-1. LCCN 53010161.

- Orgel, Leslie E. (1987). "Evolution of the Genetic Apparatus: A Review". *Evolution of Catalytic Function*. Cold Spring Harbor Symposia on Quantitative Biology **52**. Cold Spring Harbor, NY: Cold Spring Harbor Laboratory Press. doi:10.1101/SQB.1987.052.01.004. ISBN 0-87969-054-2. OCLC 19850881. "Proceedings of a symposium held at Cold Spring Harbor Laboratory in 1987"

- Raven, Peter H.; Johnson, George B. (2002). *Biology* (6th ed.). Boston, MA: McGraw-Hill. ISBN 0-07-112261-3. LCCN 2001030052. OCLC 45806501.

- Ross, Alexander (1652). *Arcana Microcosmi*. Book II. London. Retrieved 2015-07-07.

- Shapiro, Robert (1987). *Origins: A Skeptic's Guide to the Creation of Life on Earth*. Toronto; New York: Bantam Books. ISBN 0-553-34355-6.

- Sheldon, Robert B. (22 September 2005). Hoover, Richard B.; Levin, Gilbert V.; Rozanov, Alexei Y.; Gladstone, G. Randall, eds. *Historical Development of the Distinction between Bio- and Abiogenesis* (PDF). Astrobiology and Planetary Missions. Bellingham, WA: SPIE. doi:10.1117/12.663480. ISBN 978-0-8194-5911-4. LCCN 2005284378. Retrieved 2015-04-13. Proceedings of the SPIE held at San Diego, CA, 31 July–2 August 2005

- Stearns, Beverly Peterson; Stearns, Stephen C. (1999). *Watching, from the Edge of Extinction*. New Haven, CT: Yale University Press. ISBN 0-300-07606-1. LCCN 98034087. OCLC 47011675.

- Tyndall, John (1905) [Originally published 1871; London; New York: Longmans, Green & Co.; D. Appleton and Company]. *Fragments of Science* **2** (6th ed.). New York: P.F. Collier & Sons. OCLC 726998155. Retrieved 2015-06-06.

- Vartanian, Aram (1973). "Spontaneous Generation". In Wiener, Philip P. *Dictionary of the History of Ideas* **IV**. New York: Charles Scribner's Sons. ISBN 0-684-13293-1. LCCN 72007943. Retrieved 2015-06-05.

- Voet, Donald; Voet, Judith G. (2004). *Biochemistry* **1** (3rd ed.). New York: John Wiley & Sons. ISBN 0-471-19350-X. LCCN 2003269978.

- Woodward, Robert J., ed. (1969). *Our Amazing World of Nature: Its Marvels & Mysteries*. Pleasantville, NY: Reader's Digest Association. ISBN 0-340-13000-8. LCCN 69010418.

- Yarus, Michael (2010). *Life from an RNA World: The Ancestor Within*. Cambridge, MA: Harvard University Press. ISBN 978-0-674-05075-4. LCCN 2009044011.

## 1.12 Further reading

- Arrhenius, Gustaf O.; Sales, Brian C.; Mojzsis, Stephen J.; et al. (21 August 1997). "Entropy and Charge in Molecular Evolution—the Case of Phosphate" (PDF). *Journal of Theoretical Biology* (Amsterdam, the Netherlands: Elsevier) **187** (4): 503–522. doi:10.1006/jtbi.1996.0385. ISSN 0022-5193. PMID 9299295.

- Cavalier-Smith, Thomas (June 2006). "Cell evolution and Earth history: stasis and revolution". *Philosophical Transactions of the Royal Society B* (London: Royal Society) **361** (1470): 969–1006. doi:10.1098/rstb.2006.1842. ISSN 0962-8436. PMC 1578732. PMID 16754610.

- de Duve, Christian (1995). *Vital Dust: Life As A Cosmic Imperative* (1st ed.). New York: Basic Books. ISBN 0-465-09044-3. LCCN 94012964. OCLC 30624716.

- Fernando, Chrisantha T.; Rowe, Jonathan (7 July 2007). "Natural selection in chemical evolution". *Journal of Theoretical Biology* (Amsterdam, the Netherlands) **247** (1): 152–167. doi:10.1016/j.jtbi.2007.01.028. ISSN 0022-5193. PMID 17399743.

- Gribbin, John (1998). *The Case of the Missing Neutrinos: And other Curious Phenomena of the Universe* (1st Fromm International ed.). New York: Fromm International. ISBN 0-88064-199-1. LCCN 98027948. OCLC 39368356.

- Harris, Henry (2002). *Things Come to Life: Spontaneous Generation Revisited*. Oxford, UK; New York: Oxford University Press. ISBN 0-19-851538-3. LCCN 2001054856. OCLC 48100507.

- Horgan, John (February 1991). "In the Beginning...". *Scientific American* (Stuttgart: Georg von Holtzbrinck Publishing Group) **264** (2): 116–125. doi:10.1038/scientificamerican0291-116. ISSN 0036-8733.

- Ignatov, Ignat; Mosin, Oleg V. (2013). "Modeling of Possible Processes for Origin of Life and Living Matter in Hot Mineral and Seawater with Deuterium". *Journal of Environment and Earth Science* (New York: International Institute for Science, Technology and Education) **3** (14): 103–118. ISSN 2224-3216. Retrieved 2015-06-29.

- Jortner, Joshua (October 2006). "Conditions for the emergence of life on the early Earth: summary and reflections". *Philosophical Transactions of the Royal Society B* (London: Royal Society) **361** (1474): 1877–1891. doi:10.1098/rstb.2006.1909. ISSN 0962-8436. PMC 1664691. PMID 17008225.

- Klotz, Irene (24 February 2012). "Did Life Start in a Pond, Not Oceans?". *Discovery News* (Silver Spring, MD: Discovery Communications). Retrieved 2015-06-29.

- Knoll, Andrew H. (2003). *Life on a Young Planet: The First Three Billion Years of Evolution on Earth*. Princeton, NJ: Princeton University Press. ISBN 0-691-00978-3. LCCN 2002035484. OCLC 50604948.

- Luisi, Pier Luigi (2006). *The Emergence of Life: From Chemical Origins to Synthetic Biology*. Cambridge, UK: Cambridge University Press. ISBN 978-0-521-82117-9. LCCN 2006285720. OCLC 173609999.

- Maynard Smith, John; Szathmáry, Eörs (1999). *The Origins of Life: From the Birth of Life to the Origin of Language*. Oxford, UK; New York: Oxford University Press. ISBN 0-19-850493-4. LCCN 99230990. OCLC 40980149.

- Morowitz, Harold J. (1992). *Beginnings of Cellular Life: Metabolism Recapitulates Biogenesis*. New Haven, CT: Yale University Press. ISBN 0-300-05483-1. LCCN 92006849. OCLC 25316379.

- NASA Astrobiology Institute: Harrison, T. Mark; McKeegan, Kevin D.; Mojzsis, Stephen J. "Earth's Early Environment and Life: When did Earth become suitable for habitation?". Archived from the original on 2012-02-17. Retrieved 2015-06-30.

- NASA Specialized Center of Research and Training in Exobiology: Arrhenius, Gustaf O. (11 September 2002). "Arrhenius". Archived from the original on 2007-12-21. Retrieved 2015-06-30.

- "The physico-chemical basis of life". *What is Life*. Spring Valley, CA: Lukas K. Buehler. Retrieved 27 October 2005.

- Pitsch, Stefan; Krishnamurthy, Ramanarayanan; Arrhenius, Gustaf O. (6 September 2000). "Concentration of Simple Aldehydes by Sulfite-Containing Double-Layer Hydroxide Minerals: Implications for Biopoesis". *Helvetica Chimica Acta* (Hoboken, NJ: John Wiley & Sons) **83** (9): 2398–2411. doi:10.1002/1522-2675(20000906)83:9<2398::AID-HLCA2398>3.0.CO;2-5. ISSN 0018-019X. PMID 11543578.

- Pons, Marie-Laure; Quitté, Ghylaine; Fujii, Toshiyuki; et al. (25 October 2011). "Early Archean Serpentine Mud Volcanoes at Isua, Greenland, as a Niche for Early Life". *Proc. Natl. Acad. Sci. U.S.A.* (Washington, D.C.: National Academy of Sciences) **108** (43): 17639–17643. Bibcode:2011PNAS..10817639P. doi:10.1073/pnas.1108061108. ISSN 0027-8424. PMC 3203773. PMID 22006301.

- Pross, Addy (2012). *What is Life?: How Chemistry Becomes Biology* (1st ed.). Oxford, UK: Oxford University Press. ISBN 978-0-19-964101-7. LCCN 2012538842. OCLC 812020290.

- Roy, Debjani; Schleyer, Paul von Ragué (2010). "Chemical Origin of Life: How do Five HCN Molecules Combine to form Adenine under Prebiotic and Interstellar Conditions". In Matta, Chérif F. *Quantum Biochemistry*. Weinheim, Germany: Wiley-VCH. doi:10.1002/9783527629213.ch6. ISBN 978-3-527-62921-3. LCCN 2011499476. OCLC 905973537.

- Russell, Michael J.; Hall, A. J.; Cairns-Smith, Alexander Graham; et al. (10 November 1988). "Submarine hot springs and the origin of life". *Nature* (London: Nature Publishing Group) **336** (6195): 117. Bibcode:1988Natur.336..117R. doi:10.1038/336117a0. ISSN 0028-0836. PMID 11536607.

- Shock, Everett L. (25 October 1997). "High-temperature life without photosynthesis as a model for Mars" (PDF). *Journal of Geophysical Research* (Washington, D.C.: American Geophysical Union) **102** (E10): 23687–23694. Bibcode:1997JGR...10223687S. doi:10.1029/97je01087. ISSN 0148-0227.

## 1.13 External links

- "Exploring Life's Origins: A Virtual Exhibit". *Exploring Life's Origins: A Virtual Exhibit*. Arlington County, VA: National Science Foundation. Retrieved 2015-07-02.

- Fields, Helen (October 2010). "The Origins of Life". *Smithsonian* (Washington, D.C.: Smithsonian Institution). ISSN 0037-7333. Retrieved 2015-07-02.

- Fox, Douglas (28 March 2007). "Primordial Soup's On: Scientists Repeat Evolution's Most Famous Experiment". *Scientific American* (Stuttgart: Georg von Holtzbrinck Publishing Group). ISSN 0036-8733. Retrieved 2015-07-02.

- "The Geochemical Origins of Life by Michael J. Russell & Allan J. Hall". Glasgow, Scotland: University of Glasgow. 13 December 2008. Retrieved 2015-07-02.

- Kauffman, Stuart (8 August 1996). "Even peptides do it". *Nature* (London: Nature Publishing Group) **382** (6591): 496–497. Bibcode:1996Natur.382..496K.

- doi:10.1038/382496a0. ISSN 0028-0836. PMID 8700218. Archived from the original on 2006-10-15. Retrieved 2015-07-02.

- Malory, Marcia. "How life began on Earth". *Earth Facts*. Retrieved 2015-07-02.

- Nowak, Martin A.; Ohtsuki, Hisashi (30 September 2008). "Prevolutionary dynamics and the origin of evolution" (PDF). *Proc. Natl. Acad. Sci. U.S.A.* (Washington, D.C.: National Academy of Sciences) **105** (39): 14924–14927. Bibcode:2008PNAS..10514924N. doi:10.1073/pnas.0806714105. ISSN 0027-8424. PMC 2567469. PMID 18791073.

- "Possible Connections Between Interstellar Chemistry and the Origin of Life on the Earth". *Space Science and Astrobiology at Ames*. NASA. Archived from the original on 2009-07-31. Retrieved 2015-07-02.

- "Research Spotlight: Jack Szostak: Making Life from Scratch". *Origins of Life Initiative*. Cambridge, MA: Harvard University. Retrieved 2015-07-02.

- Schirber, Michael (9 June 2006). "How Life Began: New Research Suggests Simple Approach". *LiveScience* (Ogden, UT: Purch). Retrieved 2015-07-02.

- "Scientists Find Clues That Life Began in Deep Space". *NASA Astrobiology Institute*. Mountain View, CA: NASA. 30 January 2001. Archived from the original on 2013-04-29. Retrieved 2015-07-02.

- "Simple Artificial Cell Created From Scratch To Study Cell Complexity". *Science Daily* (Rockville, MD: ScienceDaily, LLC). 16 May 2008. Retrieved 2015-07-02. Post is reprinted from materials provided by Pennsylvania State University.

- Singer, Emily (19 July 2015). "Chemists Invent New Letters for Nature's Genetic Alphabet". *Wired*. New York: Condé Nast. Retrieved 2015-07-20.

- Swaminathan, Nikhil (10 June 2008). "Scientists Close to Reconstructing First Living Cell". *Scientific American* (News) (Stuttgart: Georg von Holtzbrinck Publishing Group). ISSN 0036-8733. Retrieved 2015-07-02.

- Vasas, Vera; Fernando, Chrisantha; Santos, Mauro; et al. (5 January 2012). "Evolution before genes" (PDF). *Biology Direct* (London: BioMed Central) **7**: 1. doi:10.1186/1745-6150-7-1. ISSN 1745-6150.

- Zlobin, Andrei E. (2013). "Tunguska similar impacts and origin of life". *Modern Scientific Researches and Innovations* (Moscow: International Centre of Science and Innovations Ltd.) (12). Retrieved 2015-07-02.

- Zlobin, Andrei E. (2014). "Symmetry infringement in mathematical metrics of hydrogen atom as illustration of ideas by V.I.Vernadsky concerning origin of life and biosphere" (PDF). *Acta Naturae* (Moscow: Park Media Ltd.) (Special Issue 1): 48. ISSN 2075-8251. Retrieved 2015-07-02.

### 1.13.1 Video resources

- Hazen, Robert M. (29 April 2014). *The Origins of Life* (Webcast). Baltimore, MD: Space Telescope Science Institute. Retrieved 2015-07-03. — A 2014 Spring Symposium webcast (video; 38 m)

- "The Origin of Life" on YouTube — A Royal Institution Discourse lecture given by John Maynard Smith in 1995 (video; 58 m)

- "Space Experts Discuss the Search for Life in the Universe at NASA" on YouTube — Panel discussion at NASA headquarters on 14 July 2014 (video; 87 m)

# Chapter 2

# Spontaneous generation

This article is about historical theories on the ongoing emergence of life. For the origin of life, see Abiogenesis.

**Spontaneous generation** or **anomalous generation** is an obsolete body of thought on the ordinary formation of living organisms without descent from similar organisms. Typically, the idea was that certain forms such as fleas could arise from inanimate matter such as dust, or that maggots could arise from dead flesh. A variant idea was that of **equivocal generation**, in which species such as tapeworms arose from unrelated living organisms, now understood to be their hosts. Doctrines supporting such processes of generation held that these processes are commonplace and regular. Such ideas are in contradiction to that of **univocal generation**: effectively exclusive reproduction from genetically related parent(s), generally of the same species.

The doctrine of spontaneous generation was coherently synthesized by Aristotle,[1] who compiled and expanded the work of prior natural philosophers and the various ancient explanations of the appearance of organisms; it held sway for two millennia. Today it is generally accepted to have been decisively dispelled during the 19th century by the experiments of Louis Pasteur. He expanded upon the investigations of predecessors (such as Francesco Redi who, in the 17th century, had performed experiments based on the same principles). However, some experimental difficulties were still there and objections from persons holding the traditional views persisted. Many of these residual objections were dealt with by the work of John Tyndall, succeeding the work of Pasteur.[2]

Pasteur's experiment is generally known to have refuted the theory of spontaneous generation in 1859.[3] Disproof of the traditional ideas of spontaneous generation is no longer controversial among professional biologists. By the middle of the 19th century, the theory of biogenesis had accumulated so much evidential support, due to the work of Louis Pasteur and others, that the alternative theory of spontaneous generation had been effectively disproven. John Desmond Bernal suggests that earlier theories such as spontaneous generation were based upon an explanation that life was continuously created as a result of chance events.[4][5][6]

## 2.1 Description

Spontaneous generation refers both to the supposed processes in which different types of life might repeatedly emerge from specific sources other than seeds, eggs or parents, and also to the theoretical principles which were presented in support of any such phenomena. Crucial to this doctrine is the idea that life comes from non-life, with the conditions, and that no causal agent is needed (i.e. Parent). Such hypothetical processes sometimes are referred to as *abiogenesis*, in which life routinely emerges from non-living matter on a time scale of anything from minutes to weeks, or perhaps a season or so. An example would be the supposed seasonal generation of mice and other animals from the mud of the Nile.[7] Such ideas have no operative principles in common with the modern hypothesis of abiogenesis, in which life emerged in the early ages of the planet, over a time span of at least millions of years, and subsequently diversified without evidence that there ever has been any subsequent repetition of the event.

Another version of spontaneous generation is variously termed univocal generation, *heterogenesis* or *xenogenesis*, in which one form of life has been supposed to arise from a different form, such as tapeworms from the bodies of their hosts.[8]

In the years following Louis Pasteur's experiment in 1862, the term "spontaneous generation" fell into increasing disfavor. Experimentalists used a variety of terms for the study of the origin of life from non-living materials. Heterogenesis was applied to once-living materials such as boiled broths, and Henry Charlton Bastian proposed the term *archebiosis* for life originating from inorganic materials. The two were lumped together as "spontaneous generation", but disliking the term as sounding too random, Bastian proposed *biogenesis*. In an 1870 address titled, "Spontaneous Generation", Thomas Henry Huxley defined *biogenesis* as life originating from other life and coined the negative of the term, *abiogenesis*, which was the term that became dominant.[9]

## 2.2 Pre-Aristotelian philosophers

As part of his overall attempt to give natural explanations of things that had previously been ascribed to the agency of the gods, Anaximander believed that everything arose out of the elemental nature of the universe, which he called the "apeiron" or "unbounded". According to Hippolytus of Rome in the third century CE, Anaximander claimed that living creatures were first formed in the "wet" when acted on by the Sun, and that they were different then than they are now. For example, he claimed humans, in a different form, must have earlier been born mature like other animals, or they would not have survived. Anaximander also claimed that spontaneous generation continued to this day, with aquatic forms being produced directly from lifeless matter.[10]

Anaximenes, a pupil of Anaximander, thought that air was the element that imparted life, motion and thought, and speculated that there was a *primordial terrestrial slime*, a mixture of earth and water, which when combined with the sun's heat formed plants, animals and human beings directly.[10]

Xenophanes traced the origin of man back to the transitional period between the fluid stage of the earth and the formation of land. He too held to a spontaneous generation of fully formed plants and animals under the influence of the sun.[10]

Empedocles accepted the spontaneous generation of life, but held that there had to be trials of combinations of parts of animals that spontaneously arose. Successful combinations formed the species we now see, unsuccessful forms failed to reproduce.[10]

Anaxagoras also adopted a terrestrial slime account, although he thought that the seeds of plants existed in the air from the beginning, and of animals in the aether.[10]

## 2.3 Aristotle

Aristotle laid the foundations of Western natural philosophy. In his book, The History of Animals, he stated in no uncertain terms:

> Now there is one property that animals are found to have in common with plants. For some plants are generated from the seed of plants, whilst other plants are self-generated through the formation of some elemental principle similar to a seed; and of these latter plants some derive their nutriment from the ground, whilst others grow inside other plants, as is mentioned, by the way, in my treatise on Botany. So with animals, some spring from parent animals according to their kind, whilst others grow spontaneously and not from kindred stock; and of these instances of spontaneous generation some come from putrefying earth or vegetable matter, as is the case with a number of insects, while others are spontaneously generated in the inside of animals out of the secretions of their several organs.[11]
> — Aristotle, History of Animals, Book V, Part 1

According to this theory, living things came forth from nonliving things because the nonliving material contained *pneuma*, or "vital heat". The creature generated was dependent on the proportions of this pneuma and the five elements he believed comprised all matter.[10] While Aristotle recognized that many living things emerged from putrefying matter, he pointed out that the putrefaction was not the source of life, but the byproduct of the action of the "sweet" element of water.[12]

> Animals and plants come into being in earth and in liquid because there is water in earth, and air in water, and in all air is vital heat so that in a sense all things are full of soul. Therefore living things form quickly whenever this air and vital heat are enclosed in anything. When they are so enclosed, the corporeal liquids being heated, there arises as it were a frothy bubble.
> — Aristotle, On the Generation of Animals, Book III, Part 11

*Scallop*

Numerous forms were attributed to various sources. The testaceans (shelled molluscs) are characterized by forming by spontaneous generation in mud, but differ based upon the material they grow in — for example, clams and scallops in sand, oysters in slime, and the barnacle and the limpet in the hollows of rocks. Some reddish worms form from long-standing snow which has turned reddish. Another grub was said to grow out of fire.[11]

Concerning sexual reproduction, Aristotle argued that the male parent provided the "form," or soul, that guided development through semen, and the female parent contributed unorganized matter, allowing the embryo to grow.[13]

## 2.4 Classical writers after Aristotle

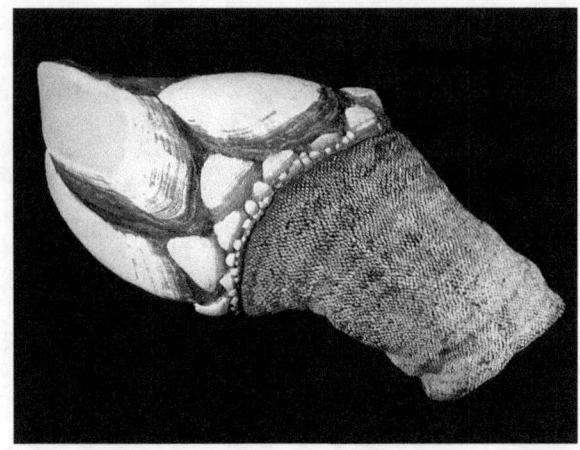

*The goose barnacle:*
Pollicipes cornucopia

Vitruvius, a Roman architect and writer of the 1st century BCE, advised that libraries be placed facing eastwards to benefit from morning light, but not towards the south or the west as those winds generate bookworms.[14]

Aristotle claimed that eels were lacking in sex and lacking milt, spawn and the passages for either.[15] Rather, he asserted eels emerged from earthworms.[16] Later philosophers dissented. Pliny the Elder did not argue against the anatomic limits of eels, but stated that eels reproduce by budding, scraping themselves against rocks, liberating particles that become eels.[17] Athenaeus described eels as entwining and discharging a fluid which would settle on mud and generate life. On the other hand, Athenaeus also dissented towards spontaneous generation, claiming that a variety of anchovy did not generate from roe, as Aristotle stated, but rather, from sea foam.[18]

*The barnacle goose:*
Branta leucopsis

## 2.5 Adoption in Christianity

As the dominant view of philosophers and thinkers continued to be in favour of spontaneous generation, some Christian theologians accepted the view. Augustine of Hippo discussed spontaneous generation in *The City of God* and *The Literal Meaning of Genesis*, citing Biblical passages such as "Let the waters bring forth abundantly the moving creature that hath life" (Genesis 1:20) as decrees that would enable ongoing creation.[19]

From the fall of the Roman Empire in 5th century to the East-West Schism in 1054, the influence of Greek science declined, although spontaneous generation generally went unchallenged. New descriptions were made. Of the numerous beliefs, some had doctrinal implications outside of the Book of Genesis. For example, the idea that a variety of bird known as the *barnacle goose* emerged from a crustacean known as the *goose barnacle*, had implications on the practice of fasting during Lent. In 1188, Gerald of Wales, after having traveled in Ireland, argued that the "unnatural" generation of barnacle geese was evidence for the virgin birth.[20] Where the practice of fasting during Lent allowed fish, but prohibited fowl, the idea that the goose was in fact a fish suggested that its consumption be permitted during Lent. The practice was eventually prohibited by decree of Pope Innocent III in 1215.[21]

Aristotle, in Arabic translation, was reintroduced to Western Europe. During the 13th century, Aristotle reached his greatest acceptance. With the availability of Latin translations Saint Albertus Magnus and his student, Saint Thomas Aquinas, raised Aristotelianism to its greatest prominence. Albert wrote a paraphrase of Aristotle, *De causis et processu universitatis*, in which he removed some and incorporated other commentaries by Arabic scholars.[22] The influential writings of Aquinas, on both the physical and metaphysical, are predominantly Aristotelian, but show numerous other influences.[23]

Spontaneous generation is discussed as a fact in literature well into the Renaissance. Where, in passing, Shakespeare discusses snakes and crocodiles forming from the mud of the Nile (Ant 2.7 F1), Izaak Walton again raises the question of the origin of eels "as rats and mice, and many other living creatures, are bred in Egypt, by the sun's heat when it shines upon the overflowing of the river...". While the ancient question of the origin of eels remained unanswered and the additional idea that eels reproduced from corruption of age was mentioned, the spontaneous generation of rats and mice engendered no debate.[24]

The Dutch biologist and microscopist Jan Swammerdam (1637 - 1680) rejected the concept that one animal could arise from another or from putrification by chance because it was impious and like others found the concept of spontaneous generation irreligious, and he associated it with atheism and Godless opinion.[25]

## 2.6 Scientific method

Jan Baptist van Helmont (1580–1644) used experimental techniques, such as growing a willow for five years and showing it increased mass while the soil showed a trivial decrease in comparison. As the process of photosynthesis was not understood, he attributed the increase of mass to the absorption of water.[26] His notes also describe a recipe for mice (a piece of soiled cloth plus wheat for 21 days) and scorpions (basil, placed between two bricks and left in sunlight). His notes suggest he may even have done these things.[27]

Where Aristotle held that the embryo was formed by a coagulation in the uterus, William Harvey (1578 – 1657) by way of dissection of deer, showed that there was no visible embryo during the first month.[13] Although his work predated the microscope, this led him to suggest that life came from invisible eggs. In the frontispiece of his book *Exercitationes de Generatione Animalium* (*Essays on the Generation of Animals*), he made an expression of biogenesis: "omnia ex ovo" (everything from eggs).[19]

The ancient beliefs were subjected to testing. In 1668, Francesco Redi challenged the idea that maggots arose spontaneously from rotting meat. In the first major experiment to challenge spontaneous generation, he placed meat in a variety of sealed, open, and partially covered containers.[28] Realizing that the sealed containers were deprived of air, he used "fine Naples veil", and observed no worm on the meat, but they appeared on the cloth.[29] Redi used his experiments to support the preexistence theory put forth by the Church at that time, which maintained that living things originated from parents.[30] In scientific circles Redi's work very soon had great influence, as evidenced in

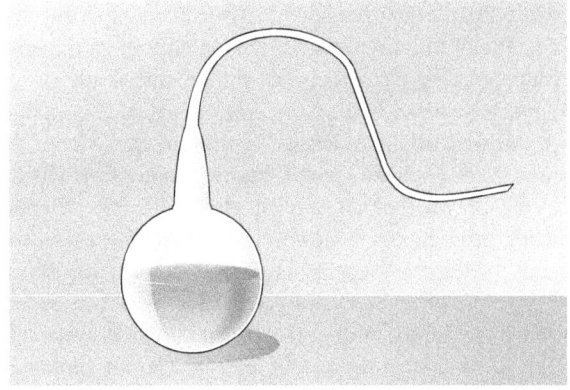

*Illustration of the swan-necked bottle used in Pasteur's experiments to disprove spontaneous generation*

a letter from John Ray in 1671 to members of the Royal Society of London:

> Whether there be any spontaneous or anomalous generation of animals, as has been the constant opinion of naturalists heretofore, I think there is good reason to question. It seems to me at present most probable, that there is no such thing; but that even all insects are the natural issue of parents of the same species with themselves. F. Redi has gone a good way in proving this, having cleared the point concerning generation *ex materia putrida*. But still there remain two great difficulties. The first is, to give an account of the production of insects bred in the by-fruits and excrescencies of vegetables, which the said Redi doubts not to ascribe to the vegetative soul of the plant that yields those excrescencies. But for this I refer you to Mr. Lister. The second, to render an account of insects bred in the bodies of other animals. I hope shortly to be able to give you an account of the generation of some of those insects which have been thought to be spontaneous, and which seem as unlikely as any to be after the ordinary and usual way.[31]

Pier Antonio Micheli, around 1729, observed that when fungal spores were placed on slices of melon the same type of fungi were produced that the spores came from, and from this observation he noted that fungi did not arise from spontaneous generation.[32]

In 1745, John Needham performed a series of experiments on boiled broths. Believing that boiling would kill all living things, he showed that when sealed right after boiling, the broths would cloud, allowing the belief in spontaneous generation to persist. His studies were rigorously scrutinized by his peers and many of them agreed.[28]

## 2.6. SCIENTIFIC METHOD

Lazzaro Spallanzani modified the Needham experiment in 1768, attempting to exclude the possibility of introducing a contaminating factor between boiling and sealing. His technique involved boiling the broth in a sealed container with the air partially evacuated to prevent explosions. Although he did not see growth, the exclusion of air left the question of whether air was an essential factor in spontaneous generation.[28] However, by that time there was already widespread scepticism among major scientists, to the principle of spontaneous generation. Observation was increasingly demonstrating that whenever there was sufficiently careful investigation of mechanisms of biological reproduction, it was plain that processes involved basing of new structures on existing complex structures, rather from chaotic muds or dead materials. Joseph Priestley, after he had fled to America and not long before his death, wrote a letter that was read to the American Philosophical Society in 1803. It said in part:

> There is nothing in modern philosophy that appears to me so extraordinary, as the revival of what has long been considered as the exploded doctrine of equivocal, or, as Dr. Darwin calls it, spontaneous generation; by which is meant the production of organized bodies from substances that have no organization, as plants and animals from no pre-existing germs of the same kinds, plants without seeds, and animals without sexual intercourse.
>
> The germ of an organized body, the seed of a plant, or the embrio of an animal, in its first discoverable state, is now found to be the future plant or animal in miniature, containing every thing essential to it when full grown, only requiring to have the several organs enlarged, and the interstices filled with extraneous nutritious matter. When the external form undergoes the greatest change, as from an aquatic insect to a flying gnat, a caterpillar to a crysalis, a crysalis to a butterfly, or a tadpole to a frog, there is nothing new in the organization; all the parts of the gnat, the butterfly, and the frog, having really existed, though not appearing to the common observer in the forms in which they are first seen. In like manner, every thing essential to the oak is found in the acorn.[33]

In 1837, Charles Cagniard de la Tour, a physicist, and Theodor Schwann, one of the founders of cell theory, published their independent discovery of yeast in alcoholic fermentation. They used the microscope to examine foam left over from the process of brewing beer. Where Leeuwenhoek described "small spheroid globules", they observed yeast cells undergo cell division. Fermentation would not occur when sterile air or pure oxygen was introduced if yeast were not present. This suggested that airborne microorganisms, not spontaneous generation, was responsible.[34]

*Before making the "swan neck flasks" Pasteur sealed flasks like this one from air. It remains sterile and on view at the Science Museum, London.*

However, although the idea of spontaneous generation had been in decline for nearly a century, its supporters did not abandon it all at once. As James Rennie wrote:

> "...inability to trace the origin of minute plants and insects led to the doctrine of what is called spontaneous or equivocal generation, of which the fancies above-mentioned are some 'of the prominent branches. The experiments of Redi on the hatching of insects from eggs, which were published at Florence in 1668, first brought discredit upon this doctrine, though it had always a few eminent disciples. At present it is maintained by a considerable number -of distinguished naturalists, such as Blumenbach, Cuvier, Bory de St. Vincent, R. Brown, &c. "The notion or spontaneous generation," says Bory, " is at first revolting to a rational mind, but it is, notwithstanding, demonstrable by the microscope. The fact is averred : Willer has seen it, I have seen it, and twenty other observers have seen it: the pandorinia exhibit it every instant. "These pan-

dorinia he elsewhere describes as probably nothing more than " animated scions of Zoocarpae". It would be unprofitable to go into any lengthened discussion upon this mysterious subject; and we have great doubts whether the ocular demonstration by the microscope would succeed except in the hands of a disciple of the school. Even with naturalists, whose business it is to deal with facts, the reason is often wonderfully influenced by the imagination..."[35]

Louis Pasteur's 1859 experiment is widely seen as having settled the question of spontaneous generation. He boiled a meat broth in a flask that had a long neck that curved downward, like a goose. The idea was that the bend in the neck prevented falling particles from reaching the broth, while still allowing the free flow of air. The flask remained free of growth for an extended period. When the flask was turned so that particles could fall down the bends, the broth quickly became clouded.[28] However, minority objections were persistent and not always unreasonable, given that the experimental difficulties were far more challenging than the popular accounts suggest. The investigations of John Tyndall, a correspondent of Pasteur and a great admirer of Pasteur's work, were decisive in disproving spontaneous generation with dealing with lingering issues. Still, even Tyndall encountered difficulties in dealing with the effects of microbial spores, which were not well understood in his day. Like Pasteur, he boiled his cultures to sterilize them, and some types of bacterial spores can survive boiling. The autoclave, which eventually came into universal application in medical practice and microbiology to sterilise equipment, was not an instrument that had come into use at the time of Tyndall's experiments, let alone those of Pasteur.[2]

## 2.7 See also

- Bugonia or bougonia

## 2.8 References

[1] André Brack (1998). "Introduction" (PDF). In André Brack. *The Molecular Origins of Life*. Cambridge University Press. p. 1. ISBN 978-0-521-56475-5. Retrieved 2009-01-07. Aristotle gathered the different claims into a real theory.

[2] Tyndall, John; Fragments of Science, Vol 2, chapters IV, XII (1876), XIII(1878); Pub. P. F. Collier, New York 1905; (Available at: https://archive.org/details/fragmenoscien02tyndrich )

[3] Levine, Russell; Evers, Chris. "The Slow Death of Spontaneous Generation (1668-1859)". *North Carolina State University*. National Health Museum.

[4] Bernal, J. D. (1967) [Reprinted work by A. I. Oparin originally published 1924; Moscow: The Moscow Worker]. *The Origin of Life*. The Weidenfeld and Nicolson Natural History. Translation of Oparin by Ann Synge. London: Weidenfeld & Nicolson. LCCN 67098482.

[5] Zubay, Geoffrey. Origins of Life, Second Edition: On Earth and in the Cosmos. Academic Press 2000. ISBN 978-0127819105

[6] Smith, John Maynard; Szathmary, Eors (1997). *The Major Transitions in Evolution*. Oxford Oxfordshire: Oxford University Press. ISBN 9780198502944.

[7] Stillingfleet, Edward. Origines Sacrae. Cambridge University Press 1697. May be downloaded from

[8] Philip P. Wiener, ed. (1973). "Spontaneous Generation". *Dictionary of the History of Ideas*. New York: Charles Scribner's Sons. Retrieved 2009-01-22.

[9] Strick, James (April 15, 2001). "Introduction". *Evolution & The Spontaneous Generation*. Continuum International Publishing Group. pp. xi–xxiv. ISBN 978-1-85506-872-8. Retrieved August 27, 2012.

[10] Wilkins, John S. (April 2004). "Spontaneous Generation and the Origin of Life". The Talk.Origins Archive. Retrieved 3 December 2008.

[11] Aristotle (1910) [c. 343 BCE]. "Book V". *The History of Animals*. translated by D'Arcy Wentworth Thompson. Oxford: Clarendon Press. ISBN 90-6186-973-0. Retrieved 2008-12-20.

[12] Aristotle (1912) [c. 350 BCE]. "Book III". *On the Generation of Animals*. translated by Arthur Platt. Oxford: Clarendon Press. ISBN 90-04-09603-5. Retrieved 2009-01-09.

[13] Lois Magner; K. Lee Lerner. "Embryology - History Of Embryology As A Science". Retrieved 2009-01-09.

[14] Marcus Vitruvius Pollio (1826) [c. 25 BCE]. "Part 4". In Joseph Gwilt (translator). *On Architecture (de Architectura)*. Book VI. electronic format by Bill Thayer. London: Priestley and Weale. Retrieved 2009-02-03.

[15] Aristotle (1910) [c. 343 BCE]. "Book IV". *The History of Animals*. translated by D'Arcy Wentworth Thompson. Oxford: Clarendon Press. ISBN 90-6186-973-0. Retrieved 2008-12-20.

[16] Aristotle (1910) [c. 343 BCE]. "Book VI". *The History of Animals*. translated by D'Arcy Wentworth Thompson. Oxford: Clarendon Press. ISBN 90-6186-973-0. Retrieved 2008-12-20.

## 2.8. REFERENCES

[17] Gaius Plinius Secundus (1855) [c. 77]. "74. (50.) — The generation of fishes". In John Bostock, Henry Thomas Riley. *Natural History*. BOOK IX. The natural history of fishes. Retrieved 2009-02-03.

[18] Athenaeus of Naucratis. "Book VII". In Yonge, C.D. *The deipnosophists, or, Banquet of the learned of Athenæus*. University of Wisconsin Digital Collection **I**. London: Henry G. Bohn. pp. 433–521. Retrieved 2009-02-03.

[19] Fry, Iris (2000). "Chapter 2: Spontaneous Generation — Ups and Downs". *The Emergence of Life on Earth*. Rutgers University Press. ISBN 978-0-8135-2740-6. Retrieved 2009-01-21.

[20] Giraldus Cambrensis (1188). *Topographia Hiberniae*. ISBN 0-85105-386-6. Retrieved 2009-02-01.

[21] Lankester, Sir Edwin Ray (1970) [1915]. "XIV. The History of the Barnacle and the Goose". *Diversions of a Naturalist* (illustrated ed.). Ayer Publishing. pp. 117–128. ISBN 978-0-8369-1471-9. Retrieved 2009-02-01.

[22] Zalta, Edward N., ed. (March 20, 2006). "Albert the Great". *Stanford Encyclopedia of Philosophy* (Winter 2009 ed.). Stanford, CA: The Metaphysics Research Lab. ISBN 1-158-37777-0. ISSN 1095-5054. OCLC 179833493. Retrieved 2009-01-23.

[23] Zalta, Edward N., ed. (July 12, 1999). "Saint Thomas Aquinas". *Stanford Encyclopedia of Philosophy* (Winter 2009 ed.). Stanford, CA: The Metaphysics Research Lab (published January 9, 2005). ISBN 1-158-37777-0. ISSN 1095-5054. OCLC 179833493. Retrieved 2009-01-23.

[24] Walton, Izaak (1903) [1653]. "XIII. Observations of the eel, and other fish that want for scales, and how to fish for them". *The Compleat Angler or the Contemplative Man's Recreation*. transcribed by Risa Bear. George Bell & Sons. ISBN 0-929309-00-6. Retrieved 2009-02-05.

[25] Margaret J. Osler; Paul Lawrence Farber (22 August 2002). *Religion, Science, and Worldview: Essays in Honor of Richard S. Westfall*. Cambridge University Press. pp. 230–. ISBN 978-0-521-52493-3. Retrieved 15 August 2012.

[26] Ducheyne, Steffen (2006). "Joan Baptista van Helmont and the Question of Experimental Modernism" (PDF). pp. 305–332. Retrieved 2009-01-07.

[27] Pasteur, Louis (April 23, 1864). Latour, Bruno, ed. "Des générations spontanées" (PDF) **1**. Conférences faite aux "soirées scientifiques de la Sorbonne" (published 1993 (English translation)): 257–265. Archived from the original (PDF) on March 26, 2009. Retrieved 2009-01-07. Check date values in: |publication-date= (help)

[28] Russell Levine; Chris Evers (1999). "The Slow Death of Spontaneous Generation (1668-1859)". Washington, D.C.: National Health Museum. Retrieved 2008-12-19.

[29] Francesco Redi of Arezzo (1909) [1669]. Mab Bigelow (translator), ed. *Experiments on the Generation of Insects*. Chicago: Open Court. Retrieved 2008-12-19.

[30] Iris Fry (1 February 2000). *Emergence of Life on Earth: A Historical and Scientific Overview*. Rutgers University Press. pp. 27–. ISBN 978-0-8135-2740-6. Retrieved 14 October 2012.

[31] Royal Society (Great Britain); Hutton, Charles, 1737-1823; Shaw, George, 1751-1813; Pearson, Richard, 1765-1836. The Extract of a Letter written by Mr. JOHN RAY, to the Editor, from Middleton, July 3, 1671, concerning Spontaneous Generation;... Number 73, p. 2219. The Philosophical transactions of the Royal Society of London, from their commencement in 1665, in the year 1800. Page 617-618. May be downloaded from:

[32] George N. Agrios (2005). *Plant Pathology*. Academic Press. pp. 17–. ISBN 978-0-12-044565-3. Retrieved 14 August 2012.

[33] Priestley, Joseph. Observations and Experiments relating to equivocal, or spontaneous, Generation. Transactions of the American Philosophical Society, Volume VI, page 119-129, 1809. Download from:

[34] Springer, Alfred (October 13, 1892). "The Microorganisms of the Soil". *Nature* (Nature Publishing Group) **46** (1198): 576–579. Bibcode:1892Natur..46R.576.. doi:10.1038/046576b0. ISSN 0028-0836.

[35] Rennie, James. Insect Transformations. Page 10. Pub: Charles Knight 1838 Download from:

# Chapter 3

# Self-organization

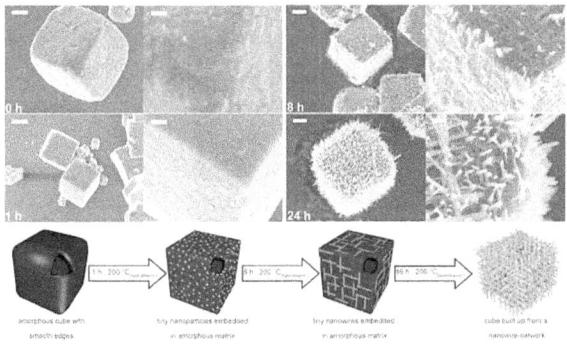

*Self-organization in micron-sized $Nb_3O_7(OH)$ cubes during a hydrothermal treatment at 200 °C. Initially amorphous cubes gradually transform into ordered 3D meshes of crystalline nanowires as summarized in the model below.*[1]

**Self-organization** is a process where some form of overall order or coordination arises out of the local interactions between smaller component parts of an initially disordered system. The process of self-organization can be spontaneous, and it is not necessarily controlled by any auxiliary agent outside of the system. It is often triggered by random fluctuations that are amplified by positive feedback. The resulting organization is wholly decentralized or distributed over all the components of the system. As such, the organization is typically robust and able to survive and, even, self-repair substantial damage or perturbations. Chaos theory discusses self-organization in terms of islands of predictability in a sea of chaotic unpredictability. Self-organization occurs in a variety of physical, chemical, biological, robotic, social, and cognitive systems. Examples of its realization can be found in crystallization, thermal convection of fluids, chemical oscillation, animal swarming, and artificial and biological neural networks.

## 3.1 Overview

Self-organization is realized[2] in the physics of non-equilibrium processes, and in chemical reactions, where it is often described as self-assembly. The concept of self-organization has proven useful in the description of biological systems,[3] from the subcellular to the ecosystem level.[4] Cited examples of self-organizing behaviour also appear in the literature of many other disciplines, both in the natural sciences and in the social sciences such as economics or anthropology. Self-organization has also been observed in mathematical systems such as cellular automata.[5] Sometimes the notion of self-organization becomes conflated with that of the related concept of emergence.[6] Properly defined, however, there may be instances of self-organization without emergence and emergence without self-organization.

Self-organization usually relies on three basic ingredients:[7]

1. strong dynamical non-linearity, often though not necessarily involving positive and negative feedback
2. balance of exploitation and exploration
3. multiple interactions

### 3.1.1 Principles of self-organization

The cybernetician William Ross Ashby formulated the original principle of self-organization in 1947.[8][9] It states that any deterministic dynamic system will automatically evolve towards a state of equilibrium that can be described in terms of an attractor in a basin of surrounding states. Once there, the further evolution of the system is constrained to remain in the attractor. This constraint on the system as a whole implies a form of mutual dependency or coordination between its constituent components or "subsystems". In Ashby's terms, each subsystem has adapted to the environment formed by all other subsystems.

The cybernetician Heinz von Foerster formulated the principle of "order from noise" in 1960.[10] It notes that self-organization is facilitated by random perturbations ("noise") that let the system explore a variety of states in its state space. This increases the chance that the system would arrive into the basin of a "strong" or "deep" attractor, from which it would then quickly enter the attractor itself. The thermodynamicist Ilya Prigogine formulated a similar principle as "order through fluctuations"[11] or "order out of chaos".[12] It is applied in the method of simulated annealing that is used in problem solving and in machine learning.

## 3.2 History of the idea

The idea that the dynamics of a system can lead to an increase of the system's organization has a long history. One of the earliest statements of this idea was by the philosopher Descartes, in the fifth part of his *Discourse on Method*, where he presents it hypothetically. Descartes further elaborated on the idea at great length in his unpublished work *The World*.

The ancient atomists believed that a designing intelligence is unnecessary to effect natural order, arguing that given enough time and space and matter, organization is ultimately inevitable, although there is no preferred tendency for this to happen. What Descartes introduced was the idea that the ordinary laws of nature *tend* to produce organization (For related history, see Aram Vartanian, *Diderot and Descartes*).

The economic concept of the "invisible hand" due to Adam Smith can be understood as an attempt to describe the influence of the market as a spontaneous order on people's actions.

Beginning with the 18th century, natural scientists sought to understand the "universal laws of form" in order to explain the observed forms of living organisms. Because of its association with Lamarckism, their ideas fell into disrepute until the early 20th century, when pioneers such as D'Arcy Wentworth Thompson revived them. The modern understanding is that there are indeed universal laws, arising from fundamental physics and chemistry, that govern growth and form in biological systems.

Sadi Carnot and Rudolf Clausius discovered the Second Law of Thermodynamics in the 19th century. It states that total entropy, sometimes understood as disorder, will always increase over time in an isolated system. This means that a system cannot spontaneously increase its order, without an external relationship that decreases order elsewhere in the system (e.g. through consuming the low-entropy energy of a battery and diffusing high-entropy heat).

Originally, the term "self-organizing" was used by Immanuel Kant in his *Critique of Judgment*, where he argued that teleology is a meaningful concept only if there exists such an entity whose parts or "organs" are simultaneously ends and means. Such a system of organs must be able to behave as if it has a mind of its own, that is, it is capable of governing itself.

The term "self-organizing" was introduced to contemporary science in 1947 by the psychiatrist and engineer W. Ross Ashby.[8] It was taken up by the cyberneticians Heinz von Foerster, Gordon Pask, Stafford Beer, and von Foerster organized a conference on "The Principles of Self-Organization" at the University of Illinois' Allerton Park in June, 1960 which led to a series of conferences on Self-Organizing Systems.[13] Norbert Wiener also took up the idea in the second edition of his *Cybernetics: or Control and Communication in the Animal and the Machine* (1961).

Self-organization as a word and concept was used by those associated with general systems theory in the 1960s, but did not become commonplace in the scientific literature until its adoption by physicists and researchers in the field of complex systems in the 1970s and 1980s.[14] After Ilya Prigogine's 1977 Nobel Prize, the *thermodynamic concept of self-organization* received some attention of the public, and scientific researchers started to migrate from the *cybernetic view* to the *thermodynamic view*.[15]

### 3.2.1 Developing views

Other views of self-organization in physical systems interpret it as a strictly accumulative construction process, commonly displaying an "S" curve history of development. As discussed somewhat differently by different researchers, local complex systems for exploiting energy gradients evolve from seeds of organization, through a succession of natural starting and ending phases for inverting their directions of development. The accumulation of working processes which their exploratory parts construct as they exploit their gradient becomes the "learning", "organization" or "design" of the system as a physical artifact, such for an ecology or economy. For example, A. Bejan's books and papers describe his approach as "Constructal Theory".[16][17][18] P. F. Henshaw's work on decoding net-energy system construction processes termed "Natural Systems Theory", uses various analytical methods to quantify and map them such as System Energy Assessment[19] for taking true quantitative measures of whole complex energy using systems, and for anticipating their successions, such as Models Learning Change[20] to permit adapting models to their emerging inverted designs. G. Y. Georgiev's work is utilizing the principle of least (stationary) action in Physics, to define organization of a complex system as the state of the constraints determining the total action of the elements in a system. Organization is then defined numerically as the reciprocal of the average action per one element and one edge crossing, if the system is described as a network. The elementary quantum of action, Planck's constant, is used to make the measure dimensionless and to define it as inversely proportional to the number of quanta of action expended by the elements for one edge crossing. The mechanism of self-organization is the interaction between the elements and the constraints, which leads to constraint minimization. This is consistent with the Gauss principle of least constraint. More elements minimize the constraints faster, another aspect of the mechanism, which is through quantity accumulation. As a result, the paths of the elements are straightened, which is consistent with Hertz's principle of least curvature. The state of a system with least average sum of actions of its elements is defined as its attractor. In open systems, where there is constant inflow and outflow of energy and elements, this final state is never reached, but the system always tends toward it.[15] This method can help describe, quantify, manage, design and predict future behavior of complex systems, to achieve the highest rates of self-organization to improve their quality, which is the numerical value of their organization. It can be applied to complex systems in physics, chemistry, biology, ecology, economics, cities, network theory and others, where they are present.[15][21][22]

## 3.3 Examples

The following list summarizes and classifies the instances of self-organization found in different disciplines. As the list grows, it becomes increasingly difficult to determine whether these phenomena are all fundamentally the same process, or the same label applied to several different processes. Self-organization, despite its intuitive simplicity as a concept, has proven notoriously difficult to define and pin down formally or mathematically, and it is entirely possible that any precise definition might not include all the phenomena to which the label has been applied.

The farther a phenomenon is removed from physics, the more controversial the idea of self-organization *as understood by physicists* becomes. Also, even when self-organization is clearly present, attempts at explaining it through physics or statistics are usually criticized as reductionistic.

Similarly, when ideas about self-organization originate in, say, biology or social science, the farther one tries to take the concept into chemistry, physics or mathematics, the more resistance is encountered, usually on the grounds that it implies direction in fundamental physical processes. However the tendency of hot bodies to get cold (see Thermodynamics) and by Le Chatelier's Principle—the statistical mechanics extension of Newton's Third Law—to oppose this tendency should be noted.

### 3.3.1 Physics

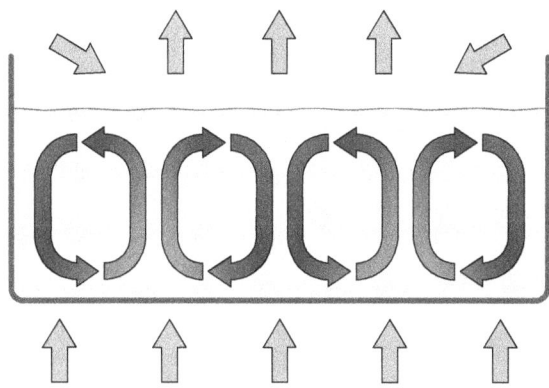

*Convection cells in a gravity field*

There are several broad classes of physical processes that can be described as self-organization. Such examples from physics include:

- structural (order-disorder, first-order) phase transitions, and spontaneous symmetry breaking such as
  - spontaneous magnetization, crystallization (see crystal growth, and liquid crystal) in the classical domain and
  - the laser, superconductivity and Bose–Einstein condensation, in the quantum domain (but with macroscopic manifestations)
- second-order phase transition, associated with "critical points" at which the system exhibits scale-invariant structures. Examples of these include:
  - critical opalescence of fluids at the critical point
  - percolation in random media
- structure formation in thermodynamic systems away from equilibrium. The theory of dissipative structures of Prigogine and Hermann Haken's Synergetics were developed to unify the understanding of these phenomena, which include lasers, turbulence and convective instabilities (e.g., Bénard cells) in fluid dynamics,
  - structure formation in astrophysics and cosmology (including star formation, planetary systems formation, galaxy formation)
  - self-similar expansion
  - Diffusion-limited aggregation
  - percolation
  - reaction-diffusion systems, such as Belousov–Zhabotinsky reaction

## 3.3. EXAMPLES

- self-organizing dynamical systems: complex systems made up of small, simple units connected to each other usually exhibit self-organization

  - Self-organized criticality (SOC)

- In tribology, friction coupled with other simultaneous effects, such as heat transfer, wear, and material diffusion. can lead to self-organized patterns at the frictional interface, ranging from stick-slip patterns to in-situ formed tribofilms and surface roughness adjustment of two materials in contact.

- In spin foam system and loop quantum gravity that was proposed by Lee Smolin. The main idea is that the evolution of space in time should be robust in general. Any fine-tuning of cosmological parameters weaken the independency of the fundamental theory. Philosophically, it can be assumed that in the early time, there has not been any agent to tune the cosmological parameters. Smolin and his colleagues in a series of works show that, based on the loop quantization of spacetime, in the very early time, a simple evolutionary model (similar to the sand pile model) behaves as a power law distribution on both the size and area of avalanche.

  - Although, this model, which is restricted only on the frozen spin networks, exhibits a non-stationary expansion of the universe. However, it is the first serious attempt toward the final ambitious goal of determining the cosmic expansion and inflation based on a self-organized criticality theory in which the parameters are not tuned, but instead are determined from within the complex system.[23]

- A laser can also be characterized as a self organized system to the extent that normal states of thermal equilibrium characterized by electromagnetic energy absorption are stimulated out of equilibrium in a reverse of the absorption process. "If the matter can be forced out of thermal equilibrium to a sufficient degree, so that the upper state has a higher population than the lower state (population inversion), then more stimulated emission than absorption occurs, leading to coherent growth (amplification or gain) of the electromagnetic wave at the transition frequency."[24]

### 3.3.2 Chemistry

Self-organization in chemistry includes:

1. molecular self-assembly

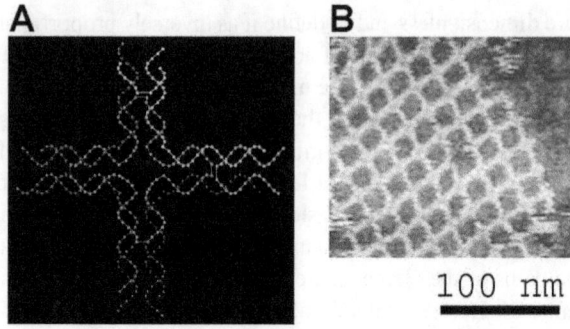

*The DNA structure at left (schematic shown) will self-assemble into the structure visualized by atomic force microscopy at right. Image from Strong.*[25]

2. reaction-diffusion systems and oscillating chemical reactions
3. autocatalytic networks (see: autocatalytic set)
4. liquid crystals
5. grid complexes
6. colloidal crystals
7. self-assembled monolayers
8. micelles
9. microphase separation of block copolymers
10. Langmuir-Blodgett films

### 3.3.3 Biology

*Birds flocking, an example of self-organization in biology*

Main article: Biological organisation

According to *Scott Camazine.. [et al.]*:

The following is an incomplete list of the diverse phenomena which have been described as self-organizing in biology.

1. spontaneous folding of proteins and other biomacromolecules

2. formation of lipid bilayer membranes

3. homeostasis (the self-maintaining nature of systems from the cell to the whole organism)

4. pattern formation and morphogenesis, or how the living organism develops and grows. See also embryology.

5. the coordination of human movement, e.g. seminal studies of bimanual coordination by Kelso

6. the creation of structures by social animals, such as social insects (bees, ants, termites), and many mammals

7. flocking behaviour (such as the formation of flocks by birds, schools of fish, etc.)

8. the origin of life itself from self-organizing chemical systems, in the theories of hypercycles and autocatalytic networks

9. the organization of Earth's biosphere in a way that is broadly conducive to life (according to the controversial Gaia hypothesis)

### 3.3.4 Computer Science

*Gosper's Glider Gun creating "gliders" in the cellular automaton Conway's Game of Life.*[27]

As mentioned above, phenomena from mathematics and computer science such as cellular automata, random graphs, and some instances of evolutionary computation and artificial life exhibit features of self-organization. In swarm robotics, self-organization is used to produce emergent behavior. In particular the theory of random graphs has been used as a justification for self-organization as a general principle of complex systems. In the field of multi-agent systems, understanding how to engineer systems that are capable of presenting self-organized behavior is a very active research area.

**Algorithms**

Many optimization algorithms can be considered as a self-organization system because the aim of the optimization is to find the optimal solution to a problem. If the solution is considered as a state of the iterative system, the optimal solution is essentially the selected, converged state or structure of the system, driven by the algorithm based on the system landscape.[28][29] In fact, one can view all optimization algorithms as a self-organization system.

**Networks**

Self-organization is an important component for a successful ability to establish networking whenever needed. Such mechanisms are also referred to as Self-organizing networks. Intensified work in the latter half of the first decade of the 21st century was mainly due to interest from the wireless communications industry. It is driven by the plug and play paradigm, and that wireless networks need to be relatively simpler to manage than they used to be.

Only certain kinds of networks are self-organizing. The best known examples are small-world networks and scale-free networks. These emerge from bottom-up interactions, and appear to be limitless in size. In contrast, there are top-down hierarchical networks, which are not self-organizing. These are typical of organizations, and have severe size limits.

In many natural systems, self-organization results from repeated phase shifts in their underlying network of connections. Such phase shifts alter the balance between internal processes (e.g. selection and variation). They give rise to the phenomenon of dual-phase evolution.

### 3.3.5 Cybernetics

Wiener regarded the automatic serial identification of a black box and its subsequent reproduction as sufficient to meet the condition of self-organization.[30] The importance of phase locking or the "attraction of frequencies", as he called it, is discussed in the 2nd edition of his

"Cybernetics".[31] Drexler sees self-replication as a key step in nano and universal assembly.

By contrast, the four concurrently connected galvanometers of W. Ross Ashby's Homeostat hunt, when perturbed, to converge on one of many possible stable states.[32] Ashby used his state counting measure of variety[33] to describe stable states and produced the "Good Regulator"[34] theorem which requires internal models for self-organized endurance and stability (e.g. Nyquist stability criterion).

Warren McCulloch proposed "Redundancy of Potential Command"[35] as characteristic of the organization of the brain and human nervous system and the necessary condition for self-organization.

Heinz von Foerster proposed Redundancy, $R = 1 - H/H_{max}$, where $H$ is entropy.[36][37] In essence this states that unused potential communication bandwidth is a measure of self-organization.

In the 1970s Stafford Beer considered this condition as necessary for autonomy which identifies self-organization in persisting and living systems. Using Variety analyses he applied his neurophysiologically derived recursive Viable System Model to management. It consists of five parts: the monitoring of performance of the survival processes (1), their management by recursive application of regulation (2), homeostatic operational control (3) and development (4) which produce maintenance of identity (5) under environmental perturbation. Focus is prioritized by an alerting "algedonic loop" feedback: a sensitivity to both pain and pleasure produced from under-performance or over-performance relative to a standard capability.[38]

In the 1990s Gordon Pask pointed out von Foerster's H and Hmax were not independent and interacted via countably infinite recursive concurrent spin processes[39] (he favoured the Bohm interpretation) which he called concepts (liberally defined in *any* medium, "productive and, incidentally reproductive"). His strict definition of concept "a procedure to bring about a relation"[40] permitted his theorem "Like concepts repel, unlike concepts attract"[41] to state a general spin based **Principle of Self-organization**. His edict, an exclusion principle, "There are No Doppelgangers"[42][39] means no two concepts can be the same (all interactions occur with different perspectives making time incommensurable for actors). This means, after sufficient duration as differences assert, all concepts will attract and coalesce as pink noise and entropy increases (and see Big Crunch, self-organized criticality). The theory is applicable to all organizationally closed or homeostatic processes that produce enduring and coherent products (where spins have a fixed average phase relationship and also in the sense of Rescher Coherence Theory of Truth with the proviso that the sets and their members exert repulsive forces at their boundaries) through interactions: evolving, learning and adapting.

Pask's Interactions of Actors "hard carapace" model is reflected in some of the ideas of emergence and coherence. It requires a knot emergence topology that produces radiation during interaction with a unit cell that has a prismatic tensegrity structure. Laughlin's contribution to emergence reflects some of these constraints.

### 3.3.6 Human society

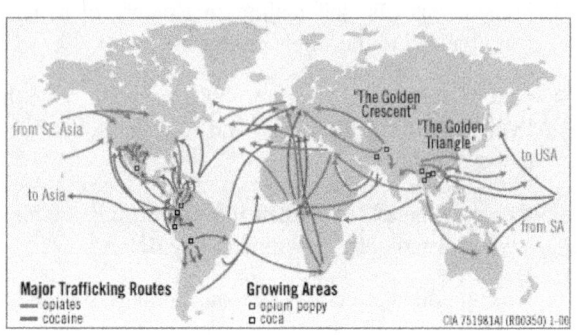

*Social self-organization in international drug routes*

The self-organizing behaviour of social animals and the self-organization of simple mathematical structures both suggest that self-organization should be expected in human society. Tell-tale signs of self-organization are usually statistical properties shared with self-organizing physical systems (see Zipf's law, power law, Pareto principle). Examples such as critical mass, herd behaviour, groupthink and others, abound in sociology, economics, behavioral finance and anthropology.[43] The theory of human social self-organization is also known as spontaneous order theory.

In social theory the concept of self-referentiality has been introduced as a sociological application of self-organization theory by Niklas Luhmann (1984). For Luhmann the elements of a social system are self-producing communications, i.e. a communication produces further communications and hence a social system can reproduce itself as long as there is dynamic communication. For Luhmann human beings are sensors in the environment of the system. Luhmann developed an evolutionary theory of Society and its subsytems, using functional *analyses* and systems *theory*.[44]

Self-organization in human and computer networks can give rise to a decentralized, distributed, self-healing system, protecting the security of the actors in the network by limiting the scope of knowledge of the entire system held by each individual actor. The Underground Railroad is a good example of this sort of network. The networks that arise from drug trafficking exhibit similar self-organizing properties. The Sphere College Project seeks to apply self-organization to adult education. Parallel examples exist in the world of privacy-preserving computer networks such as Tor. In each case, the network as a whole exhibits distinctive synergistic behavior through the combination of the behaviors of individual actors in the network. Usually the growth of such networks is fueled by an ideology or sociological force that is adhered to or shared by all participants in the network.[15]

**Economics**

In economics, a market economy is sometimes said to be self-organizing. Paul Krugman has written on the role that market self-organization plays in the business cycle in his book "The Self Organizing Economy".[45] Friedrich Hayek coined the term *catallaxy*[46] to describe a "self-organizing system of voluntary co-operation", in regards to the spontaneous order of the free market economy. Neo-classical economists hold that imposing central planning usually makes the self-organized economic system less efficient. On the other end of the spectrum, economists consider that market failures are so significant that self-organization produces bad results and that the state should direct production and pricing. Most economists adopt an intermediate position and recommend a mixture of market economy and command economy characteristics (sometimes called a mixed economy). When applied to economics, the concept of self-organization can quickly become ideologically imbued.[15][47]

**Collective intelligence**

Non-thermodynamic concepts of entropy and self-organization have been explored by many theorists. Cliff Joslyn and colleagues and their so-called "global brain" projects. Marvin Minsky's "Society of Mind" and the no-central editor in charge policy of the open sourced internet encyclopedia, called Wikipedia, are examples of applications of these principles – see collective intelligence.

Donella Meadows, who codified twelve leverage points that a self-organizing system could exploit to organize itself, was one of a school of theorists who saw human creativity as part of a general process of adapting human lifeways to the planet and taking humans out of conflict with natural processes. See Gaia philosophy, deep ecology, ecology movement and Green movement for similar self-organizing ideals. (The connections between self-organisation and Gaia theory and the environmental movement are explored in the book *The Unity of Nature* by Alan Marshall).

### 3.3.7 Psychology and education

**Self-organised learning**

Enabling others to "learn how to learn"[48] is usually misconstrued as instructing them[49] how to successfully submit to being taught. Whilst fully accepting that we can always learn from others, particularly those with more and/or different experience than ourselves; self-organised learning (SOL) repudiates any idea[50] that this reduces to accepting that "the expert knows best" or that there is ever "the one best method." It offers an alternative definition of learning as "the construction of personally significant, relevant and viable meaning."

This more democratic 'bottom up' approach to learning is to be frequently tested experientially[51] by the learner(s) as being more "meaningful, constructive and creatively effective for me or us."

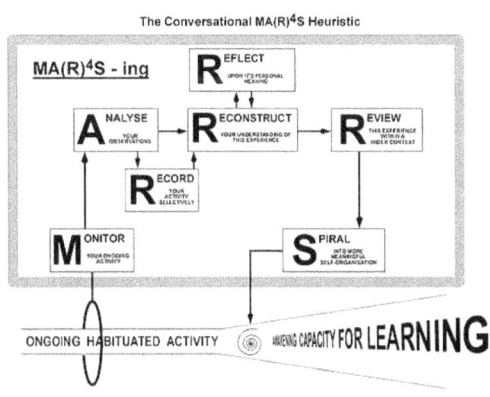
Cybernetic algorithm

*Visualization of links between pages on a wiki. This is an example of collective intelligence through collaborative editing.*

## 3.3. EXAMPLES

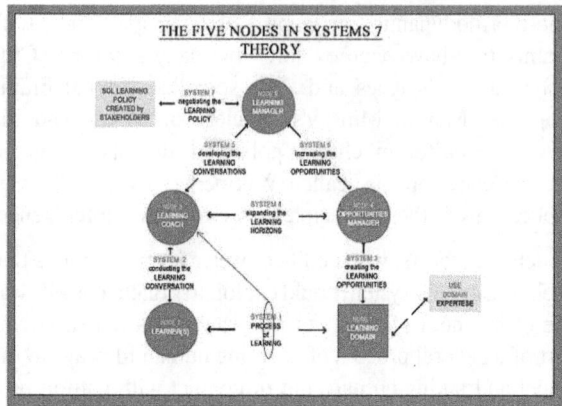

*Systems algorithm*

Since human learning may be achieved by one person,[52] or groups of learners working together;[53] SOL is not only a more rewarding and effective way of living one's personal life; it is also applicable in any group of people living, playing and/or working together.

As many young children, pupils, students and lifelong learners eventually become ruefully aware, this 'testing out of what I have learned' needs to be carried out in each learner(s) whole process of living, and so it extends well beyond the confines of specific learning environments (home, school, university, etc.), and eventually beyond the reaches of the controllers of these environments (parents, teachers, employers, etc.)[54]

SOL needs to be tested, and intermittently revised, through the ongoing personal experience[55] of the learner(s) themselves in their ever-expanding outer and inner lives.

Whilst internal life may cease to expand, the external environment does not. If a learner allows themselves to become progressively more other-organised, they become less able to recognise and respond to varying needs for change. Unfortunately this is often the current reported experience of many during, and hence after their parenting, schooling and/or higher education.

But, this SOL way of understanding the learning process need not be restricted by either consciousness or language.[56] Nor is it restricted to humans, since analogous directional self-organizing (learning?) processes are reported variously within the life sciences and even within the less-living sciences, for example, of physics and chemistry: (as is clearly articulated in other sections of this 'Self-organization' Section).

Since SOL is as yet only very superficially recognised within psychology and education, it is useful to place it more firmly within the human public mind-pool[57] of achievement, knowledge, experience and understanding. SOL can also be placed within a hierarchy of scientific explanatory concepts, for example:

1. Cause and Effect (requires "other things being equal")

2. Cybernetics[40] (incorporates item 1 in this list) with greater complexity, providing internal feedback and feed-forward controls: but still implying a sealed boundary. (i.e. other things being equal)

3. Systems Theory[58] (incorporates item 2 in this list, and opens the boundaries)

4. Self-organized System (incorporates item 3 in this list) and attributes this property to the interaction, patterning and coordination among the sub-systems of the system in question; in response to flow across its boundaries

5. Self-Organised Learning (SOL)[59] (incorporates item 4 in this list) but also requires that the parts each systematically respond, change and develop in the light of their experience, whilst self-organizing in the developing experiential interest of the whole).
SOL not only involves self-organization of the first order, i.e. what is mostly experienced as learning from experience without much conscious awareness of the process. At a second level of SOL consciousness enables us, (possibly uniquely among living beings) to reflect upon and thus self-organise the very process of self-organisation itself, (See 'Cybernetic algorithm' figure). It also enables organisations small and large to self-organise themselves, (see 'System algorithm' figure).
Once this approach to human learning is acknowledged, then we can re-set science into its place within the total human mind-pool. A mind-pool of human know-how and feel-how as an ever expanding and hopefully self-organizing resource.

6. Learning Conversation (incorporates item 5 in this list) and yet is at the same time its major tool. The Learning Conversation is a two-way process between SOLers, even within one person (conversing with oneself). Whilst not necessarily requiring language i.e. dialogue; it does require that the each participant really attempts to represent their meaning to the other(s), and that they all attempt to create personally significant, relevant and viable meaning in themselves in response to the others representations. So art, drama, music, computer programs, maths problems, ???, etc., can all create different, if limited, forms of Learning Conversation which really only become fully functional when at least two humans really attempt to fully communicate, and effectively share their understanding. That is achieve shared meaning in an event that approximates to what Maslow called a creative encounter[60]

7. Conversational Science[61] (will require item 6 in this list, the main method of SOL) among all seekers after significant, relevant and viable shared meaning. Science and many other human activities still need major paradigm shifts if we are to achieve Self-Organised Living. It also requires equal stakeholder-ship for each converser. Thus SOL can be seen as necessary but not sufficient for science to contribute positively to the benefit of the society, within which it may have only spasmodically been conversing successfully (SOL wise). Until, perhaps, both science and society as a whole will become Self-Organised Learners (SOLers) continually learning from their own shared experience and using what they learn in the shared interest of all concerned.

### 3.3.8 Traffic flow

The self-organizing behaviour of drivers in traffic flow determines almost all traffic spatiotemporal phenomena observed in real traffic data like traffic breakdown at a highway bottleneck, highway capacity, the emergence of moving traffic jams, etc. Self-organization in traffic flow is extremely complex spatiotemporal dynamic process. For this reason, only in 1996–2002 spatiotemporal self-organization effects in traffic have been understood in real measured traffic data and explained by Boris Kerner's three-phase traffic theory.

### 3.3.9 Methodology

In many complex systems in nature, there are global phenomena that are the irreducible result of local interactions between components whose individual study would not allow us to see the global properties of the whole combined system. Thus, a growing number of researchers think that many properties of language are not directly encoded by any of the components involved, but are the self-organized outcomes of the interactions of the components.

Building mathematical models in the context of research into language origins and the evolution of languages is enjoying growing popularity in the scientific community, because it is a crucial tool for studying the phenomena of language in relation to the complex interactions of its components. These systems are put to two main types of use: 1) they serve to evaluate the internal coherence of verbally expressed theories already proposed by clarifying all their hypotheses and verifying that they do indeed lead to the proposed conclusions ; 2) they serve to explore and generate new theories, which themselves often appear when one simply tries to build an artificial system reproducing the verbal behavior of humans.

As it were, the construction of operational models to test proposed hypotheses in linguistics is gaining much contemporary attention. An operational model is one which defines the set of its assumptions explicitly and above all shows how to calculate their consequences, that is, to prove that they lead to a certain set of conclusions.

**In the emergence of language**

Investigators have examined the emergence of language in the human species in a game-theoretic framework[62] based on a model of senders and receivers of information. The evolution of certain properties of language such as inference follow from this sort of framework (with the parameters stating that information transmitted can be partial or redundant, and the underlying assumption that the sender and receiver each want to take the action in their own best interest). Likewise, models have shown that compositionality, a central component of human language, emerges dynamically during linguistic evolution, and need not be introduced by biological evolution. Tomasello (1999) argues that one evolutionary step, the ability to sustain culture, laid the groundwork for the evolution of human language.[63] The ability to ratchet cultural advances cumulatively allowed for the complex development of human cognition not seen in other animals.

**In language acquisition**

Within a species' ontogeny, the acquisition of language has also been shown to self-organize. Through the ability to see others as intentional agents (theory of mind), and actions such as 'joint attention,' human children have the scaffolding they need to learn the language of those around them.

**In articulatory phonology**

Articulatory phonology takes the approach that speech production consists of a coordinated series of gestures, called 'constellations,' which are themselves dynamical systems. In this theory, linguistic contrast comes from the distinction between such gestural units, which can be described on a low-dimensional level in the abstract. However, these structures are necessarily context-dependent in real-time production. Thus the context-dependence emerges naturally from the dynamical systems themselves. This statement is controversial, however, as it suggests a universal phonetics which is not evident across languages.[64] Cross-linguistic patterns show that what can be treated as the same gestural units produce different contextualised patterns in different languages.[65] Articulatory Phonology fails to attend to the acoustic output of the gestures themselves (meaning that many typological patterns remain unexplained).[66] Freedom among listeners in the weighting of perceptual cues in the acoustic signal has a more fundamental role to play in the emergence of structure.[67] The realization of the perceptual contrasts by means of articulatory movements means that articulatory considerations do play a role,[68] but these are purely secondary.

**In diachrony and synchrony**

Several mathematical models of language change rely on self-organizing or dynamical systems. Abrams and Strogatz (2003) produced a model of language change that focused on "language death" – the process by which a speech community merges into the surrounding speech communities. Nakamura et al. (2008) proposed a variant of this model that incorporates spatial dynamics into language contact transactions in order to describe the emergence of creoles. Both of these models proceed from the assumption that language change, like any self-organizing system, is a large-scale act or entity (in this case the creation or death of a language, or changes in its boundaries) that emerges from many actions on a micro-level. The microlevel in this example is the everyday production and comprehension of language by speakers in areas of language contact.

## 3.4 Criticism

Heinz Pagels, in a balanced, but ultimately negative 1985 book review of Ilya Prigogine and Isabelle Stengers' *Order Out of Chaos* in *Physics Today*, appeals to authority:[69]

In theology, Thomas Aquinas (1225–1274) in his *Summa Theologica* assumes a teleological created universe in rejecting the idea that something can be a self-sufficient cause of its own organization:[70]

("The body of the Article" consists of the *quinque viae*.)

## 3.5 See also

- Ant mill
- Autowave
- Biology concepts: Bow tie (biology) – evolution – morphogenesis – homeostasis – Gaia Hypothesis
- Causality
- Chemistry concepts: reaction-diffusion – autocatalysis
- Complex systems concepts: emergence – evolutionary computation – artificial life – self-organized criticality – "edge of chaos" – spontaneous order – metastability – Chaos theory – Butterfly effect
- Computer science concepts: swarm intelligence
- Constructal law
- Dual-phase evolution
- Self-organized criticality control
- Free energy principle
- Free will
- Information theory
- Language – Operator grammar
- Mathematics concepts: fractal – random graph – power law – small world phenomenon – cellular automata
- Organization of the artist
- Philosophical concepts: tectology – Religious naturalism
- Physics concepts: thermodynamics – non-equilibrium thermodynamics – constructal theory – statistical mechanics – phase transition – dissipative structures – turbulence
- Social concepts: participatory organization
- Spontaneous order
- Stigmergy
- Systems theory concepts: cybernetics – autopoiesis – polytely
- Santiago theory of cognition
- Thermodynamics concepts: Second Law of Thermodynamics – Heat death of the Universe

## 3.6 References

[1] Betzler, S. B.; Wisnet, A.; Breitbach, B.; Mitterbauer, C.; Weickert, J.; Schmidt-Mende, L.; Scheu, C. (2014). "Template-free synthesis of novel, highly-ordered 3D hierarchical $Nb_3O_7(OH)$ superstructures with semiconductive and photoactive properties". *Journal of Materials Chemistry A* **2** (30): 12005. doi:10.1039/C4TA02202E.

[2] Glansdorff, P., Prigogine, I. (1971). *Thermodynamic Theory of Structure, Stability and Fluctuations*, Wiley-Interscience, London. ISBN 0-471-30280-5

[3] Witzany G (2014). Biological Self-Organization. International Journal of Signs and Semiotic Systems 3: 1-11.

[4] Compare: Camazine, Scott (2003). *Self-organization in Biological Systems*. Princeton studies in complexity (reprint ed.). Princeton University Press. ISBN 9780691116242. Retrieved 2016-04-05.

[5] Ilachinski, Andrew (2001). *Cellular Automata: A Discrete Universe*. World Scientific. p. 247. ISBN 9789812381835. Retrieved 2016-04-05. We have already seen ample evidence for what is arguably the single most impressive general property of CA, namely their capacity for self-organization.

[6] Bernard Feltz et al (2006). *Self-organization and Emergence in Life Sciences*. ISBN 9781402039164. p. 1.

[7] Bonabeau, Eric; Dorigo, Marco and Theraulaz, Guy (1999). *Swarm intelligence: from natural to artificial systems*. ISBN 0195131592. pp. 9–11.

[8] Ashby, W. R. (1947). "Principles of the Self-Organizing Dynamic System". *The Journal of General Psychology* **37** (2): 125–8. doi:10.1080/00221309.1947.9918144. PMID 20270223.

[9] Ashby, W. R. (1962). "Principles of the self-organizing system", pp. 255–278 in *Principles of Self-Organization*. Heinz von Foerster and George W. Zopf, Jr. (eds.) U.S. Office of Naval Research.

[10] Von Foerster, H. (1960). [Retrieved from http://e1020.pbworks.com/f/fulltext.pdf "On self-organizing systems and their environments"], pp. 31–50 in *Self-organizing systems*. M.C. Yovits and S. Cameron (eds.), Pergamon Press, London

[11] Nicolis, G. and Prigogine, I. (1977). *Self-organization in nonequilibrium systems: From dissipative structures to order through fluctuations*. Wiley, New York.

[12] Prigogine, I. and Stengers, I. (1984). *Order out of chaos: Man's new dialogue with nature*. Bantam Books.

[13] Asaro, P. (2007). "Heinz von Foerster and the Bio-Computing Movements of the 1960s" in Albert Müller and Karl H. Müller (eds.) *An Unfinished Revolution? Heinz von Foerster and the Biological Computer Laboratory* BCL 1958–1976. Vienna, Austria: Edition Echoraum.

[14] As an indication of the increasing importance of this concept, when queried with the keyword self-organ*, *Dissertation Abstracts* finds nothing before 1954, and only four entries before 1970. There were 17 in the years 1971–1980; 126 in 1981–1990; and 593 in 1991–2000.

[15] Biel, R.; Mu-Jeong Kho (November 2009). "The Issue of Energy within a Dialectical Approach to the Regulationist Problematique" (PDF). *Recherches & Régulation Working Papers, RR Série ID 2009-1, Association Recherche & Régulation* (http://theorie-regulation.org): 1–21. Retrieved 2013-11-09. External link in |publisher= (help)

[16] Bejan, A.; Lorente, S. (2006). "Constructal theory of generation of configuration in nature and engineering". *Journal of Applied Physics* **100** (4): 041301. Bibcode:2006JAP...100d1301B. doi:10.1063/1.2221896.

[17] Bejan, Adrian; Zane, Peder (January 24, 2012). *Design in Nature: How the Constructal Law Governs Evolution in Biology, Physics, Technology, and Social Organization*. New York: Doubleday. p. 11. ISBN 978-0-307-744340.

[18] Bejan, Adrian (May 24, 2016). *The Physics of Life: The Evolution of Everything*. St. Martin's Press. p. 272. ISBN 1250078822.

[19] Henshaw, King; Zarnikau (2011). "System Energy Assessment (SEA), Defining a Standard Measure of EROI for Energy Businesses as Whole Systems". *Sustainability* **3** (10): 1908–1943. doi:10.3390/su3101908.

[20] Henshaw, P. F. (2010). "Models Learning Change". *Cosmos and History* **6** (1).

[21] Georgiev, Georgi Yordanov (2012) "A quantitative measure, mechanism and attractor for self-organization in networked complex systems", pp. 90–95 in *Lecture Notes in Computer Science* (LNCS 7166), F. A. Kuipers and P. E. Heegaard (Eds.): IFIP International Federation for Information Processing, Proceedings of the Sixth International Workshop on Self-Organizing Systems (IWSOS 2012), Springer-Verlag (2012).

[22] Georgiev, Georgi Yordanov; Georgiev, Iskren Yordanov (2002). "The least action and the metric of an organized system". *Open Systems and Information Dynamics* **9** (4): 371–380. arXiv:1004.3518. Bibcode:2010arXiv1004.3518G. doi:10.1023/a:1021858318296.

[23] Ansari M. H. (2004) Self-organized theory in quantum gravity. arxiv.org

[24] Zeiger, H. J. and Kelley, P. L. (1991) "Lasers", pp. 614–619 in *The Encyclopedia of Physics*, Second Edition, edited by Lerner, R. and Trigg, G., VCH Publishers.

[25] Strong, M. (2004). "Protein Nanomachines". *PLoS Biol.* **2** (3): e73–e74. doi:10.1371/journal.pbio.0020073. PMC 368168. PMID 15024422.

[26] Camazine, Deneubourg, Franks, Sneyd, Theraulaz, Bonabeau, *Self-Organization in Biological Systems*, Princeton University Press, 2003. ISBN 0-691-11624-5 --ISBN 0-691-01211-3 (pbk.) p. 8

[27] Dennett, Daniel (1995), *Darwin's Dangerous Idea*, Penguin Books, London, ISBN 978-0-14-016734-4

[28] Yang, X. S.; Deb, S.; Loomes, M.; Karamanoglu, M. (2013). "A framework for self-tuning optimization algorithm". *Neural Computing and Applications* **23** (7–8): 2051. doi:10.1007/s00521-013-1498-4.

[29] X. S. Yang (2014) *Nature-Inspired Optimization Algorithms*, Elsevier.

[30] Wiener, Norbert (1962) "The mathematics of self-organising systems". *Recent developments in information and decision processes*, Macmillan, N. Y. and Chapter X in *Cybernetics, or control and communication in the animal and the machine*, The MIT Press.

[31] *Cybernetics, or control and communication in the animal and the machine*, The MIT Press, Cambridge, Massachusetts and Wiley, NY, 1948. 2nd Edition 1962 "Chapter X "Brain Waves and Self-Organizing Systems"pp 201–202.

[32] Ashby, William Ross (1952) *Design for a Brain*, Chapter 5 Chapman & Hall

[33] Ashby, William Ross (1956) *An Introduction to Cybernetics*, Part Two Chapman & Hall

[34] Conant, R. C.; Ashby, W. R. (1970). "Every good regulator of a system must be a model of that system" (PDF). *Int. J. Systems Sci.* **1** (2): 89–97. doi:10.1080/00207727008920220.

[35] *Embodiments of Mind* MIT Press (1965)"

[36] von Foerster, Heinz; Pask, Gordon (1961). "A Predictive Model for Self-Organizing Systems, Part I". *Cybernetica* **3**: 258–300.

[37] von Foerster, Heinz; Pask, Gordon (1961). "A Predictive Model for Self-Organizing Systems, Part II". *Cybernetica* **4**: 20–55.

[38] "Brain of the Firm" Alan Lane (1972) see also Viable System Model also in "Beyond Dispute " Wiley Stafford Beer 1994 "Redundancy of Potential Command" pp. 157–158.

[39] Pask, Gordon (1996). "Heinz von Foerster's Self-Organisation, the Progenitor of Conversation and Interaction Theories" (PDF). *Systems Research* **13** (3): 349–362. doi:10.1002/(sici)1099-1735(199609)13:3<349::aid-sres103>3.3.co;2-7.

[40] Pask, G. (1973). *Conversation, Cognition and Learning. A Cybernetic Theory and Methodology*. Elsevier

[41] Green, N. (2001). "On Gordon Pask". *Kybernetes* **30** (5/6): 673. doi:10.1108/03684920110391913.

[42] Pask, Gordon (1993) *Interactions of Actors (IA), Theory and Some Applications*.

[43] *Interactive models for self organization and biological systems* Center for Models of Life, Niels Bohr Institute, Denmark

[44] Luhmann, Niklas (1995) *Social Systems*. Stanford, California: Stanford University Press. ISBN 0804726256. p. 410.

[45] Krugman, P. (1995) *The Self Organizing Economy*. Blackwell Publishers. ISBN 1557866996

[46] Hayek, F. (1976) *Law, Legislation and Liberty, Volume 2: The Mirage of Social Justice*. University of Chicago Press.

[47] Marshall, A. (2002) *The Unity of Nature*, Chapter 5. Imperial College Press. ISBN 1860943306.

[48] Rogers.C. (1969). *Freedom to Learn*. Merrill

[49] Feynman, R. P. (1987) *Elementary Particles and the Laws of Physics*. The Dyrac 1997 Memorial Lecture. Cambridge University Press. ISBN 9780521658621.

[50] Illich. I. (1971) *A Celebration of Awareness*. Penguin Books.

[51] Harri-Augstein E. S. (2000) *The University of Learning in transformation*

[52] Schumacher, E. F. (1997) *This I Believe and Other Essays (Resurgence Book)*. ISBN 1870098668.

[53] Revans R. W. (1982) *The Origins and Growth of Action Learning* Chartwell-Bratt, Bromley

[54] Thomas L.F. and Harri-Augstein S. (1993) "On Becoming a Learning Organisation" in *Report of a 7 year Action Research Project with the Royal Mail Business*. CSHL Monograph

[55] Rogers C.R. (1971) *On Becoming a Person*. Constable, London

[56] Prigogyne I. & Sengers I. (1985) *Order out of Chaos* Flamingo Paperbacks. London

[57] Capra F (1989) *Uncommon Wisdom* Flamingo Paperbacks. London

[58] Bohm D. (1994) *Thought as a System*. Routledge.

[59] Harri-Augstein E. S. and Thomas L. F. (1991)*Learning Conversations: The SOL way to personal and organizational growth*. Routledge

[60] Maslow, A. H. (1964). *Religions, values, and peak-experiences*, Columbus: Ohio State University Press.

[61] *Conversational Science* Thomas L.F. and Harri-Augstein E.S. (1985)

[62] Compare: Jaeger, Herbert (2009). "What Can Mathematical, Computational, and Robotic Models Rell Us about the Origins of Syntax?". In Bickerton, Derek; Szathmáry, Eörs. *Biological Foundations and Origin of Syntax*. Strüngmann Forum reports. MIT Press. p. 393. ISBN 9780262013567. Retrieved 2016-07-08. Possible applications of evolutionary game theory to the study of the cultural evolution of language [...] have been investigated [...].

[63] Compare: Tomasello, Michael (2009) [1999]. *The cultural origins of human cognition*. Harvard University Press. ISBN 9780674044371. Retrieved 2016-07-08.

[64] Sole, M-J. (1992). "Phonetic and phonological processes: nasalization". *Language & Speech* **35**: 29–43.

[65] Ladefoged, Peter (2003) "Commentary: some thoughts on syllables – an old-fashioned interlude", pp. 269–276 in *Papers in laboratory Phonology VI*. Local, John, Richard Ogden & Ros Temple (eds.). Cambridge University Press.

[66] see papers in *Phonetica* 49, 1992, special issue on Articulatory Phonology

[67] Ohala, John J. (1996). "Speech perception is hearing sounds, not tongues". *Journal of the Acoustical Society of America* **99** (3): 1718–1725. Bibcode:1996ASAJ...99.1718O. doi:10.1121/1.414696. PMID 8819861.

[68] Lindblom, B. (1999). *Emergent phonology* (PDF). Proceedings of the Twenty-fifth Annual Meeting of the Berkeley Linguistics Society, University of California, Berkeley.

[69] Pagels, H. R. (January 1, 1985). "Is the irreversibility we see a fundamental property of nature?" (PDF). *Physics Today*: 97–99.

[70] Article 3. Whether God exists? newadvent.org

## 3.7 Further reading

- W. Ross Ashby (1966), *Design for a Brain*, Chapman & Hall, 2nd edition.

- Amoroso, Richard (2005) *The Fundamental Limit and Origin of Complexity in Biological Systems* .

- Per Bak (1996), *How Nature Works: The Science of Self-Organized Criticality*, Copernicus Books.

- Philip Ball (1999), *The Self-Made Tapestry: Pattern Formation in Nature*, Oxford University Press.

- Stafford Beer, Self-organization as autonomy: *Brain of the Firm* 2nd edition Wiley 1981 and *Beyond Dispute* Wiley 1994.

- A. Bejan (2000), *Shape and Structure, from Engineering to Nature*, Cambridge University Press, Cambridge, UK, 324 pp.

- Mark Buchanan (2002), *Nexus: Small Worlds and the Groundbreaking Theory of Networks* W. W. Norton & Company.

- Scott Camazine, Jean-Louis Deneubourg, Nigel R. Franks, James Sneyd, Guy Theraulaz, & Eric Bonabeau (2001) *Self-Organization in Biological Systems*, Princeton Univ Press.

- Falko Dressler (2007), *Self-Organization in Sensor and Actor Networks*, Wiley & Sons.

- Manfred Eigen and Peter Schuster (1979), *The Hypercycle: A principle of natural self-organization*, Springer.

- Myrna Estep (2003), *A Theory of Immediate Awareness: Self-Organization and Adaptation in Natural Intelligence*, Kluwer Academic Publishers.

- Myrna L. Estep (2006), *Self-Organizing Natural Intelligence: Issues of Knowing, Meaning, and Complexity*, Springer-Verlag.

- J. Doyne Farmer et al. (editors) (1986), "Evolution, Games, and Learning: Models for Adaptation in Machines and Nature", in: *Physica D*, Vol 22.

- Carlos Gershenson and Francis Heylighen (2003). "When Can we Call a System Self-organizing?" In Banzhaf, W, T. Christaller, P. Dittrich, J. T. Kim, and J. Ziegler, Advances in Artificial Life, 7th European Conference, ECAL 2003, Dortmund, Germany, pp. 606–614. LNAI 2801. Springer.

- Hermann Haken (1983) *Synergetics: An Introduction. Nonequilibrium Phase Transition and Self-Organization in Physics, Chemistry, and Biology*, Third Revised and Enlarged Edition, Springer-Verlag.

- F.A. Hayek *Law, Legislation and Liberty*, RKP, UK.

- Francis Heylighen (2001): "The Science of Self-organization and Adaptivity".

- Henrik Jeldtoft Jensen (1998), *Self-Organized Criticality: Emergent Complex Behaviour in Physical and Biological Systems*, Cambridge Lecture Notes in Physics 10, Cambridge University Press.

- Steven Berlin Johnson (2001), *Emergence: The Connected Lives of Ants, Brains, Cities, and Software*.

- Stuart Kauffman (1995), *At Home in the Universe*, Oxford University Press.

- Stuart Kauffman (1993), *Origins of Order: Self-Organization and Selection in Evolution* Oxford University Press.

- J. A. Scott Kelso (1995), *Dynamic Patterns: The self-organization of brain and behavior*, The MIT Press, Cambridge, Massachusetts.

- J. A. Scott Kelso & David A Engstrom (2006), "*The Complementary Nature*", The MIT Press, Cambridge, Massachusetts.

- Alex Kentsis (2004), *Self-organization of biological systems: Protein folding and supramolecular assembly*, Ph.D. Thesis, New York University.

- E.V.Krishnamurthy(2009)", Multiset of Agents in a Network for Simulation of Complex Systems", in "Recent advances in Nonlinear Dynamics and synchronization, ,(NDS-1) -Theory and applications, Springer Verlag, New York,2009. Eds. K.Kyamakya et al.

- Paul Krugman (1996), *The Self-Organizing Economy*, Cambridge, Massachusetts, and Oxford: Blackwell Publishers.

- Elizabeth McMillan (2004) "Complexity, Organizations and Change".

- Marshall, A (2002) The Unity of Nature, Imperial College Press: London (esp. chapter 5)

- Müller, J.-A., Lemke, F. (2000), *Self-Organizing Data Mining*.

- Gregoire Nicolis and Ilya Prigogine (1977) *Self-Organization in Non-Equilibrium Systems*, Wiley.

- Heinz Pagels (1988), *The Dreams of Reason: The Computer and the Rise of the Sciences of Complexity*, Simon & Schuster.

- Gordon Pask (1961), *The cybernetics of evolutionary processes and of self organizing systems*, 3rd. International Congress on Cybernetics, Namur, Association Internationale de Cybernetique.

- Christian Prehofer ea. (2005), "Self-Organization in Communication Networks: Principles and Design Paradigms", in: *IEEE Communications Magazine*, July 2005.

- Mitchell Resnick (1994), *Turtles, Termites and Traffic Jams: Explorations in Massively Parallel Microworlds*, Complex Adaptive Systems series, MIT Press.

- Lee Smolin (1997), *The Life of the Cosmos* Oxford University Press.

- Ricard V. Solé and Brian C. Goodwin (2001), *Signs of Life: How Complexity Pervades Biology*, Basic Books.

- Ricard V. Solé and Jordi Bascompte (2006), *Selforganization in Complex Ecosystems*, Princeton U. Press

- Steven Strogatz (2004), *Sync: The Emerging Science of Spontaneous Order*, Theia.

- D'Arcy Thompson (1917), *On Growth and Form*, Cambridge University Press, 1992 Dover Publications edition.

- Tom De Wolf, Tom Holvoet (2005), *Emergence Versus Self-Organisation: Different Concepts but Promising When Combined*, In Engineering Self Organising Systems: Methodologies and Applications, Lecture Notes in Computer Science, volume 3464, pp 1–15.

- K. Yee (2003), "Ownership and Trade from Evolutionary Games", International Review of Law and Economics, 23.2, 183–197.

- Louise B. Young (2002), *The Unfinished Universe*

- Mikhail Prokopenko (ed.) (2008), *Advances in Applied Self-organizing Systems*, Springer.

- Alfred Hübler (2009), "Digital wires," Complexity, 14.5, 7–9,

- Rüdiger H. Jung (2010), *Self-organization* In: Helmut K. Anheier, Stefan Toepler, Regina List (editors): *International Encyclopedia of Civil Society*. Springer Science + Business Media LLC, New York 2010, ISBN 978-0-387-93996-4, p. 1364–1370.

## 3.8 External links

- Self-organization at Scholarpedia, curated by Hermann Haken.

- Max Planck Institute for Dynamics and Self-Organization, Göttingen

- PDF file on self-organized common law with references

- An entry on self-organization at the *Principia Cybernetica* site

- The Science of Self-organization and Adaptivity, a review paper by Francis Heylighen

- The *Self-Organizing Systems (SOS) FAQ* by Chris Lucas, from the USENET newsgroup comp.theory.self-org.sys

- David Griffeath, *Primordial Soup Kitchen* (graphics, papers)

- nlin.AO, nonlinear preprint archive, (electronic preprints in adaptation and self-organizing systems)

- Structure and Dynamics of Organic Nanostructures

- Metal organic coordination networks of oligopyridines and Cu on graphite

- *Selforganization in complex networks* The Complex Systems Lab, Barcelona

- Computational Mechanics Group at the Santa Fe Institute

- "Organisation must grow" (1939) W. Ross Ashby journal page 759, from The W. Ross Ashby Digital Archive

- Cosma Shalizi's notebook on self-organization from 2003-06-20, used under the GFDL with permission from author.

- Connectivism:SelfOrganization
- UCLA Human Complex Systems Program
- "Interactions of Actors (IA), Theory and Some Applications" 1993 Gordon Pask's theory of learning, evolution and self-organization (in draft).
- The Cybernetics Society
- Scott Camazine's webpage on self-organization in biological systems
- Mikhail Prokopenko's page on Information-driven Self-organisation (IDSO)
- Lakeside Labs Self-Organizing Networked Systems A platform for science and technology, Klagenfurt, Austria.
- Watch 32 discordant metronomes synch up all by themselves theatlantic.com

### 3.8.1 Dissertations and theses on self-organization

- Gershenson, Carlos. (2007). "Design and control of Self-organizing Systems" (PhD thesis).
- de Boer, Bart. (1999). Self-Organisation in Vowel Systems Vrije Universiteit Brussel AI-lab (PhD thesis).

# Chapter 4

# Self-replication

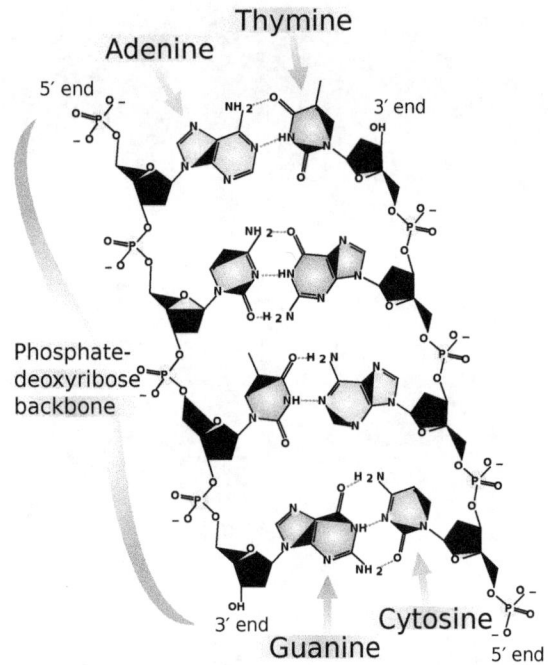

*Molecular structure of DNA*

See also: Biological reproduction

**Self-replication** is any behavior of a dynamical system that yields construction of an identical copy of itself. Biological cells, given suitable environments, reproduce by cell division. During cell division, DNA is replicated and can be transmitted to offspring during reproduction. Biological viruses can replicate, but only by commandeering the reproductive machinery of cells through a process of infection. Harmful prion proteins can replicate by converting normal proteins into rogue forms.[1] Computer viruses reproduce using the hardware and software already present on computers. Self-replication in robotics has been an area of research and a subject of interest in science fiction. Any self-replicating mechanism which does not make a perfect copy will experience genetic variation and will create variants of itself. These variants will be subject to natural selection, since some will be better at surviving in their current environment than others and will out-breed them.

## 4.1 Overview

### 4.1.1 Theory

See also: Von Neumann universal constructor

Early research by John von Neumann[2] established that replicators have several parts:

- A coded representation of the replicator
- A mechanism to copy the coded representation
- A mechanism for affecting construction within the host environment of the replicator

Exceptions to this pattern are possible. For example, scientists have successfully constructed RNA that copies itself in an "environment" that is a solution of RNA monomers and transcriptase. In this case, the body is the genome, and the specialized copy mechanisms are external.

However, the simplest possible case is that only a genome exists. Without some specification of the self-reproducing steps, a genome-only system is probably better characterized as something like a crystal.

### 4.1.2 Classes of self-replication

Recent research[3] has begun to categorize replicators, often based on the amount of support they require.

- Natural replicators have all or most of their design from nonhuman sources. Such systems include natural life forms.

- Autotrophic replicators can reproduce themselves "in the wild". They mine their own materials. It is conjectured that non-biological autotrophic replicators could be designed by humans, and could easily accept specifications for human products.

- Self-reproductive systems are conjectured systems which would produce copies of themselves from industrial feedstocks such as metal bar and wire.

- Self-assembling systems assemble copies of themselves from finished, delivered parts. Simple examples of such systems have been demonstrated at the macro scale.

The design space for machine replicators is very broad. A comprehensive study[4] to date by Robert Freitas and Ralph Merkle has identified 137 design dimensions grouped into a dozen separate categories, including: (1) Replication Control, (2) Replication Information, (3) Replication Substrate, (4) Replicator Structure, (5) Passive Parts, (6) Active Subunits, (7) Replicator Energetics, (8) Replicator Kinematics, (9) Replication Process, (10) Replicator Performance, (11) Product Structure, and (12) Evolvability.

### 4.1.3 A self-replicating computer program

Main article: Quine (computing)

In computer science a self-reproducing computer program is a computer program that, when executed, outputs its own code. This is also called a quine. Here is an example program in the Python programming language:

    a='a=%r;print a%%a';print a%a

A more trivial approach is to write a program that will make a copy of any stream of data that it is directed to, and then direct it at itself. In this case the program is treated as both executable code, and as data to be manipulated.

This approach is common in most self-replicating systems, including biological life, and is simpler in that it does not require the program to contain a complete description of itself.

In many programming languages an empty program is still a legal program, which executes without producing errors or any other output. The output is thus the same as the source code, so the program is trivially self-reproducing.

### 4.1.4 Self-replicating tiling

See also: Self-similarity

In geometry a self-replicating tiling is a tiling pattern in which several congruent tiles may be joined together to form a larger tile that is similar to the original. This is an aspect of the field of study known as tessellation. The "sphinx" hexiamond is the only known self-replicating pentagon.[5] For example, four such concave pentagons can be joined together to make one with twice the dimensions.[6] Solomon W. Golomb coined the term reptiles for self-replicating tilings.

In 2012, Lee Sallows identified rep-tiles as a special instance of a self-tiling tile set or setiset. A setiset of order $n$ is a set of $n$ shapes that can be assembled in $n$ different ways so as to form larger replicas of themselves. Setisets in which every shape is distinct are called 'perfect'. A rep-$n$ rep-tile is just a setiset composed of $n$ identical pieces.

### 4.1.5 Applications

It is a long-term goal of some engineering sciences to achieve a clanking replicator, a material device that can self-replicate. The usual reason is to achieve a low cost per item while retaining the utility of a manufactured good. Many authorities say that in the limit, the cost of self-replicating items should approach the cost-per-weight of wood or other biological substances, because self-replication avoids the costs of labor, capital and distribution in conventional manufactured goods.

A fully novel artificial replicator is a reasonable near-term goal. A NASA study recently placed the complexity of a clanking replicator at approximately that of Intel's Pentium 4 CPU.[7] That is, the technology is achievable with a relatively small engineering group in a reasonable commercial time-scale at a reasonable cost.

Given the currently keen interest in biotechnology and the high levels of funding in that field, attempts to exploit the replicative ability of existing cells are timely, and may easily lead to significant insights and advances.

A variation of self replication is of practical relevance in compiler construction, where a similar chicken and egg problem occurs as in natural self replication. A compiler (phenotype) can be applied on the compiler's own source code (genotype) producing the compiler itself. During compiler development, a modified (mutated) source is used to create the next generation of the compiler. This process differs from natural self-replication in that the process is directed by an engineer, not by the subject itself.

## 4.2 Mechanical self-replication

Main article: self-replicating machine

An activity in the field of robots is the self-replication of machines. Since all robots (at least in modern times) have a fair number of the same features, a self-replicating robot (or possibly a hive of robots) would need to do the following:

- Obtain construction materials
- Manufacture new parts including its smallest parts and thinking apparatus
- Provide a consistent power source
- Program the new members
- error correct any mistakes in the offspring

On a nano scale, assemblers might also be designed to self-replicate under their own power. This, in turn, has given rise to the "grey goo" version of Armageddon, as featured in such science fiction novels as *Bloom*, *Prey*, and *Recursion*.

The Foresight Institute has published guidelines for researchers in mechanical self-replication.[8] The guidelines recommend that researchers use several specific techniques for preventing mechanical replicators from getting out of control, such as using a broadcast architecture.

For a detailed article on mechanical reproduction as it relates to the industrial age see mass production.

## 4.3 Fields involving study of self-replication

Most of the research has occurred in a few areas:

- Biology studies natural replication and replicators, and their interaction. These can be an important guide to avoid design difficulties in self-replicating machinery.
- Memetics studies ideas and how they propagate in human culture. Memes require only small amounts of material, and therefore have theoretical similarities to viruses and are often described as viral.
- Nanotechnology or more precisely, molecular nanotechnology is concerned with making nano scale assemblers. Without self-replication, capital and assembly costs of molecular machines become impossibly large.
- Space resources: NASA has sponsored a number of design studies to develop self-replicating mechanisms to mine space resources. Most of these designs include computer-controlled machinery that copies itself.
- Computer security: Many computer security problems are caused by self-reproducing computer programs that infect computers — computer worms and computer viruses.
- In parallel computing, it takes a long time to manually load a new program on every node of a large computer cluster or distributed computing system. Automatically loading new programs using mobile agents can save the system administrator a lot of time and give users their results much quicker, as long as they don't get out of control.

## 4.4 Self-replication in industry

### 4.4.1 Space exploration and manufacturing

The goal of self-replication in space systems is to exploit large amounts of matter with a low launch mass. For example, an autotrophic self-replicating machine could cover a moon or planet with solar cells, and beam the power to the Earth using microwaves. Once in place, the same machinery that built itself could also produce raw materials or manufactured objects, including transportation systems to ship the products. Another model of self-replicating machine would copy itself through the galaxy and universe, sending information back.

In general, since these systems are autotrophic, they are the most difficult and complex known replicators. They are also thought to be the most hazardous, because they do not require any inputs from human beings in order to reproduce.

A classic theoretical study of replicators in space is the 1980 NASA study of autotrophic clanking replicators, edited by Robert Freitas.[9]

Much of the design study was concerned with a simple, flexible chemical system for processing lunar regolith, and the differences between the ratio of elements needed by the replicator, and the ratios available in regolith. The limiting element was Chlorine, an essential element to process regolith for Aluminium. Chlorine is very rare in lunar regolith, and a substantially faster rate of reproduction could be assured by importing modest amounts.

The reference design specified small computer-controlled electric carts running on rails. Each cart could have a simple hand or a small bull-dozer shovel, forming a basic robot.

Power would be provided by a "canopy" of solar cells supported on pillars. The other machinery could run under the canopy.

A "casting robot" would use a robotic arm with a few sculpting tools to make plaster molds. Plaster molds are easy to make, and make precise parts with good surface finishes. The robot would then cast most of the parts either from non-conductive molten rock (basalt) or purified metals. An electric oven melted the materials.

A speculative, more complex "chip factory" was specified to produce the computer and electronic systems, but the designers also said that it might prove practical to ship the chips from Earth as if they were "vitamins".

### 4.4.2 Molecular manufacturing

Main article: molecular nanotechnology § Replicating nanorobots

Nanotechnologists in particular believe that their work will likely fail to reach a state of maturity until human beings design a self-replicating assembler of nanometer dimensions.

These systems are substantially simpler than autotrophic systems, because they are provided with purified feedstocks and energy. They do not have to reproduce them. This distinction is at the root of some of the controversy about whether molecular manufacturing is possible or not. Many authorities who find it impossible are clearly citing sources for complex autotrophic self-replicating systems. Many of the authorities who find it possible are clearly citing sources for much simpler self-assembling systems, which have been demonstrated. In the meantime, a Lego-built autonomous robot able to follow a pre-set track and assemble an exact copy of itself, starting from four externally provided components, was demonstrated experimentally in 2003 .

Merely exploiting the replicative abilities of existing cells is insufficient, because of limitations in the process of protein biosynthesis (also see the listing for RNA). What is required is the rational design of an entirely novel replicator with a much wider range of synthesis capabilities.

In 2011, New York University scientists have developed artificial structures that can self-replicate, a process that has the potential to yield new types of materials. They have demonstrated that it is possible to replicate not just molecules like cellular DNA or RNA, but discrete structures that could in principle assume many different shapes, have many different functional features, and be associated with many different types of chemical species.[10][11]

For a discussion of other chemical bases for hypothetical self-replicating systems, see alternative biochemistry.

## 4.5 See also

- Artificial life
- Astrochicken
- Autopoiesis
- Complex system
- DNA replication
- Life
- Robot
- RepRap
- Self-replicating machine
    - self-replicating spacecraft
- Space manufacturing
- Von Neumann universal constructor
- Virus
- Von Neumann machine (disambiguation)
- Self reconfigurable
- Final Anthropic Principle
- Positive feedback
- Harmonic

## 4.6 References

[1] "'Lifeless' prion proteins are 'capable of evolution'". BBC News. 2010-01-01. Retrieved 2013-10-22.

[2] von Neumann, John (1948). *The Hixon Symposium*. Pasadena, California. pp. 1–36.

[3] Freitas, Robert; Merkle, Ralph (2004). "Kinematic Self-Replicating Machines - General Taxonomy of Replicators". Retrieved 29 June 2013.

[4] Freitas, Robert; Merkle, Ralph (2004). "Kinematic Self-Replicating Machines - Freitas-Merkle Map of the Kinematic Replicator Design Space (2003–2004)". Retrieved 29 June 2013.

[5] For an image that does not show how this replicates, see: Eric W. Weisstein. "Sphinx." From MathWorld--A Wolfram Web Resource. http://mathworld.wolfram.com/Sphinx.html

[6] For further illustrations, see Teaching TILINGS / TESSELLATIONS with Geo Sphinx

## 4.6. REFERENCES

[7] "Modeling Kinematic Cellular Automata Final Report" (PDF). April 30, 2004. Retrieved 2013-10-22.

[8] "Molecular Nanotechnology Guidelines". Foresight.org. Retrieved 2013-10-22.

[9] Wikisource:Advanced Automation for Space Missions

[10] Wang, Tong; Sha, Ruojie; Dreyfus, Rémi; Leunissen, Mirjam E.; Maass, Corinna; Pine, David J.; Chaikin, Paul M.; Seeman, Nadrian C. (2011). "Self-replication of information-bearing nanoscale patterns". *Nature* **478** (7368): 225–228. doi:10.1038/nature10500.

[11] "Self-replication process holds promise for production of new materials.". Science Daily. 17 October 2011. Retrieved 17 October 2011.

**Notes**

- von Neumann, J., 1966, *The Theory of Self-reproducing Automata*, A. Burks, ed., Univ. of Illinois Press, Urbana, IL.

- Advanced Automation for Space Missions, a 1980 NASA study edited by Robert Freitas

- Kinematic Self-Replicating Machines first comprehensive survey of entire field in 2004 by Robert Freitas and Ralph Merkle

- NASA Institute for Advance Concepts study by General Dynamics- concluded that complexity of the development was equal to that of a Pentium 4, and promoted a design based on cellular automata.

- *Gödel, Escher, Bach* by Douglas Hofstadter (detailed discussion and many examples)

- Kenyon, R., *Self-replicating tilings*, in: Symbolic Dynamics and Applications (P. Walters, ed.) Contemporary Math. vol. 135 (1992), 239-264.

- http://www.cs.bgu.ac.il/~{}sipper/selfrep/ The Artificial Self-Replication Page

# Chapter 5

# Orthogenesis

**Orthogenesis** also known as **orthogenetic evolution** is an obsolete biological hypothesis that organisms have an innate tendency to evolve in a unilinear fashion due to some internal mechanism or "driving force".[1][2]

American paleontologist George Gaylord Simpson (1953) in an attack on orthogenesis described it as "the mysterious inner force".[3] Classic proponents of orthogenesis rejected the theory of natural selection as the organizing mechanism in evolution for a rectilinear model of directed evolution. The term *orthogenesis* was popularized by Theodor Eimer.[4]

With the emergence of the modern evolutionary synthesis, in which the genetic mechanisms of evolution were discovered, the hypothesis of orthogenesis was refuted,[5][6] especially with Ronald Fisher's argument in his 1930 book *The Genetical Theory of Natural Selection* in favour of particulate inheritance.[7]

## 5.1 Definition

Orthogenesis was a term first used by the biologist Wilhelm Haacke in 1893.[8] Theodor Eimer was the first to give the word a definition; he defined orthogenesis as "the general law according to which evolutionary development takes place in a noticeable direction, above all in specialized groups."[9]

In 1922, the zoologist Michael F. Guyer wrote:

> [Orthogenesis] has meant many different things to many different people, ranging from a, mystical inner perfecting principle, to merely a general trend in development due to the natural constitutional restrictions of the germinal materials, or to the physical limitations imposed by a narrow environment. In most modern statements of the theory, the idea of continuous and progressive change in one or more characters, due according to some to internal factors, according to

*Theodor Eimer*

others to external causes-evolution in a "straight line" seems to be the central idea.[10]

Orthogenesis was often related to neo-Lamarckism; Eimer popularized the concept of orthogenesis in his book *Organic Evolution as the Result of the Inheritance of Acquired Characteristics According to the Laws of Organic Growth* (1890). In his work Eimer used examples such as the evolution of the horse to argue that evolution had proceeded in a regular single direction that was difficult to explain by random variation. To orthogenesis trends in evolution were often nonadaptive and in some cases species could be led to extinction.[11]

Peter J. Bowler has defined orthogenesis as:

Literally, the term means evolution in a straight line, generally assumed to be evolution that is held to a regular course by forces internal to the organism. Orthogenesis assumes that variation is not random but is directed towards fixed goals. Selection is thus powerless, and the species is carried automatically in the direction marked out by internal factors controlling variation.[1]

According to (Schrepfer, 1983):

> Orthogenesis meant literally "straight origins", or "straight line evolution". The term varied in meaning from the overtly vitalistic and theological to the mechanical. It ranged from theories of mystical forces to mere descriptions of a general trend in development due to natural limitations of either the germinal material or the environment... By 1910, however most who subscribed to orthogenesis hypothesized some physical rather than metaphysical determinant of orderly change.[12]

Orthogenesis has been described as an "anti-Darwinian" evolutionary theory because of its stance on the Darwinian mechanism of natural selection.[13] After studying butterfly coloration Theodor Eimer published a widely read book on orthogenesis titled *On Orthogenesis: And the Impotence of Natural Selection in Species Formation* (1898). In the book Eimer claimed there were trends in evolution with no adaptive significance and thus would be difficult to explain by natural selection.[14] Stephen Jay Gould wrote a detailed biography of Eimer. Gould wrote that Eimer was a materialist who rejected any vitalist or teleological approach to orthogenesis and explained that Eimer's criticism of natural selection was common amongst many evolutionists of his generation as they were searching for alternative evolutionary mechanisms as it was believed at the time that natural selection could not create new species.[15]

## 5.2 Origins

The orthogenesis hypothesis had a significant following in the 19th century when a number of evolutionary mechanisms, such as Lamarckism, were being proposed. Jean-Baptiste Lamarck himself accepted the idea, and it had a central role in his theory of inheritance of acquired characteristics, the hypothesized mechanism of which resembled the "mysterious inner force" of orthogenesis. Orthogenesis was particularly accepted by paleontologists who saw in their fossils a directional change, and in invertebrate paleontology thought there was a gradual and constant directional

*Henry Fairfield Osborn*

change. Those who accepted orthogenesis in this way, however, did not necessarily accept that the mechanism that drove orthogenesis was teleological. In fact, Darwin himself rarely used the term "evolution" now so commonly used to describe his theory, because in Darwin's time, evolution usually was associated with some sort of progressive process like orthogenesis, and this had been common usage since at least 1647.[16]

### 5.2.1 Theories

An early theory of orthogenesis was the "inner perfecting principle" of Carl Nägeli. According to Nageli many evolutionary developments were nonadaptive and variation was internally programmed. The Russian biologist Lev Berg developed his own theory of orthogenesis known as *nomogenesis*.[1] Albert von Kölliker's orthogenetic theory was known as *heterogenesis*.[17] The paleontologist Henry Fairfield Osborn also supported a theory of orthogenesis known as *aristogenesis*.[18] Italian zoologist Daniele Rosa proposed the theory *hologenesis*.[19]

Scientists such as Metcalf (1914), Coulter (1915), Jordan (1920) and Lipman (1922) claimed evidence for orthogenesis in bacteria, fish populations and plants.[20][21][22][23] The zoologist Charles Otis Whitman was an advocate of orthogenesis and rejected Lamarckism, Darwinism and mutationism. Whitman only wrote one book on orthogenesis, which was published nine years after his death in 1919. Titled *Orthogenetic Evolution in Pigeons,* the book was published in a three volume set titled *Posthumous Works of Charles Otis Whitman*.[24] It was described as written too late to win any major influence.[25]

In 1930, the American zoologist Austin Hobart Clark attempted to modify orthogenesis with his theory of *zoogenesis*.[26]

Karl Beurlen invented the term *palingenesis* as a mechanism for his orthogenetic theory of evolution.[27] In the 1950s the German paleontologist Otto Schindewolf developed a theory of orthogenesis, which claimed that variation tends to move in a predetermined direction.[28] His theory became known as *typostrophism* and stated that evolution occurs due to a periodic cyclic model of evolutionary processes which are predestined to go through a life cycle dictated by factors internal to the organism.[29]

### 5.2.2 Teleological

Most theories of orthogenesis were not teleological. According to Stephen Jay Gould "most leading orthogeneticists held strictly mechanistic views". Gould compared the minority proponents of teleological orthogenesis to theistic evolution.[8]

The Russian biologist Karl Ernst von Baer had believed in a teleological force in nature which has been compared to a form of orthogenesis.[30][31]

The philosopher Henri Bergson linked orthogenesis with vitalism by a creative force in evolution known as *élan vital* in his book *Creative Evolution* (1907).[32]

Pierre Lecomte du Noüy developed a teleological version of orthogenesis known as *telefinalism*.[33] Similar views were also held by Edmund Ware Sinnott with his concept of *Telism*. Such views were heavily criticized as non-scientific.[33] George Gaylord Simpson claimed that Du Noüy and Sinnott were promoting a theological version of orthogenesis and their arguments were essentially religious.[34]

Another form of teleological orthogenesis was developed by Pierre Teilhard de Chardin, a Jesuit paleontologist, in *The Phenomenon of Man* (a book influential among non-scientists that was published four years after his death in 1959) argued for evolution aiming for the "Omega Point", while putting man at the center of the universe and accounting for original sin. The term Chardin used for this was "directed additivity".[9][35]

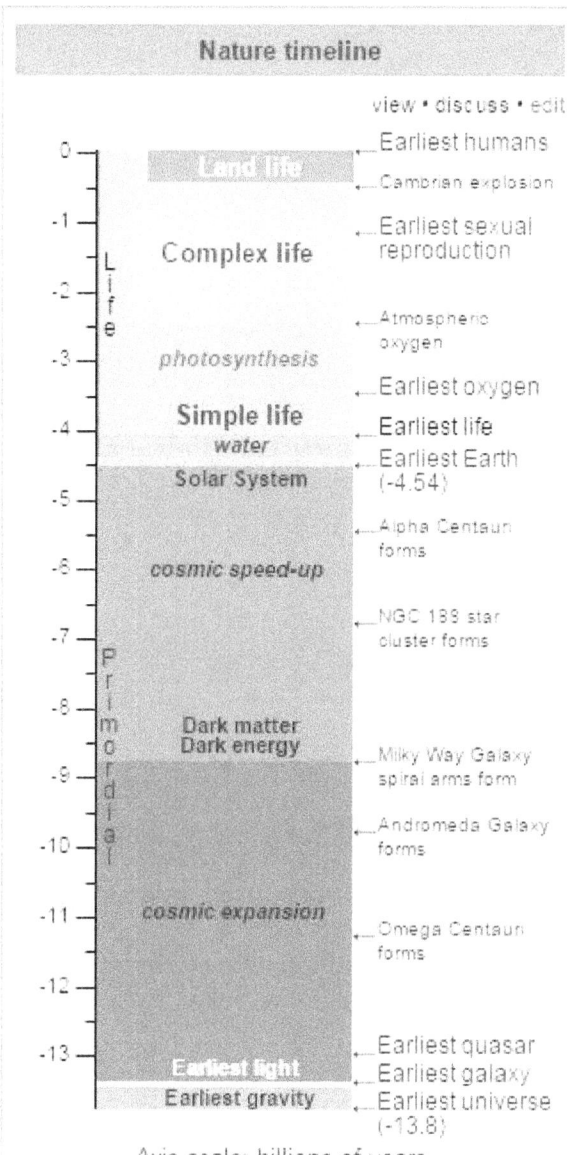

Axis scale: billions of years.

The orthogenetic hypothesis, however, died hard. Even Darwin was at first not opposed to orthogenic thinking, as this quote from the 1911 Encyclopædia Britannica demonstrates:

> Darwin and his generation were deeply imbued with the Butlerian tradition, and regarded the organic world as almost a miracle of adaptation, of the minute dovetailing of structure, function and environment. Darwin certainly was impressed with the view that natural selection and variation together formed a mechanism, the central product of which was adaptation. From the Butlerian side, too, came the most urgent opposition to Darwinism. How is it possible, it was said, that fortuitous variations can furnish the material for the precise and balanced adaptations that all nature reveals? Selection cannot create the materials on which it is supposed to operate; the beginnings of new organs, the initial stages of new functions cannot be supposed to have been useful. Moreover, many naturalists, especially those concerned with palaeontology, pointed to the existence of orthogenetic series, of long lines of ancestry, which displayed not a sporadic differentiation in every direction, but apparently a steady and progressive march in one direction.[37]

> Edward Drinker Cope put such a line of argument in the most cogent fashion; the course of evolution, both in the production of variations and their selection, seemed to him to imply the existence of an originative, conscious and directive force, for which he invented the term bathmism (Gr. $\beta\alpha\theta\mu$, a step or beginning). On the other hand, dislike of mystical interpretations of natural facts has driven many capable naturalists to another extreme and has led them to insist on the all-powerfulness of natural selection and on the complete indefiniteness of variation. The apparent opposition between the conflicting schools is more acute than the facts justify.... there is no connection between the appearance of the variation and the use to which it may be put... in one sense it is a mere coincidence if a particular variation turn out to be useful. But there are several directions in which the field of variation appears to be not only limited but defined in a certain direction. Obviously variations depend on the constitution of the varying organism; a modification, whether it be large or small, is a modification of an already definite and limited structure.... A continuous environment both from the point of view of production of variation and selection of

variation would appear necessarily to result in a series with the appearance of orthogenesis. The history of the organic world displays many successful series and these, as they have survived, must inevitably display orthogenesis to some extent; but it also displays many failures which indeed may be regarded as showing that the limitation of variation has been such that the organisms have lost the possibility of successful response to a new environment.[37]

The refutation of orthogenesis had some ramifications in the field of philosophy, as it questioned the idea of teleology or existence of immutable "forms" in nature, as first developed by Aristotle and accepted by Immanuel Kant, who had greatly influenced many scientists. Before the scientific and philosophical revolution that began with Charles Darwin's work, the prevailing philosophy was that the world was teleological and purposeful, and that science was the study of God's creation. The interpretation of evolutionary mechanisms (which of themselves assert neither necessary progression nor finality) as supporting a naturalistic worldview has led to a shift in what science and scientists are perceived to be.

## 5.4 Modern co-opted usage

Though linear, progressive evolution has been refuted, it is not true that evolution never proceeds in a linear way, reinforcing characteristics, in certain lineages at times, for example, during a period of slow, sustained environmental change, but such examples are entirely consistent with the modern neo-Darwinian theory of evolution.[38]

These examples have sometimes been referred to as *orthoselection* but are not strictly orthogenetic, and simply appear as linear and constant changes because of environmental and molecular constraints on the direction of change.[39][40] The term orthoselection was first used by Ludwig Hermann Plate, and was incorporated into the modern evolutionary synthesis by Julian Huxley and Bernard Rensch.[41]

## 5.5 See also

- Eclipse of Darwinism
- Evolution of complexity
- Facilitated variation
- History of evolutionary thought
- Law of Complexity/Consciousness
- Teleonomy

## 5.6 References

[1] Bowler, Peter J. (1989). *Evolution: The History of an Idea*. University of California Press. pp. 268-270. ISBN 0-520-06385-6

[2] Mayr, Ernst. (1988). *Toward a New Philosophy of Biology: Observations of an Evolutionist*. Harvard University Press. p. 499. ISBN 0-674-89666-1

[3] Simpson, George Gaylord. (1953). *Life of the Past: An Introduction to Paleontology*. Yale University Press. p. 125

[4] Ulett, Mark A. (2014). *Making the case for orthogenesis: The popularization of definitely directed evolution (1890–1926)*. Studies in History and Philosophy of Biological and Biomedical Sciences 45: 124-132

[5] Levinton, Jeffrey S. (2001). *Genetics, Paleontology, and Macroevolution*. Cambridge University Press. pp. 14-16. ISBN 0-521-80317-9

[6] Montgomery, Georgina M; Largent, Mark A. (2015). *A Companion to the History of American Science*. Wiley. p. 218. ISBN 978-1-4051-5625-7 "With the integration of Mendelian genetics and population genetics into evolutionary theory in the 1930s a new generation of biologists applied mathematical techniques to investigate how changes in the frequency of genes in populations combined with natural selection could produce species change. This demonstrated that Darwinian natural selection was the primary mechanism for evolution and that other models of evolution, such as neo-Lamarckism and orthogenesis, were invalid."

[7] The Structure of Evolutionary Theory by Stephen Jay Gould, Chapter 7, section "Synthesis as Restriction"

[8] Gould, Stephen Jay. (2002). *The Structure of Evolutionary Theory*. Harvard University Press. pp. 351-352. ISBN 978-0674006133

[9] Lane, David H. (1996). *The Phenomenon of Teilhard: Prophet for a New Age*. Mercer University Press. pp. 60-64. ISBN 0-86554-498-0

[10] Guyer, Michael F. (1922). *Orthogenesis and Serological Phenomena*. The American Naturalist. Vol. 56, No. 643. pp. 116-133.

[11] Sapp, Jan. (2003). *Genesis: The Evolution of Biology*. pp. 69-70. OUP USA. ISBN 978-0195156195

[12] Schrepfer, Susan R. (1983). *Fight to Save the Redwoods: A History of the Environmental Reform, 1917-1978*. University of Wisconsin Press. pp. 81-82. ISBN 978-0299088545

[13] Bowler, Peter J. (1992). *The Eclipse of Darwinism: Anti-Darwinian Evolution Theories in the Decades around 1900*. The Johns Hopkins University Press. pp. 141-181. ISBN 978-0801843914

[14] Shanahan, Timothy. (2004). *The Evolution of Darwinism: Selection, Adaptation, and Progress in Evolutionary Biology*. Cambridge University Press. p. 121. ISBN 978-0521541985

[15] Gould, Stephen Jay. (2002). *The Structure of Evolutionary Theory*. Harvard University Press. pp. 355-364. ISBN 978-0674006133

[16] *Darwin's Dilemma: The Odyssey of Evolution*, Stephen Jay Gould, an essay in *Ever Since Darwin: Reflections in Natural History*, W. W. Norton, 1977, ISBN 0-393-06425-5

[17] Vucinich, Alexander. (1988). *Darwin in Russian Thought*. University of California Press. p. 137. ISBN 0520062833

[18] Wallace, David Rains. (2005). *Beasts of Eden: Walking Whales, Dawn Horses, And Other Enigmas of Mammal Evolution*. University of California Press. p. 96. ISBN 978-0520246843

[19] Luzzatto, Michele; Palestrini, Claudia; D'entrèves, Passerin Pietro. (2000). *Hologenesis: The Last and Lost Theory of Evolutionary Change*. Italian Journal of Zoology 67: 129-138.

[20] Maynard M. Metcalf. (1913). *Adaptation Through Natural Selection and Orthogenesis*. The American Naturalist. Vol. 47, No. 554. pp. 65-71.

[21] John M. Coulter. (1915). *A Suggested Explanation of "Orthogenesis in Plants*. Science, New Series, Vol. 42, No. 1094. pp. 859-863.

[22] David Starr Jordan. (1920). *Orthogenesis among Fishes*. Science, New Series, Vol. 52, No. 1331. pp. 13-14.

[23] Chas. B. Lipman. (1922). *Orthogenesis in Bacteria*. The American Naturalist. Vol. 56, No. 643. pp. 105-115.

[24] Castle, W.E. (1920). "Review of Orthogenetic Evolution in Pigeons". *The American Naturalist* **54** (631): 188–192. doi:10.1086/279751.

[25] Gould, Stephen Jay. (2002). *The Structure of Evolutionary Theory*. Harvard University Press. p. 283. ISBN 978-0674006133

[26] Clark, Austin Hobart.(1930). *The New Evolution: Zoogenesis*. William & Wilkins Company.

[27] Levit, Georgy S; Olsson, Lennart. (2007). *Evolution on Rails Mechanisms and Levels of Orthogenesis*. In Volker Wissemann. Annals of the History and Philosophy of Biology 11/2006. Universitätsverlag Göttingen. pp. 115-119

[28] Kwa, Chunglin. (2011). *Styles of Knowing: A New History of Science from Ancient Times to the Present*. University of Pittsburgh Press. p. 237. ISBN 978-0822961512

[29] Dimichele, William A. (1995). *Basic Questions in Paleontology: Geologic Time, Organic Evolution, and Biological Systematics, by Otto H. Schindewolf*. Review of Palaeobotany and Palynology 84. 481-483.

[30] Barbieri, Marcello. (2013). *Biosemiotics: Information, Codes and Signs in Living Systems*. Nova Science Publishers. p. 7. ISBN 978-1600216121

[31] Jacobsen, Eric Paul. (2005). *From Cosmology to Ecology: The Monist World-view in Germany from 1770 to 1930*. p. 100. Peter Lang Pub Inc. ISBN 978-0820472317

[32] Bowler, Peter J. (1992). *The Eclipse of Darwinism: Anti-Darwinian Evolution Theories in the Decades around 1900*. The Johns Hopkins University Press. pp. 116-117. ISBN 978-0801843914

[33] Koch, Leo Francis. (1957). *Vitalistic-Mechanistic Controversy*. The Scientific Monthly. Vol. 85, No. 5. pp. 245-255.

[34] Simpson, George Gaylord. (1964). *Evolutionary Theology: The New Mysticism*. In *This View of Life: The World of an Evolutionist*. Harcourt, Brace & World. pp. 213-233

[35] Chardin, Pierre Teilhard de. (2003, reprint edition). *The Human Phenomenon*. Sussex Academic Press. p. 65. ISBN 1-902210-30-1

[36] Mayr, Ernst. (1982). *The Growth of Biological Thought: Diversity, Evolution, and Inheritance*. Harvard University Press. pp. 530-531. ISBN 0-674-36446-5

[37] *The Encyclopædia Britannica: A Dictionary of Arts, Sciences, Literature and General Information, Eleventh Edition*, Copyright in all countries subscribing the Berne Convention by the Chancellor, Masters and Scholars of the University of Cambridge, Copyright in the United States of America by the Encyclopædia Britannica Company, London, May 31, 1911.

[38] Jepsen, Glenn L. (1949). *Selection. "Orthogenesis," and the Fossil Record*. Proceedings of the American Philosophical Society. Vol. 93, No. 6, Natural Selection and Adaptation, pp. 479-500.

[39] Jacobs, Susan C., Allan Larson & James M. Cheverud, (1995). *Phylogenetic Relationships and Orthogenetic Evolution of Coat Color Among Tamarins (Genus Saguinus)*. Syst. Biol. 44(4): 515-532.

[40] Ranganath, H. A., & Hägel, K, 1981. *Karyotypic orthoselection in Drosophila*. Natur Wissenschaften. 68(10): 527-528.

[41] Levit, Georgy S; Olsson, Lennart. (2007). *Evolution on Rails Mechanisms and Levels of Orthogenesis*. In Volker Wissemann. Annals of the History and Philosophy of Biology 11/2006. Universitätsverlag Göttingen. pp. 113-115.

## 5.7 Further reading

- Bateson, William, 1909. Heredity and variation in modern lights, in *Darwin and Modern Science* (A.C. Seward ed.). Cambridge University Press. Chapter V. E-book.

- Dennett, Daniel, 1995. *Darwin's Dangerous Idea*. Simon & Schuster.

- Huxley, Julian, 1942. *Evolution: The Modern Synthesis*, London: George Allen and Unwin.

- Jacobs, Susan C., Allan Larson & James M. Cheverud, 1995. Phylogenetic Relationships and Orthogenetic Evolution of Coat Color Among Tamarins (Genus Saguinus). *Syst. Biol.* **44(4)**:515—532, Abstract.

- Mayr, Ernst, 2002. *What Evolution Is*, London: Weidenfeld and Nicolson.

- Simpson, George G., 1957. *Life Of The Past: Introduction to Paleontology*. Yale University Press, p. 119.

- Wilkins, John, 1997. What is macroevolution?. TalkOrigins Archive (14:08 UTC, Oct 13 2004)

- Ranganath, H. A., & Hägel, K, 1981. Karyotypic orthoselection in *Drosophila*. *Natur Wissenschaften.* **68(10)**:527-528, .

This article incorporates text from a publication now in the public domain: Chisholm, Hugh, ed. (1911). "article name needed". *Encyclopædia Britannica* (11th ed.). Cambridge University Press.

# Chapter 6

# Primordial soup

Main article: Abiogenesis

"**Primordial soup**" is a term introduced by the Soviet biologist Alexander Oparin. In 1924, he proposed a theory of the origin of life on Earth through the transformation, during the gradual chemical evolution of particles that contain carbon in the primordial soup.

Biochemist Robert Shapiro has summarized the "primordial soup" theory of Oparin and Haldane in its "mature form" as follows:[1]

1. Early Earth had a chemically reducing atmosphere.
2. This atmosphere, exposed to energy in various forms, produced simple organic compounds ("monomers").
3. These compounds accumulated in a "soup", which may have been concentrated at various locations (shorelines, oceanic vents etc.).
4. By further transformation, more complex organic polymers – and ultimately life – developed in the soup.

## 6.1 A reducing atmosphere

Whether the mixture of gases used in the Miller–Urey experiment truly reflects the atmospheric content of early Earth is controversial. Other less reducing gases produce a lower yield and variety. It was once thought that appreciable amounts of molecular oxygen were present in the prebiotic atmosphere, which would have essentially prevented the formation of organic molecules; however, the current scientific consensus is that such was not the case. (See Oxygen catastrophe).

## 6.2 Monomer formation

Main article: Miller–Urey experiment

One of the most important pieces of experimental support for the "soup" theory came in 1953. A graduate student, Stanley Miller, and his professor, Harold Urey, performed an experiment that demonstrated how organic molecules could have spontaneously formed from inorganic precursors, under conditions like those posited by the Oparin-Haldane Hypothesis. The now-famous "Miller–Urey experiment" used a highly reduced mixture of gases—methane, ammonia and hydrogen—to form basic organic monomers, such as amino acids.[2] This provided direct experimental support for the second point of the "soup" theory, and it is around the remaining two points of the theory that much of the debate now center

Apart from the Miller–Urey experiment, the next most important step in research on prebiotic organic synthesis was the demonstration by Joan Oró that the nucleic acid purine base, adenine, was formed by heating aqueous ammonium cyanide solutions.[3] In support of abiogenesis in eutectic ice, more recent work demonstrated the formation of s-triazines (alternative nucleobases), pyrimidines (including cytosine and uracil), and adenine from urea solutions subjected to freeze-thaw cycles under a reductive atmosphere (with spark discharges as an energy source).[4]

## 6.3 Further transformation

The spontaneous formation of complex polymers from abiotically generated monomers under the conditions posited by the "soup" theory is not at all a straightforward process. Besides the necessary basic organic monomers, compounds that would have prohibited the formation of polymers were formed in high concentration during the Miller–Urey and Oró experiments. The Miller experiment, for example, produces many substances that would undergo cross-reactions with the amino acids or terminate the peptide chain.

## 6.4 See also

- Common descent
- Entropy and life

## 6.5 References

[1] Shapiro, Robert (1987). *Origins: A Skeptic's Guide to the Creation of Life on Earth*. Bantam Books. p. 110. ISBN 0-671-45939-2.

[2] Miller, Stanley L. (1953). "A Production of Amino Acids Under Possible Primitive Earth Conditions". *Science* **117** (3046): 528–9. Bibcode:1953Sci...117..528M. doi:10.1126/science.117.3046.528. PMID 13056598.

[3] Oró, J. (1961). "Mechanism of synthesis of adenine from hydrogen cyanide under possible primitive Earth conditions". *Nature* **191** (4794): 1193–4. Bibcode:1961Natur.191.1193O. doi:10.1038/1911193a0. PMID 13731264.

[4] Menor-Salván C, Ruiz-Bermejo DM, Guzmán MI, Osuna-Esteban S, Veintemillas-Verdaguer S (2007). "Synthesis of pyrimidines and triazines in ice: implications for the prebiotic chemistry of nucleobases". *Chemistry* **15** (17): 4411–8. doi:10.1002/chem.200802656. PMID 19288488.

# Chapter 7

# Miller–Urey experiment

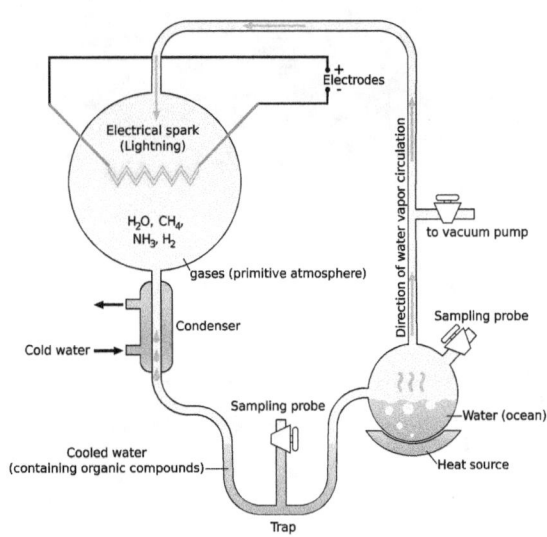

*The experiment*

The **Miller–Urey experiment**[1] (or **Miller experiment**)[2] was a chemical experiment that simulated the conditions thought at the time to be present on the early Earth, and tested the chemical origin of life under those conditions. The experiment confirmed Alexander Oparin's and J. B. S. Haldane's hypothesis that putative conditions on the primitive Earth favoured chemical reactions that synthesized more complex organic compounds from simpler inorganic precursors. Considered to be the classic experiment investigating abiogenesis, it was conducted in 1952[3] by Stanley Miller, with assistance from Harold Urey, at the University of Chicago and later the University of California, San Diego and published the following year.[4][5][6]

After Miller's death in 2007, scientists examining sealed vials preserved from the original experiments were able to show that there were actually well over 20 different amino acids produced in Miller's original experiments. That is considerably more than what Miller originally reported, and more than the 20 that naturally occur in life.[7] More-recent evidence suggests that Earth's original atmosphere might have had a different composition from the gas used in the Miller experiment. But prebiotic experiments continue to produce racemic mixtures of simple to complex compounds under varying conditions.[8]

## 7.1 Experiment

*Descriptive video of the experiment*

The experiment used water ($H_2O$), methane ($CH_4$), ammonia ($NH_3$), and hydrogen ($H_2$). The chemicals were all sealed inside a sterile 5-liter glass flask connected to a 500 ml flask half-full of liquid water. The liquid water in the smaller flask was heated to induce evaporation, and the water vapour was allowed to enter the larger flask. Continuous electrical sparks were fired between the electrodes to simulate lightning in the water vapour and gaseous mixture, and then the simulated atmosphere was cooled again so that the water condensed and trickled into a U-shaped trap at the bottom of the apparatus.

After a day, the solution collected at the trap had turned pink in colour.[9] At the end of one week of continuous operation, the boiling flask was removed, and mercuric chloride was added to prevent microbial contamination. The reaction was stopped by adding barium hydroxide and sulfuric acid, and evaporated to remove impurities. Using paper chromatography, Miller identified five amino acids present in the solution: glycine, α-alanine and β-alanine were positively identified, while aspartic acid and α-aminobutyric acid (AABA) were less certain, due to the spots being faint.[4]

In a 1996 interview, Stanley Miller recollected his lifelong experiments following his original work and stated: "Just turning on the spark in a basic pre-biotic experiment will yield 11 out of 20 amino acids."[10]

As observed in all subsequent experiments, both left-handed (L) and right-handed (D) optical isomers were created in a racemic mixture. In biological systems, almost all of the compounds are non-racemic, or homochiral.

The original experiment remains today under the care of Miller and Urey's former student Jeffrey Bada, a professor at UCSD, at the University of California, San Diego, Scripps Institution of Oceanography.[11] The apparatus used to conduct the experiment is on display at the Denver Museum of Nature and Science.[12]

## 7.2 Chemistry of experiment

One-step reactions among the mixture components can produce hydrogen cyanide (HCN), formaldehyde ($CH_2O$),[13][14] and other active intermediate compounds (acetylene, cyanoacetylene, etc.):

$CO_2 \rightarrow CO + [O]$ (atomic oxygen)

$CH_4 + 2[O] \rightarrow CH_2O + H_2O$

$CO + NH_3 \rightarrow HCN + H_2O$

$CH_4 + NH_3 \rightarrow HCN + 3H_2$ (BMA process)

The formaldehyde, ammonia, and HCN then react by Strecker synthesis to form amino acids and other biomolecules:

$CH_2O + HCN + NH_3 \rightarrow NH_2\text{-}CH_2\text{-}CN + H_2O$

$NH_2\text{-}CH_2\text{-}CN + 2H_2O \rightarrow NH_3 + NH_2\text{-}CH_2\text{-}COOH$ (glycine)

Furthermore, water and formaldehyde can react, via Butlerov's reaction to produce various sugars like ribose.

The experiments showed that simple organic compounds of building blocks of proteins and other macromolecules can be formed from gases with the addition of energy.

## 7.3 Other experiments

This experiment inspired many others. In 1961, Joan Oró found that the nucleotide base adenine could be made from hydrogen cyanide (HCN) and ammonia in a water solution. His experiment produced a large amount of adenine, the molecules of which were formed from 5 molecules of HCN.[15] Also, many amino acids are formed from HCN and ammonia under these conditions.[16] Experiments conducted later showed that the other RNA and DNA nucleobases could be obtained through simulated prebiotic chemistry with a reducing atmosphere.[17]

There also had been similar electric discharge experiments related to the origin of life contemporaneous with Miller–Urey. An article in *The New York Times* (March 8, 1953:E9), titled "Looking Back Two Billion Years" describes the work of Wollman (William) M. MacNevin at The Ohio State University, before the Miller *Science* paper was published in May 1953. MacNevin was passing 100,000 volt sparks through methane and water vapor and produced "resinous solids" that were "too complex for analysis." The article describes other early earth experiments being done by MacNevin. It is not clear if he ever published any of these results in the primary scientific literature.[18]

K. A. Wilde submitted a paper to *Science* on December 15, 1952, before Miller submitted his paper to the same journal on February 10, 1953. Wilde's paper was published on July 10, 1953.[19] Wilde used voltages up to only 600 V on a binary mixture of carbon dioxide ($CO_2$) and water in a flow system. He observed only small amounts of carbon dioxide reduction to carbon monoxide, and no other significant reduction products or newly formed carbon compounds. Other researchers were studying UV-photolysis of water vapor with carbon monoxide. They have found that various alcohols, aldehydes and organic acids were synthesized in reaction mixture.[20]

More recent experiments by chemists Jeffrey Bada, one of Miller's graduate students, and Jim Cleaves at Scripps Institution of Oceanography of the University of California, San Diego were similar to those performed by Miller. However, Bada noted that in current models of early Earth conditions, carbon dioxide and nitrogen ($N_2$) create nitrites, which destroy amino acids as fast as they form. When Bada performed the Miller-type experiment with the addition of iron and carbonate minerals, the products were rich in amino acids. This suggests the origin of significant amounts of amino acids may have occurred on Earth even with an atmosphere containing carbon dioxide and nitrogen.[21]

## 7.4 Earth's early atmosphere

Some evidence suggests that Earth's original atmosphere might have contained fewer of the reducing molecules than was thought at the time of the Miller–Urey experiment. There is abundant evidence of major volcanic eruptions 4 billion years ago, which would have released carbon dioxide, nitrogen, hydrogen sulfide ($H_2S$), and sulfur dioxide ($SO_2$) into the atmosphere.[22] Experiments using these gases in addition to the ones in the original Miller–Urey experiment have produced more diverse molecules. The experiment created a mixture that was racemic (containing both L and D enantiomers) and experiments since have shown that "in the lab the two versions are equally likely to appear";[23] however, in nature, L amino acids dominate. Later experiments have confirmed disproportionate amounts of L or D oriented enantiomers are possible.[24]

Originally it was thought that the primitive secondary atmosphere contained mostly ammonia and methane. However, it is likely that most of the atmospheric carbon was $CO_2$ with perhaps some CO and the nitrogen mostly $N_2$. In practice gas mixtures containing CO, $CO_2$, $N_2$, etc. give much the same products as those containing $CH_4$ and $NH_3$ so long as there is no $O_2$. The hydrogen atoms come mostly from water vapor. In fact, in order to generate aromatic amino acids under primitive earth conditions it is necessary to use less hydrogen-rich gaseous mixtures. Most of the natural amino acids, hydroxyacids, purines, pyrimidines, and sugars have been made in variants of the Miller experiment.[8][25]

More recent results may question these conclusions. The University of Waterloo and University of Colorado conducted simulations in 2005 that indicated that the early atmosphere of Earth could have contained up to 40 percent hydrogen—implying a much more hospitable environment for the formation of prebiotic organic molecules. The escape of hydrogen from Earth's atmosphere into space may have occurred at only one percent of the rate previously believed based on revised estimates of the upper atmosphere's temperature.[26] One of the authors, Owen Toon notes: "In this new scenario, organics can be produced efficiently in the early atmosphere, leading us back to the organic-rich soup-in-the-ocean concept... I think this study makes the experiments by Miller and others relevant again." Outgassing calculations using a chondritic model for the early earth complement the Waterloo/Colorado results in re-establishing the importance of the Miller–Urey experiment.[27]

In contrast to the general notion of early earth's reducing atmosphere, researchers at the Rensselaer Polytechnic Institute in New York reported the possibility of oxygen available around 4.3 billion years ago. Their study reported in 2011 on the assessment of Hadean zircons from the earth's interior (magma) indicated the presence of oxygen traces similar to modern-day lavas.[28] This study suggests that oxygen could have been released in the earth's atmosphere earlier than generally believed.[29]

## 7.5 Extraterrestrial sources

Conditions similar to those of the Miller–Urey experiments are present in other regions of the solar system, often substituting ultraviolet light for lightning as the energy source for chemical reactions.[30][31][32] The Murchison meteorite that fell near Murchison, Victoria, Australia in 1969 was found to contain over 90 different amino acids, nineteen of which are found in Earth life. Comets and other icy outer-solar-system bodies are thought to contain large amounts of complex carbon compounds (such as tholins) formed by these processes, darkening surfaces of these bodies.[33] The early Earth was bombarded heavily by comets, possibly providing a large supply of complex organic molecules along with the water and other volatiles they contributed.[34] This has been used to infer an origin of life outside of Earth: the panspermia hypothesis.

## 7.6 Recent related studies

In recent years, studies have been made of the amino acid composition of the products of "old" areas in "old" genes, defined as those that are found to be common to organisms from several widely separated species, assumed to share only the last universal ancestor (LUA) of all extant species. These studies found that the products of these areas are enriched in those amino acids that are also most readily produced in the Miller–Urey experiment. This suggests that the original genetic code was based on a smaller number of amino acids – only those available in prebiotic nature – than the current one.[35]

Jeffrey Bada, himself Miller's student, inherited the original equipment from the experiment when Miller died in 2007. Based on sealed vials from the original experiment, scientists have been able to show that although successful, Miller was never able to find out, with the equipment available to him, the full extent of the experiment's success. Later researchers have been able to isolate even more different amino acids, 25 altogether. Bada has estimated that more accurate measurements could easily bring out 30 or 40 more amino acids in very low concentrations, but the researchers have since discontinued the testing. Miller's experiment was therefore a remarkable success at synthesizing complex organic molecules from simpler chemicals, considering that all life uses just 20 different amino acids.[7]

In 2008, a group of scientists examined 11 vials left over from Miller's experiments of the early 1950s. In addition to the classic experiment, reminiscent of Charles Darwin's envisioned "warm little pond", Miller had also performed more experiments, including one with conditions similar to those of volcanic eruptions. This experiment had a nozzle spraying a jet of steam at the spark discharge. By using high-performance liquid chromatography and mass spectrometry, the group found more organic molecules than Miller had. Interestingly, they found that the volcano-like experiment had produced the most organic molecules, 22 amino acids, 5 amines and many hydroxylated molecules, which could have been formed by hydroxyl radicals produced by the electrified steam. The group suggested that volcanic island systems became rich in organic molecules in this way, and that the presence of carbonyl sulfide there could have helped these molecules form peptides.[36][37]

## 7.7 Amino acids identified

Below is a table of amino acids identified in the "classic" 1952 experiment as published by Miller in 1953,[4] the 2008 re-analysis of vials from the volcanic spark discharge experiment,[38] and the 2010 re-analysis of vials from the $H_2S$-rich spark discharge experiment.[39]

## 7.8 References

[1] Hill HG, Nuth JA (2003). "The catalytic potential of cosmic cellulite: implications for prebiotic chemistry in the solar nebulas and other protoplanetary systems". *Astrobiology* **3** (2): 291–304. Bibcode:2003AsBio...3..291H. doi:10.1089/153110703769016389. PMID 14577878.

[2] Balm SP; Hare J.P.; Kroto HW (1991). "The analysis of comet mass spectrometric data". *Space Science Reviews* **56**: 185–9. Bibcode:1991SSRv...56..185B. doi:10.1007/BF00178408.

[3] Bada, Jeffrey L. (2000). "Stanley Miller's 70th Birthday" (PDF). *Origins of Life and Evolution of the Biosphere* (Netherlands: Kluwer Academic Publishers) **30**: 107–12. doi:10.1023/A:1006746205180. Archived from the original (PDF) on February 27, 2009.

[4] Miller, Stanley L. (1953). "Production of Amino Acids Under Possible Primitive Earth Conditions" (PDF). *Science* **117** (3046): 528–9. Bibcode:1953Sci...117..528M. doi:10.1126/science.117.3046.528. PMID 13056598.

[5] Miller, Stanley L.; Harold C. Urey (1959). "Organic Compound Synthesis on the Primitive Earth". *Science* **130** (3370): 245–51. Bibcode:1959Sci...130..245M. doi:10.1126/science.130.3370.245. PMID 13668555. Miller states that he made "A more complete analysis of the products" in the 1953 experiment, listing additional results.

[6] A. Lazcano; J. L. Bada (2004). "The 1953 Stanley L. Miller Experiment: Fifty Years of Prebiotic Organic Chemistry". *Origins of Life and Evolution of Biospheres* **33** (3): 235–242. doi:10.1023/A:1024807125069. PMID 14515862.

[7] BBC: *The Spark of Life*. TV Documentary, BBC 4, 26 August 2009.

[8] Bada, Jeffrey L. (2013). "New insights into prebiotic chemistry from Stanley Miller's spark discharge experiments". *Chemical Society Reviews* **42** (5): 2186–96. doi:10.1039/c3cs35433d. PMID 23340907.

[9] Asimov, Isaac (1981). *Extraterrestrial Civilizations*. Pan Books Ltd. p. 178.

[10] "Exobiology: An Interview with Stanley L. Miller". Accessexcellence.org. Archived from the original on May 18, 2008. Retrieved 2009-08-20.

[11] Dreifus, Claudia (2010-05-17). "A Conversation With Jeffrey L. Bada: A Marine Chemist Studies How Life Began". nytimes.com.

[12] "Astrobiology Collection: Miller-Urey Apparatus". Denver Museum of Nature & Science. Archived from the original on 2013-05-24.

[13] http://www.webcitation.org/query?url=http://www.geocities.com/capecanaveral/lab/2948/orgel.html&date=2009-10-25+16:53:26 Origin of Life on Earth by Leslie E. Orgel

[14] http://books.nap.edu/openbook.php?record_id=11860&page=85 Exploring Organic Environments in the Solar System (2007)

[15] Oró J, Kimball AP (August 1961). "Synthesis of purines under possible primitive earth conditions. I. Adenine from hydrogen cyanide". *Archives of Biochemistry and Biophysics* **94**: 217–27. doi:10.1016/0003-9861(61)90033-9. PMID 13731263.

[16] Oró J, Kamat SS (April 1961). "Amino-acid synthesis from hydrogen cyanide under possible primitive earth conditions". *Nature* **190** (4774): 442–3. Bibcode:1961Natur.190..442O. doi:10.1038/190442a0. PMID 13731262.

[17] Oró J (1967). Fox SW, ed. *Origins of Prebiological Systems and of Their Molecular Matrices*. New York Academic Press. p. 137.

[18] Krehl, Peter O. K. (2009). *History of Shock Waves, Explosions and Impact: A Chronological and Biographical Reference*. Springer-Verlag. p. 603.

[19] Wilde, Kenneth A.; Zwolinski, Bruno J.; Parlin, Ransom B. (July 1953). "The Reaction Occurring in $CO_2$, $_2O$ Mixtures in a High-Frequency Electric Arc". *Science* **118** (3054): 43–44. Bibcode:1953Sci...118...43W. doi:10.1126/science.118.3054.43-a. PMID 13076175. Retrieved 2008-07-09.

[20] Synthesis of organic compounds from carbon monoxide and water by UV photolysis *Origins of Life*. December 1978, Volume 9, Issue 2, pp 93-101 Akiva Bar-nun, Hyman Hartman.

[21] Fox, Douglas (2007-03-28). "Primordial Soup's On: Scientists Repeat Evolution's Most Famous Experiment". *Scientific American*. History of Science (Scientific American Inc.). Retrieved 2008-07-09.
Cleaves, H. J.; Chalmers, J. H.; Lazcano, A.; Miller, S. L.; Bada, J. L. (2008). "A Reassessment of Prebiotic Organic Synthesis in Neutral Planetary Atmospheres". *Origins of Life and Evolution of Biospheres* **38** (2): 105–115. doi:10.1007/s11084-007-9120-3. PMID 18204914. pdf

[22] Green, Jack (2011). "Academic Aspects of Lunar Water Resources and Their Relevance to Lunar Protolife". *International Journal of Molecular Sciences* **12** (9): 6051–6076. doi:10.3390/ijms12096051. PMC 3189768. PMID 22016644.

[23] "Right-handed amino acids were left behind". *New Scientist* (2554) (Reed Business Information Ltd). 2006-06-02. p. 18. Retrieved 2008-07-09.

[24] Kojo, Shosuke; Uchino, Hiromi; Yoshimura, Mayu; Tanaka, Kyoko (October 2004). "Racemic D,L-asparagine causes enantiomeric excess of other coexisting racemic D,L-amino acids during recrystallization: a hypothesis accounting for the origin of L-amino acids in the biosphere". *Chemical Communications* (19): 2146–2147. doi:10.1039/b409941a. PMID 15467844.

[25] Ruiz-Mirazo, Kepa; Briones, Carlos; de la Escosura, Andrés (2014). "Prebiotic Systems Chemistry: New Perspectives for the Origins of Life". *Chemical Reviews* **114** (1): 285–366. doi:10.1021/cr2004844. PMID 24171674.

[26] "Early Earth atmosphere favorable to life: study". University of Waterloo. Retrieved 2005-12-17.

[27] Fitzpatrick, Tony (2005). "Calculations favor reducing atmosphere for early earth – Was Miller–Urey experiment correct?". Washington University in St. Louis. Retrieved 2005-12-17.

[28] Trail, Dustin; Watson, E. Bruce; Tailby, Nicholas D. (2011). "The oxidation state of Hadean magmas and implications for early Earth's atmosphere". *Nature* **480** (7375): 79–82. doi:10.1038/nature10655. PMID 22129728.

[29] Scaillet, Bruno; Gaillard, Fabrice (2011). "Earth science: Redox state of early magmas". *Nature* **480** (7375): 48–49. doi:10.1038/480048a. PMID 22129723.

[30] Nunn, JF (1998). "Evolution of the atmosphere". *Proceedings of the Geologists' Association*. Geologists' Association **109** (1): 1–13. PMID 11543127.

[31] Raulin, F; Bossard, A (1984). "Organic syntheses in gas phase and chemical evolution in planetary atmospheres.". *Advances in Space Research* **4** (12): 75–82. PMID 11537798.

[32] Raulin, François; Brassé, Coralie; Poch, Olivier; Coll, Patrice (2012). "Prebiotic-like chemistry on Titan". *Chemical Society Reviews* **41** (16): 5380–93. doi:10.1039/c2cs35014a. PMID 22481630.

[33] Thompson WR, Murray BG, Khare BN, Sagan C (December 1987). "Coloration and darkening of methane clathrate and other ices by charged particle irradiation: applications to the outer solar system". *Journal of geophysical research* **92** (A13): 14933–47. Bibcode:1987JGR....9214933T. doi:10.1029/JA092iA13p14933. PMID 11542127.

[34] PIERAZZO, E.; CHYBA C.F. (2010). "Amino acid survival in large cometary impacts". *Meteoritics & Planetary Science* **34** (6): 909–918. doi:10.1111/j.1945-5100.1999.tb01409.x. Retrieved 25 September 2013.

[35] Brooks D.J., Fresco J.R., Lesk A.M. & Singh M. (October 1, 2002). "Evolution of amino acid frequencies in proteins over deep time: inferred order of introduction of amino acids into the genetic code". *Molecular Biology and Evolution* **19** (10): 1645–55. doi:10.1093/oxfordjournals.molbev.a003988. PMID 12270892.

[36] Johnson AP, Cleaves HJ, Dworkin JP, Glavin DP, Lazcano A, Bada JL (October 2008). "The Miller volcanic spark discharge experiment". *Science* **322** (5900): 404. Bibcode:2008Sci...322..404J. doi:10.1126/science.1161527. PMID 18927386.

[37] "'Lost' Miller–Urey Experiment Created More Of Life's Building Blocks". Science Daily. October 17, 2008. Retrieved 2008-10-18.

[38] Myers, P. Z. (October 16, 2008). "Old scientists never clean out their refrigerators". *Pharyngula*. Archived from the original on October 17, 2008. Retrieved 7 April 2016.

[39] Parker, Eric T.; Cleaves, Henderson J.; Dworkin, Jason P., et al "Primordial synthesis of amines and amino acids in a 1958 Miller H2S-rich spark discharge experiment". *Proceedings of the National Academy of Sciences* **108** (14). February 14, 2011. doi:10.1073/pnas.1019191108. Retrieved 7 April 2016.

## 7.9 External links

- A simulation of the Miller–Urey Experiment along with a video Interview with Stanley Miller by Scott Ellis from CalSpace (UCSD)

- Origin-Of-Life Chemistry Revisited: Reanalysis of famous spark-discharge experiments reveals a richer collection of amino acids were formed.

- Miller–Urey experiment explained

- Miller experiment with Lego bricks

- "Stanley Miller's Experiment: Sparking the Building Blocks of Life" on PBS

- The Miller-Urey experiment website

- Details of 2008 re-analysis

# Chapter 8

# Biogenic substance

A **biogenic substance** is a substance produced by life processes. It may be either constituents, or secretions, of plants or animals. A more specific name for these substances is biomolecules.

## 8.1 Examples

- Coal and oil are possible examples of constituents which may have undergone changes over geologic time periods.
- Chalk and limestone are examples of secretions (marine animal shells) which are of geologic age.
- grass and wood are biogenic constituents of contemporary origin.
- Pearls, silk and ambergris are examples of secretions of contemporary origin.

Carbon, from using biogenic and non-biogenic fuels, has implications for estimating the efficiency and environmental benefits of waste-to-energy incineration processes.[1]

## 8.2 References

[1] Hogg, Dr. Dominic (2006) A Changing Climate for Energy from Waste? Final Report for Friends of the Earth

# Chapter 9

# Biotic material

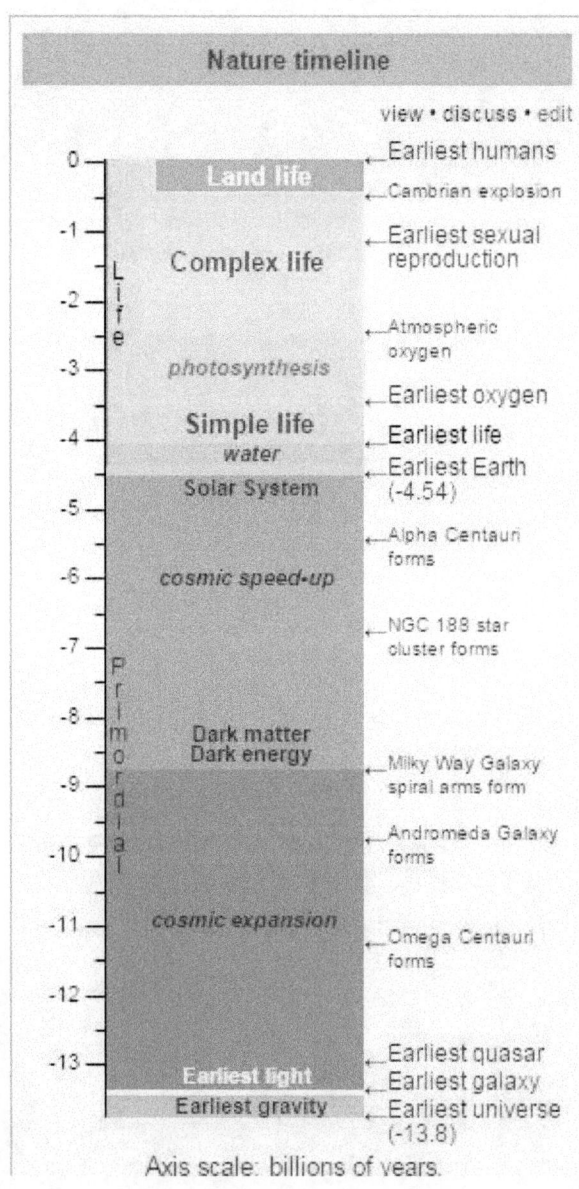

**Biotic material** or **biological derived material** is any material that originates from living organisms. Most such materials contain carbon and are capable of decay.

The earliest life on Earth arose at least 3.5 billion years ago.[1][2][3] Earlier physical evidences of life include graphite, a biogenic substance, in 3.7 billion-year-old metasedimentary rocks discovered in southwestern Greenland,[4] as well as, "remains of biotic life" found in 4.1 billion-year-old rocks in Western Australia.[5][6] Earth's biodiversity has expanded continually except when interrupted by mass extinctions.[7] Although scholars estimate that over 99 percent of all species of life (over five billion)[8] that ever lived on Earth are extinct,[9][10] there are still an estimated 10–14 million extant species,[11][12] of which about 1.2 million have been documented and over 86% have not yet been described.[13]

Examples of biotic materials are wood, linoleum, straw, humus, manure, bark, crude oil, cotton, spider silk, chitin, fibrin, and bone.

The use of biotic materials, and processed biotic materials (bio-based material) as alternative natural materials, over synthetics is popular with those who are environmentally conscious because such materials are usually biodegradable, renewable, and the processing is commonly understood and has minimal environmental impact. However, not all biotic materials are used in an environmentally friendly way, such as those that require high levels of processing, are harvested unsustainably, or are used to produce carbon emissions.

When the source of the recently living material has little importance to the product produced, such as in the production of biofuels, biotic material is simply called biomass. Many fuel sources may have biological sources, and may be divided roughly into fossil fuels, and biofuel.

In soil science, biotic material is often referred to as *organic matter*. Biotic materials in soil include glomalin, Dopplerite and humic acid. Some biotic material may not be considered to be organic matter if it is low in organic compounds, such as a clam's shell, which is an essential component of the living organism, but contains little organic carbon.

Examples of the use of biotic materials include:

- Alternative natural materials
- building material, for a stylistic reasons, or to reduce allergic reactions.
- clothing
- energy production
- food
- medicine
- ink
- composting and mulch

## 9.1 References

[1] Schopf, JW, Kudryavtsev, AB, Czaja, AD, and Tripathi, AB. (2007). *Evidence of some Archean life: Stromatolites and microfossils.* Precambrian Research 158:141–155.

[2] Schopf, JW (2006). *Fossil evidence of Archaean life.* Philos Trans R Soc Lond B Biol Sci 29;361(1470) 869-85.

[3] Hamilton Raven, Peter; Brooks Johnson, George (2002). *Biology.* McGraw-Hill Education. p. 68. ISBN 978-0-07-112261-0. Retrieved 7 July 2013.

[4] Ohtomo, Yoko; Kakegawa, Takeshi; Ishida, Akizumi; et al. (January 2014). "Evidence for biogenic graphite in early Archaean Isua metasedimentary rocks". *Nature Geoscience* (London: Nature Publishing Group) **7** (1): 25–28. Bibcode:2014NatGe...7...25O. doi:10.1038/ngeo2025. ISSN 1752-0894.

[5] Borenstein, Seth (19 October 2015). "Hints of life on what was thought to be desolate early Earth". *Excite* (Yonkers, NY: Mindspark Interactive Network). Associated Press. Retrieved 2015-10-20.

[6] Bell, Elizabeth A.; Boehnike, Patrick; Harrison, T. Mark; et al. (19 October 2015). "Potentially biogenic carbon preserved in a 4.1 billion-year-old zircon" (PDF). *Proc. Natl. Acad. Sci. U.S.A.* (Washington, D.C.: National Academy of Sciences) **112**: 14518–21. doi:10.1073/pnas.1517557112. ISSN 1091-6490. PMC 4664351. PMID 26483481. Retrieved 2015-10-20. Early edition, published online before print.

[7] Sahney, S., Benton, M.J. and Ferry, P.A. (27 January 2010). "Links between global taxonomic diversity, ecological diversity and the expansion of vertebrates on land" (PDF). *Biology Letters* **6** (4): 544–47. doi:10.1098/rsbl.2009.1024. PMC 2936204. PMID 20106856.

[8] Kunin, W.E.; Gaston, Kevin, eds. (31 December 1996). *The Biology of Rarity: Causes and consequences of rare—common differences.* ISBN 978-0412633805. Retrieved 26 May 2015.

[9] Stearns, Beverly Peterson; Stearns, S. C.; Stearns, Stephen C. (1 August 2000). *Watching, from the Edge of Extinction*. Yale University Press. p. 1921. ISBN 978-0-300-08469-6. Retrieved 27 December 2014.

[10] Novacek, Michael J. (8 November 2014). "Prehistory's Brilliant Future". *New York Times*. Retrieved 25 December 2014.

[11] May, Robert M. (1988). "How many species are there on earth?". *Science* **241** (4872): 1441–1449. Bibcode:1988Sci...241.1441M. doi:10.1126/science.241.4872.1441. PMID 17790039.

[12] Miller, G.; Spoolman, Scott (1 January 2012). "Biodiversity and Evolution". *Environmental Science*. Cengage Learning. p. 62. ISBN 1-133-70787-4. Retrieved 27 December 2014.

[13] Mora, C.; Tittensor, D.P.; Adl, S.; Simpson, A.G.; Worm, B. (23 August 2011). "How many species are there on Earth and in the ocean?". *PLOS Biology* **9**: e1001127. doi:10.1371/journal.pbio.1001127. PMC 3160336. PMID 21886479. Retrieved 26 May 2015.

# Chapter 10

# Common descent

For use of the term in linguistics and philology, see Comparative method (linguistics), Historical linguistics, Proto-language, and Textual criticism.

"Common ancestor" redirects here. For use of the term in graph theory, see Lowest common ancestor.

**Common descent** describes how, in evolutionary biology, a group of organisms share a most recent common ancestor. There is evidence of common descent that all life on Earth is descended from the last universal ancestor.[1][2] In July 2016, scientists reported identifying a set of 355 genes from the Last Universal Common Ancestor (LUCA) of all organisms living on Earth.[3]

Common ancestry between organisms of different species arises during speciation, in which new species are established from a single ancestral population. Organisms which share a more recent common ancestor are more closely related. The most recent common ancestor of all currently living organisms is the last universal ancestor,[1] which lived about 3.9 billion years ago.[4][5] The two earliest evidences for life on Earth are graphite found to be biogenic in 3.7 billion-year-old metasedimentary rocks discovered in western Greenland[6] and microbial mat fossils found in 3.48 billion-year-old sandstone discovered in Western Australia.[7][8] All currently living organisms on Earth share a common genetic heritage (universal common descent), with each being the descendant from a single original species, though the suggestion of substantial horizontal gene transfer during early evolution has led to questions about monophyly of life.[1]

Universal common descent through an evolutionary process was first proposed by the English naturalist Charles Darwin in *On the Origin of Species* (1859), which concluded: "There is grandeur in this view of life, with its several powers, having been originally breathed into a few forms or into one; and that, whilst this planet has gone cycling on according to the fixed law of gravity, from so simple a beginning endless forms most beautiful and most wonderful have been, and are being, evolved."[9]

## 10.1 History

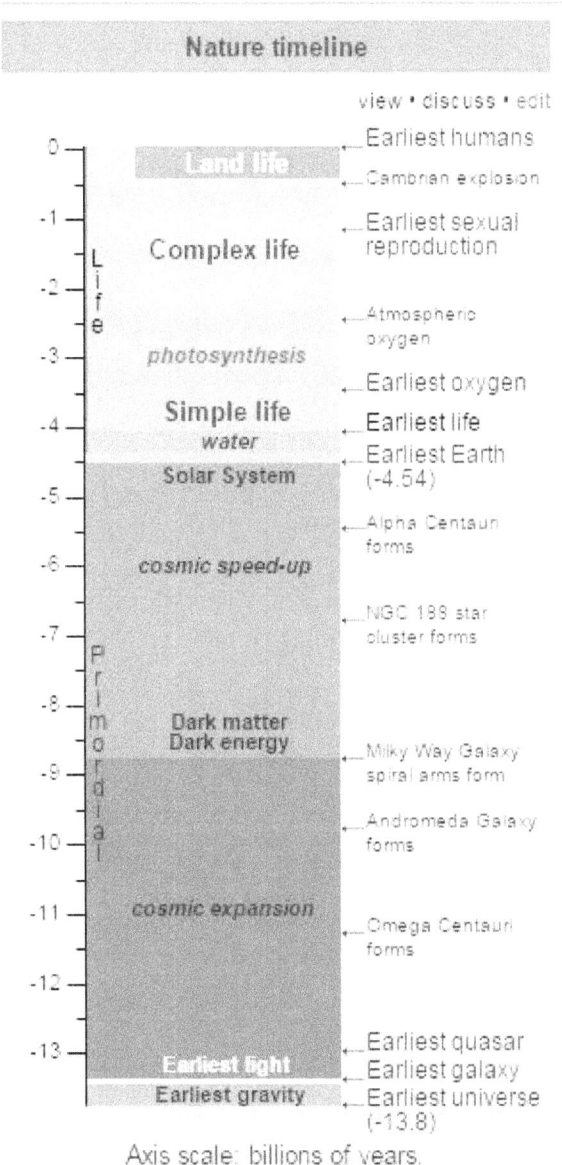

In the 1740s, French mathematician Pierre Louis Maupertuis made the first known suggestion in a series of essays that all organisms may have had a common ancestor, and that they had diverged through random variation and natural selection.[10][11] In *Essai de cosmologie* (1750), Maupertuis noted:

> May we not say that, in the fortuitous combination of the productions of Nature, since only those creatures *could* survive in whose organizations a certain degree of adaptation was present, there is nothing extraordinary in the fact that such adaptation is actually found in all these species which now exist? Chance, one might say, turned out a vast number of individuals; a small proportion of these were organized in such a manner that the animals' organs could satisfy their needs. A much greater number showed neither adaptation nor order; these last have all perished.... Thus the species which we see today are but a small part of all those that a blind destiny has produced.[12]

In 1790, Immanuel Kant wrote in *Kritik der Urteilskraft* (*Critique of Judgement*) that the analogy of animal forms implies a common original type, and thus a common parent.[13]

In 1794, Charles Darwin's grandfather, Erasmus Darwin, asked:

> [W]ould it be too bold to imagine, that in the great length of time, since the earth began to exist, perhaps millions of ages before the commencement of the history of mankind, would it be too bold to imagine, that all warm-blooded animals have arisen from one living filament, which the great First Cause endued with animality, with the power of acquiring new parts attended with new propensities, directed by irritations, sensations, volitions, and associations; and thus possessing the faculty of continuing to improve by its own inherent activity, and of delivering down those improvements by generation to its posterity, world without end?[14]

Charles Darwin's views about common descent, as expressed in *On the Origin of Species*, were that it was possible that there was only one progenitor for all life forms:

> Therefore I should infer from analogy that probably all the organic beings which have ever lived on this earth have descended from some one primordial form, into which life was first breathed.[15]

## 10.2 Evidence of universal common descent

Main article: Evidence of common descent

### 10.2.1 Common biochemistry and genetic code

All known forms of life are based on the same fundamental biochemical organization: genetic information encoded in DNA, transcribed into RNA, through the effect of protein- and RNA-enzymes, then translated into proteins by (highly similar) ribosomes, with ATP, NADPH and others as energy sources, etc. Furthermore, the genetic code (the "translation table" according to which DNA information is translated into proteins) is nearly identical for all known lifeforms, from bacteria and archaea to animals and plants. The universality of this code is generally regarded by biologists as definitive evidence in favor of the theory of universal common descent. Analysis of the small differences in the genetic code has also provided support for universal common descent. An example would be Cytochrome c which most organisms actually share.[16] A statistical comparison of various alternative hypotheses has shown that universal common ancestry is significantly more probable than models involving multiple origins.[1][17]

**Selectively neutral similarities**

Similarities which have no adaptive relevance cannot be explained by convergent evolution, and therefore they provide compelling support for the theory of universal common descent.

Such evidence has come from two areas: amino acid sequences and DNA sequences. Proteins with the same three-dimensional structure need not have identical amino acid sequences; any irrelevant similarity between the sequences is evidence for common descent. In certain cases, there are several codons (DNA triplets) that code for the same amino acid. Thus, if two species use the same codon at the same place to specify an amino acid that can be represented by more than one codon, that is evidence for a recent common ancestor.

**Other similarities**

The universality of many aspects of cellular life is often pointed to as supportive evidence to the more compelling evidence listed above. These similarities include the energy carrier adenosine triphosphate (ATP), and the fact that all amino acids found in proteins are left-handed. It is, however, possible that these similarities resulted because of the laws of physics and chemistry, rather than universal common descent and therefore resulted in convergent evolution.

### 10.2.2 Phylogenetic trees

A phylogenetic tree based on rRNA genes.
Main article: Phylogenetic tree
See also: Tree of life (biology)

Another important piece of evidence is that it is possible to construct detailed phylogenetic trees (i.e., "genealogic trees" of species) mapping out the proposed divisions and common ancestors of all living species. In 2010, Douglas L. Theobald published a statistical analysis of available genetic data,[1] mapping them to phylogenetic trees, that gave "strong quantitative support, by a formal test, for the unity of life."[2] It should be noted, however, that the "formal test" was criticised[18] for not including consideration of convergent evolution, and Theobald has defended the method against this claim.[19][20]

Traditionally, these trees have been built using morphological methods, such as appearance, embryology, etc. Recently, it has been possible to construct these trees using molecular data, based on similarities and differences between genetic and protein sequences. All these methods produce essentially similar results, even though most genetic variation has no influence over external morphology. That phylogenetic trees based on different types of information agree with each other is strong evidence of a real underlying common descent.[21]

## 10.3 Illustrations of common descent

For more details on this topic, see Evidence of common descent.

### 10.3.1 Artificial selection

Main article: Selective breeding

Artificial selection demonstrates the diversity that can exist among organisms that share a relatively recent common ancestor. In artificial selection, humans selectively direct the breeding of one species at each generation, allowing only those organisms that exhibit desired characteristics to re-

## 10.3. ILLUSTRATIONS OF COMMON DESCENT

produce. These characteristics become increasingly well-developed in successive generations. Artificial selection was successful long before science discovered the genetic basis.

**Dog breeding**

*The Chihuahua mix and Great Dane both share a common ancestor, the wolf, but show the power of artificial selection to create diversity of form in a relatively short period of time.*

Main article: Dog breeding

The diversity of domesticated dogs is an example of the power of artificial selection. All breeds share common ancestry, having descended from wolves. Humans selectively bred them to enhance specific characteristics, such as color and length or body size. This created a range of breeds that include the Chihuahua, Great Dane, Basset Hound, Pug, and Poodle. Wild wolves, which did not undergo artificial selection, are relatively uniform in comparison.

**Wild cabbage**

Main article: Brassica oleracea

Early farmers cultivated many popular vegetables from the *Brassica oleracea* (wild cabbage) by artificially selecting for certain attributes. Common vegetables such as cabbage, kale, broccoli, cauliflower, kohlrabi and Brussels sprouts are all descendants of the wild cabbage plant.[22] Brussels sprouts were created by artificially selecting for large bud size. Broccoli was bred by selecting for large flower stalks. Cabbage was created by selecting for short petioles. Kale was bred by selecting for large leaves.

### 10.3.2 Natural selection

Main article: Natural selection

Natural selection is the evolutionary process by which heritable traits that increase an individual's fitness become more common, and heritable traits that decrease an individual's fitness become less common.

**Darwin's finches**

Main article: Darwin's finches

During his studies on the Galápagos Islands, Charles Dar-

*Darwin's finches*

win observed 13 species of finches that are closely related and differ most markedly in the shape of their beaks. The beak of each species is suited to the food available in its particular environment, suggesting that beak shapes evolved by natural selection. Large beaks were found on the islands where the primary source of food for the finches are nuts and therefore the large beaks allowed the birds to be better equipped for opening the nuts and staying well nourished. Slender beaks were found on the finches which found insects to be the best source of food on the island they inhabited; their slender beaks allowed the birds to be better equipped for pulling out the insects from their tiny hiding places. The finch is also found on the main land and it is thought that they migrated to the islands and began adapting to their environment through natural selection.

## 10.4 See also

- *The Ancestor's Tale*
- Urmetazoan

## 10.5 References

[1] Theobald, Douglas L. (13 May 2010). "A formal test of the theory of universal common ancestry". *Nature* (London: Nature Publishing Group) **465** (7295): 219–222. doi:10.1038/nature09014. ISSN 0028-0836. PMID 20463738.

[2] Steel, Mike; Penny, David (13 May 2010). "Origins of life: Common ancestry put to the test". *Nature* (London: Nature Publishing Group) **465** (7295): 168–169. doi:10.1038/465168a. ISSN 0028-0836. PMID 20463725.

[3] Wade, Nicholas (25 July 2016). "Meet Luca, the Ancestor of All Living Things". *New York Times*. Retrieved 25 July 2016.

[4] Doolittle, W. Ford (February 2000). "Uprooting the Tree of Life" (PDF). *Scientific American* (Stuttgart: Georg von Holtzbrinck Publishing Group) **282** (2): 90–95. doi:10.1038/scientificamerican0200-90. ISSN 0036-8733. PMID 10710791. Archived from the original (PDF) on 2006-09-07. Retrieved 2015-11-22.

[5] Glansdorff, Nicolas; Ying Xu; Labedan, Bernard (9 July 2008). "The Last Universal Common Ancestor: emergence, constitution and genetic legacy of an elusive forerunner". *Biology Direct* (London: BioMed Central) **3**: 29. doi:10.1186/1745-6150-3-29. ISSN 1745-6150. PMC 2478661. PMID 18613974.

[6] Ohtomo, Yoko; Kakegawa, Takeshi; Ishida, Akizumi; et al. (January 2014). "Evidence for biogenic graphite in early Archaean Isua metasedimentary rocks". *Nature Geoscience* (London: Nature Publishing Group) **7** (1): 25–28. Bibcode:2014NatGe...7...25O. doi:10.1038/ngeo2025. ISSN 1752-0894.

[7] Borenstein, Seth (13 November 2013). "Oldest fossil found: Meet your microbial mom". *Excite* (Yonkers, NY: Mindspark Interactive Network). Associated Press. Retrieved 2015-11-22.

[8] Noffke, Nora; Christian, Daniel; Wacey, David; Hazen, Robert M. (16 December 2013). "Microbially Induced Sedimentary Structures Recording an Ancient Ecosystem in the *ca.* 3.48 Billion-Year-Old Dresser Formation, Pilbara, Western Australia". *Astrobiology* (New Rochelle, NY: Mary Ann Liebert, Inc.) **13** (12): 1103–1124. doi:10.1089/ast.2013.1030. ISSN 1531-1074. PMC 3870916. PMID 24205812.

[9] Darwin 1859, p. 490

[10] Crombie & Hoskin 1970, pp. 62–63

[11] Treasure 1985, p. 142

[12] Harris 1981, p. 107

[13] Kant 1987, p. 304: "Despite all the variety among these forms, they seem to have been produced according to a common archetype, and this analogy among them reinforces our suspicion that they are actually akin, produced by a common original mother."

[14] Darwin 1818, p. 397 [§ 39.4.8]

[15] Darwin 1859, p. 484

[16] Knight, Robin; Freeland, Stephen J.; Landweber, Laura F. (January 2001). "Rewiring the keyboard: evolvability of the genetic code". *Nature Reviews Genetics* (London: Nature Publishing Group) **2** (1): 49–58. doi:10.1038/35047500. ISSN 1471-0056. PMID 11253070.

[17] Than, Ker (14 May 2010). "All Species Evolved From Single Cell, Study Finds". *National Geographic News* (Washington, D.C.: National Geographic Society). Retrieved 2011-10-30.

[18] Yonezawa, Takahiro; Hasegawa, Masami (16 December 2010). "Was the universal common ancestry proved?". *Nature* (London: Nature Publishing Group) **468** (7326): E9. doi:10.1038/nature09482. ISSN 1476-4687. PMID 21164432.

[19] Theobald, Douglas L. (16 December 2010). "Theobald reply". *Nature* (London: Nature Publishing Group) **468** (7326): E10. doi:10.1038/nature09483. ISSN 1476-4687.

[20] Theobald, Douglas L. (24 November 2011). "On universal common ancestry, sequence similarity, and phylogenetic structure: The sins of P-values and the virtues of Bayesian evidence". *Biology Direct* (London: BioMed Central) **6** (1): 60. doi:10.1186/1745-6150-6-60. ISSN 1745-6150. PMC 3314578. PMID 22114984.

[21] Theobald, Douglas L. "Prediction 1.3: Consilience of independent phylogenies". *29+ Evidences for Macroevolution: The Scientific Case for Common Descent*. Version 2.89. Houston, TX: The TalkOrigins Foundation, Inc. Retrieved 2009-11-20.

[22] Raven, Evert & Eichhorn 2005, p. 200: "[These vegetables were] all produced from a single species of plant (Brassica oleracea), a member of the mustard family."

## 10.6 Bibliography

- Crombie, A. C.; Hoskin, Michael (1970). "The Scientific Movement and the Diffusion of Scientific Ideas, 1688–1751". In Bromley, J. S. *The Rise of Great Britain and Russia, 1688–1715/25*. The New

Cambridge Modern History **6**. London: Cambridge University Press. ISBN 0-521-07524-6. LCCN 57014935. OCLC 7588392.

- Darwin, Charles (1859). *On the Origin of Species by Means of Natural Selection, or the Preservation of Favoured Races in the Struggle for Life* (1st ed.). London: John Murray. LCCN 06017473. OCLC 741260650. The book is available from The Complete Work of Charles Darwin Online. Retrieved 2015-11-23.

- Darwin, Erasmus (1818) [Originally published 1794]. *Zoonomia; or the Laws of Organic Life* **1** (4nd American ed.). Philadelphia, PA: Edward Earle. *Zoonomia; or The laws of organic life: in three parts (Volume 1) (1818)* on the Internet Archive Retrieved 2015-11-23.

- Harris, C. Leon (1981). *Evolution: Genesis and Revelations: With Readings from Empedocles to Wilson*. Albany, NY: State University of New York Press. ISBN 0-87395-487-4. LCCN 81002555. OCLC 7278190.

- Kant, Immanuel (1987) [Originally published 1790 in Prussia as *Kritik der Urteilskraft*]. *Critique of Judgment*. Translated, with an introduction, by Werner S. Pluhar; foreword by Mary J. Gregor. Indianapolis, IN: Hackett Publishing Company. ISBN 0-87220-025-6. LCCN 86014852. OCLC 13796153.

- Raven, Peter H.; Evert, Ray F.; Eichhorn, Susan E. (2005). *Biology of Plants* (7th ed.). New York: W. H. Freeman and Company. ISBN 0-7167-1007-2. LCCN 2004053303. OCLC 183148564.

- Treasure, Geoffrey (1985). *The Making of Modern Europe, 1648-1780*. New York: Methuen. ISBN 0-416-72370-5. LCCN 85000255. OCLC 11623262.

## 10.7 External links

- 29+ Evidences for Macroevolution: The Scientific Case for Common Descent from the TalkOrigins Archive.

- The Tree of Life Web Project

# Chapter 11

# Last universal ancestor

For lowest common ancestors in graph theory and computer science, see lowest common ancestor.
"LUCA" redirects here. For other uses, see Luca (disambiguation).

The **last universal ancestor** (**LUA**), also called the **last universal common ancestor** (**LUCA**), **cenancestor**, or **progenote**, is the most recent organism from which all organisms now living on Earth have a common descent.[1] Thus, it is the most recent common ancestor (MRCA) of all current life on Earth. As such, it should not be assumed to be the first living organism. The LUA is estimated to have lived some 3.5 to 3.8 billion years ago (sometime in the Paleoarchean era).[2][3] The earliest evidence for life on Earth is biogenic graphite found in 3.7 billion-year-old metamorphized sedimentary rocks discovered in Western Greenland[4] and microbial mat fossils found in 3.48 billion-year-old sandstone discovered in Western Australia.[5][6] A study in 2015 found potentially biogenic carbon from 4.1 billion years ago in ancient rocks in Western Australia. Such findings would indicate the existence of different conditions on Earth during that period than what is generally assumed today and point to an earlier origination of life.[7][8] In July 2016, scientists reported identifying a set of 355 genes from the Last Universal Common Ancestor (LUCA) of all life living on Earth.[9]

Charles Darwin proposed the theory of universal common descent through an evolutionary process in his book *On the Origin of Species* in 1859, saying, "Therefore I should infer from analogy that probably all the organic beings which have ever lived on this earth have descended from some one primordial form, into which life was first breathed."[10]

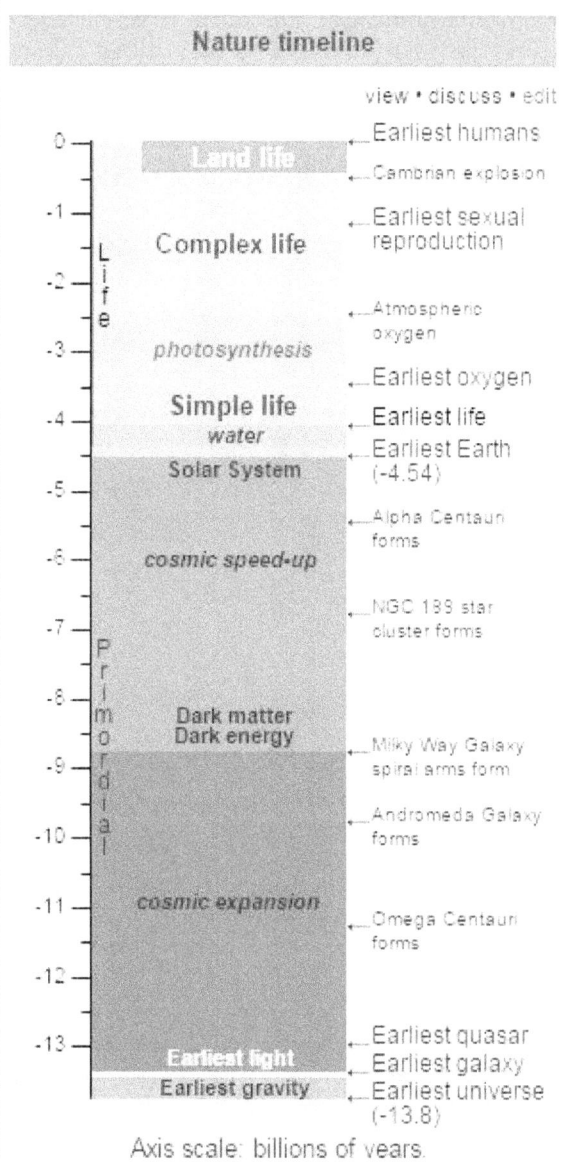

## 11.1 Features

## 11.1. FEATURES

Considering what is known of the offspring groups (see phylogenetic bracketing), the LUA is thought to have been a small, single-cell organism. It likely had a ring-shaped coil of DNA floating freely within the cell, like modern bacteria. Morphologically, it would likely not have been exceptionally distinctive among a collection of generalized, small-size, modern-day bacteria. However, Carl Woese *et al*, who first proposed the currently-used three domain system based on an analysis of the 16S rRNA sequences of bacteria, archaea, and eukaryotes, stated that the LUA would have been a "...simpler, more rudimentary entity than the individual ancestors that spawned the three [domains] (and their descendants)" regarding its genetic machinery.[11]

While the gross anatomy of the LUA must be reconstructed with much uncertainty, its internal mechanisms may be described in some detail, based on the properties currently shared by all independently living organisms on Earth:[12][13][14][15]

- The genetic code was based on DNA.[16] Note, however, that some studies suggest that the LUCA may have lacked DNA and been defined wholly through RNA.[17]
    - The DNA was composed of four nucleotides (deoxyadenosine, deoxycytidine, deoxythymidine, and deoxyguanosine), to the exclusion of other possible deoxynucleotides.
    - The genetic code was composed of three-nucleotide codons, thus producing 64 different codons. Since only 20 amino acids were used, multiple codons code for the same amino acids.
    - The DNA was kept double-stranded by a template-dependent DNA polymerase.
    - The integrity of the DNA was maintained by a group of maintenance enzymes, including DNA topoisomerase, DNA ligase and other DNA repair enzymes. The DNA was also protected by DNA-binding proteins such as histones.
- The genetic code was expressed via RNA intermediates, which were single-strand.
    - RNA was produced by a DNA-dependent RNA polymerase using nucleotides similar to those of DNA with the exception that thymidine in DNA was replaced by uridine in RNA.
- The genetic code was expressed into proteins.
- Proteins were assembled from free amino acids by translation of an mRNA by ribosomes, tRNA and a group of related proteins.
    - Ribosomes were composed of two subunits, one big 50S and one small 30S.
    - Each ribosomal subunit was composed of a core of ribosomal RNA surrounded by ribosomal proteins.
    - The RNA molecules (rRNA and tRNA) played an important role in the catalytic activity of the ribosomes.

- Only 20 amino acids were used, to the exclusion of countless other amino acids.

- Only the L-isomers of the amino acids were used.

- ATP was used as an energy intermediate.

- There were several hundred protein enzymes that catalyzed chemical reactions that extract energy from fats, sugars, and amino acids, and that synthesize fats, sugars, amino acids, and nucleic acid bases using arbitrary chemical pathways.

- The cell contained a water-based cytoplasm that was surrounded and effectively enclosed by a lipid bilayer membrane.

- Inside the cell, the concentration of sodium was lower, and potassium was higher, than outside. This gradient was maintained by specific ion transporters (also referred to as *ion pumps*).

- The cell multiplied by duplicating all its contents followed by cellular division.

- The cell used chemiosmosis to produce energy. It also reduced $CO_2$ and oxidized $H_2$ (methanogenesis or acetogenesis) via acetyl-thioesters [18][19]

- The cell had 355 genes and lived in conditions found in deep sea vents caused by ocean water interacting with magma erupting through the ocean floor.[20][21]

## 11.2 Hypotheses

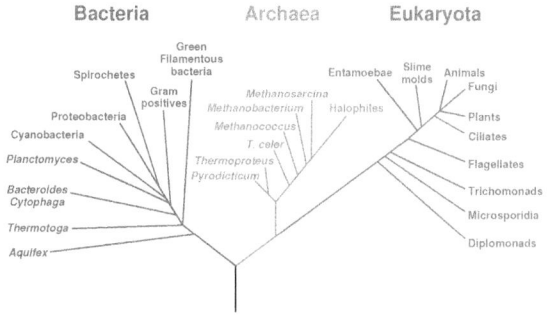

*A cladogram linking all major groups of living organisms to the LUA (the black trunk at the bottom). This graph is derived from ribosomal RNA sequence data.*[22]

In 1859, Charles Darwin published *The Origin of Species* in which he twice stated the hypothesis that there was only one progenitor for all life forms. In the summation he states, "Therefore I should infer from analogy that probably all the organic beings which have ever lived on this earth have descended from some one primordial form, into which life was first breathed."[23] The very last sentence is a restatement of the hypothesis: "There is grandeur in this view of life, with its several powers, having been originally breathed into a few forms or into one."[23]

When the LUA was hypothesized, cladograms based on genetic distance between living cells indicated that Archaea split early from the rest of life. This was inferred from the fact that all known archaeans were highly resistant to environmental extremes such as high salinity, temperature or acidity, and led some scientists to suggest that the LUA evolved in areas like the deep ocean vents, where such extremes prevail today. Archaea, however, were discovered in less hostile environments and are now believed to be more closely related to eukaryotes than bacteria, although many details are still unknown.[24][25]

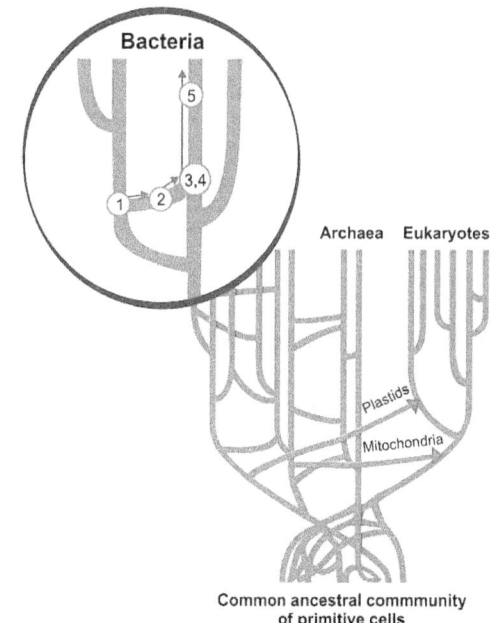

*Current (2005) tree of life showing horizontal gene transfers, giving rise to a phylogenetic network*

In 2010, based on "the vast array of molecular sequences now available from all domains of life,"[26] a formal test of universal common ancestry was published.[1] The formal test favored the existence of a universal common ancestor over a wide class of alternative hypotheses that included horizontal gene transfer. While the formal test overwhelmingly favored the existence of a single LUA, this does not imply that the LUA was ever alone. Instead, it was a member of the early microbial community.[1] Given that many other nucleotides are possible besides adenine (A), thymine (T, DNA only), guanine (G), cytosine (C), and uracil (U, RNA only), it is extremely unlikely that organisms descendent from separate abiogenesis events (that is to say separate incidents where organic molecules initially came together to form cell-like structures) would be able to complete a horizontal gene transfer without garbling each other's genes, converting them into noncoding segments. Also, many more amino acids are chemically possible than the twenty found in modern protein molecules. These lines of chemical evidence, taken into account for the formal statistical test by Theobald (2010), point to a single cell having been the LUA in that, although it was a member of the early microbial community, only its descendents survived beyond the Paleoarchean Era. With a common framework in the AT/GC rule and the standard twenty amino acids, horizontal gene transfer would have been feasible and may have been very common later on among the progeny of that single cell.

In 1998, Carl Woese proposed (1) that no individual organism can be considered a LUA, and (2) that the genetic heritage of all modern organisms derived through horizontal gene transfer among an ancient community of organisms.[27] While the results described by the later papers Theobald (2010) and Saey (2010) demonstrate the existence of a single LUCA, the argument in Woese (1998) can still be applied to Ur-organisms. At the beginnings of life, ancestry was not as linear as it is today because the genetic code took time to evolve.[28] Before high fidelity replication, organisms could not be easily mapped on a phylogenetic tree. Not to be confused with the Ur-organism, however, the LUCA lived after the genetic code and at least some rudimentary early form of molecular proofreading had already evolved. It was not the very first cell, but rather, the one whose descendents survived beyond the very early stages of microbial evolution.

## 11.3 Location of the root

For branching of Bacteria phyla, see Bacterial phyla.

The most commonly accepted location of the root of the tree of life is between a monophyletic domain Bacteria and a clade formed by Archaea and Eukaryota of what is referred to as the "traditional tree of life" based on several molecular studies starting with C. Woese.[29] A very small minority of studies have concluded differently, namely that the root is in the Domain Bacteria, either in the phylum Firmicutes[30] or that the phylum Chloroflexi is basal to a clade with Archaea+Eukaryotes and the rest of Bacteria as proposed by Thomas Cavalier-Smith.[31]

## 11.4 See also

- Abiogenesis
- Bacterial phyla
- Common descent
- Most recent common ancestor
- Origin of the first cell
- Protocell
- Panspermia
- Timeline of evolution

## 11.5 References

[1] Theobald DL (May 2010). "A formal test of the theory of universal common ancestry". *Nature* **465** (7295): 219–22. Bibcode:2010Natur.465..219T. doi:10.1038/nature09014. PMID 20463738.

[2] Doolittle WF (February 2000). "Uprooting the tree of life". *Scientific American* **282** (2): 90–5. doi:10.1038/scientificamerican0200-90. PMID 10710791.

[3] Glansdorff N, Xu Y, Labedan B (2008). "The last universal common ancestor: emergence, constitution and genetic legacy of an elusive forerunner". *Biology Direct* **3**: 29. doi:10.1186/1745-6150-3-29. PMC 2478661. PMID 18613974.

[4] Ohtomo, Yoko; Kakegawa, Takeshi; Ishida, Akizumi; Nagase, Toshiro; Rosing, Minik T. (2013). "Evidence for biogenic graphite in early Archaean Isua metasedimentary rocks". *Nature Geoscience* **7**: 25–8. Bibcode:2014NatGe...7...25O. doi:10.1038/ngeo2025.

[5] Borenstein, Seth (13 November 2013). "Oldest fossil found: Meet your microbial mom". *AP News*. Retrieved 15 November 2013.

[6] Noffke N, Christian D, Wacey D, Hazen RM (December 2013). "Microbially induced sedimentary structures recording an ancient ecosystem in the ca. 3.48 billion-year-old Dresser Formation, Pilbara, Western Australia". *Astrobiology* **13** (12): 1103–24. doi:10.1089/ast.2013.1030. PMC 3870916. PMID 24205812.

[7] "Excite News - Hints of life on what was thought to be desolate early Earth". apnews.excite.com. Retrieved 2016-06-18.

[8] Bell, Elizabeth A.; Boehnke, Patrick; Harrison, T. Mark; Mao, Wendy L. (2015-11-24). "Potentially biogenic carbon preserved in a 4.1 billion-year-old zircon". *Proceedings of the National Academy of Sciences of the United States of America* **112** (47): 14518–14521. doi:10.1073/pnas.1517557112. ISSN 1091-6490. PMC 4664351. PMID 26483481.

[9] Wade, Nicholas (25 July 2016). "Meet Luca, the Ancestor of All Living Things". *New York Times*. Retrieved 25 July 2016.

[10] Darwin, C. (1859), *The Origin of Species by Means of Natural Selection*, John Murray, p. 490

[11] Woese, C. R.; Kandler, O.; Wheelis, M. L. (1990-06-01). "Towards a natural system of organisms: proposal for the domains Archaea, Bacteria, and Eucarya.". *Proceedings of the National Academy of Sciences* **87** (12): 4576–4579. doi:10.1073/pnas.87.12.4576. ISSN 0027-8424. PMC 54159. PMID 2112744.

[12] Wächtershäuser, Günter (1998). "Towards a Reconstruction of Ancestral Genomes by Gene Cluster Alignment". *Systematic and Applied Microbiology* **21** (4): 473–4, IN1, 475–7. doi:10.1016/S0723-2020(98)80058-1.

[13] Gregory, Michael. "What is Life?". Clinton College.

[14] Pace NR (January 2001). "The universal nature of biochemistry". *Proceedings of the National Academy of Sciences of the United States of America* **98** (3): 805–8. Bibcode:2001PNAS...98..805P. doi:10.1073/pnas.98.3.805. PMC 33372. PMID 11158550.

[15] Wächtershäuser G (January 2003). "From pre-cells to Eukarya--a tale of two lipids". *Molecular Microbiology* **47** (1): 13–22. doi:10.1046/j.1365-2958.2003.03267.x. PMID 12492850.

[16] Garwood, Russell J. (2012). "Patterns In Palaeontology: The first 3 billion years of evolution". *Palaeontology Online* **2** (11): 1–14. Retrieved June 25, 2015.

[17] Marshall, Michael. "Life began with a planetary mega-organism". New Scientist.

[18] Martin W, Russell MJ (October 2007). "On the origin of biochemistry at an alkaline hydrothermal vent". *Philosophical Transactions of the Royal Society of London. Series B, Biological Sciences* **362** (1486): 1887–925. doi:10.1098/rstb.2006.1881. PMC 2442388. PMID 17255002.

[19] Lane N, Allen JF, Martin W (April 2010). "How did LUCA make a living? Chemiosmosis in the origin of life". *BioEssays* **32** (4): 271–80. doi:10.1002/bies.200900131. PMID 20108228.

[20] Wade, Nicholas (2016-07-25). "Meet Luca, the Ancestor of All Living Things". *The New York Times*. ISSN 0362-4331. Retrieved 2016-07-26.

[21] "The physiology and habitat of the last universal common ancestor".

[22] Woese CR, Kandler O, Wheelis ML (June 1990). "Towards a natural system of organisms: proposal for the domains Archaea, Bacteria, and Eucarya". *Proceedings of the National Academy of Sciences of the United States of America* **87** (12): 4576–9. doi:10.1073/pnas.87.12.4576. PMC 54159. PMID 2112744.

[23] Darwin, Charles. *On the Origin of Species*. London: John Murray, Albermarle Street. 1859. Pg. 484 and 490.

[24] Xie Q, Wang Y, Lin J, Qin Y, Wang Y, Bu W (2012). "Potential key bases of ribosomal RNA to kingdom-specific spectra of antibiotic susceptibility and the possible archaeal origin of eukaryotes". *PLoS ONE* **7** (1): e29468. doi:10.1371/journal.pone.0029468. PMC 3256160. PMID 22247777.

[25] Yutin N, Makarova KS, Mekhedov SL, Wolf YI, Koonin EV (August 2008). "The deep archaeal roots of eukaryotes". *Molecular Biology and Evolution* **25** (8): 1619–30. doi:10.1093/molbev/msn108. PMC 2464739. PMID 18463089.

[26] Steel M, Penny D (May 2010). "Origins of life: Common ancestry put to the test". *Nature* **465** (7295): 168–9. Bibcode:2010Natur.465..168S. doi:10.1038/465168a. PMID 20463725.

[27] Woese C (June 1998). "The universal ancestor". *Proceedings of the National Academy of Sciences of the United States of America* **95** (12): 6854–9. Bibcode:1998PNAS...95.6854W. doi:10.1073/pnas.95.12.6854. PMC 22660. PMID 9618502.

[28] Maynard Smith, John; Szathmáry, Eörs (1995). *The Major Transitions in Evolution*. Oxford, England: Oxford University Press. ISBN 0-19-850294-X.

[29] Boone, David R.; Castenholz, Richard W.; Garrity, George M. (eds.). *The* Archaea *and the Deeply Branching and Phototrophic* Bacteria. Bergey's Manual of Systematic Bacteriology. ISBN 978-0-387-21609-6.

[30] Valas RE, Bourne PE (2011). "The origin of a derived superkingdom: how a gram-positive bacterium crossed the desert to become an archaeon". *Biology Direct* **6**: 16. doi:10.1186/1745-6150-6-16. PMC 3056875. PMID 21356104.

[31] Cavalier-Smith T (2006). "Rooting the tree of life by transition analyses". *Biology Direct* **1**: 19. doi:10.1186/1745-6150-1-19. PMC 1586193. PMID 16834776.

# Chapter 12

# Proteinoid

**Proteinoids**, or **thermal proteins**, are protein-like, often cross-linked molecules formed abiotically from amino acids.[1] A slightly altered definition of proteinoids is from Hayakawa et al. (1967):[2] "*macromolecular preparations of mean molecular weights in the thousands, containing most of the twenty amino acids found in protein hydrolyzates. Although these polymers have other properties of contemporary protein as well, identity with the latter is not a necessary inference*".[1]

Its discoverer, Sidney W. Fox proposed the hypothesis that proteinoids were a precursor to the first living cells (protocell).[1]

## 12.1 History

In trying to uncover the intermediate stages of abiogenesis, scientist Sidney W. Fox in the 1950s and 1960s, studied the spontaneous formation of peptide structures under conditions that might plausibly have existed early in Earth's history. He demonstrated that amino acids could spontaneously form small chains called peptides. In one of his experiments, he allowed amino acids to dry out as if puddled in a warm, dry spot in prebiotic conditions. He found that, as they dried, the amino acids formed long, often cross-linked, thread-like microscopic polypeptide globules, he named "proteinoid microspheres".[3]

## 12.2 Polymerization

The abiotic polymerization of amino acids into proteins through the formation of peptide bonds was thought to occur only at temperatures over 140 °C. However, the biochemist Sidney Walter Fox and his co-workers discovered that phosphoric acid acted as a catalyst for this reaction. They were able to form protein-like chains from a mixture of 18 common amino acids at 70 °C in the presence of phosphoric acid, and dubbed these protein-like chains proteinoids. Fox later found naturally occurring proteinoids similar to those he had created in his laboratory in lava and cinders from Hawaiian volcanic vents and determined that the amino acids present polymerized due to the heat of escaping gases and lava. Other catalysts have since been found; one of them, amidinium carbodiimide, is formed in primitive Earth experiments and is effective in dilute aqueous solutions.

When present in certain concentrations in aqueous solutions, proteinoids form small microspheres. This is because some of the amino acids incorporated into proteinoid chains are more hydrophobic than others, and so proteinoids cluster together like droplets of oil in water. These structures exhibit a few characteristics of living cells:

1. an outer wall.

2. osmotic swelling and shrinking.

3. budding.

4. binary fission (dividing into two daughter microspheres).

5. streaming movement of internal particles.

Fox thought that the microspheres may have provided a cell compartment within which organic molecules could have become concentrated and protected from the outside environment during the process of chemical evolution.[1]

Proteinoid microspheres are today being considered for use in pharmaceuticals, providing microscopic biodegradable capsules in which to package and deliver oral drugs.[4]

In another experiment using a similar method to set suitable conditions for life to form, Fox collected volcanic material from a cinder cone in Hawaii. He discovered that the temperature was over 100 °C (212 °F) just 4 inches (100 mm) beneath the surface of the cinder cone, and suggested that this might have been the environment in which life was created—molecules could have formed and then been washed through the loose volcanic ash and into the sea. He placed lumps of lava over amino acids derived from methane, ammonia and water, sterilized all materials, and baked the lava over the amino acids for a few hours in a glass oven. A brown, sticky substance formed over the surface and when the lava was drenched in sterilized water a thick, brown liquid leached out. It turned out that the amino acids had combined to form proteinoids, and the proteinoids had combined to form small spheres. Fox called these "microspheres". His protobionts were not cells, although they formed clumps and chains reminiscent of bacteria. Based upon such experiments, Colin S. Pittendrigh stated in December 1967 that "laboratories will be creating a living cell within ten years," a remark that reflected the typical contemporary levels of innocence of the complexity of cell structures.[5]

## 12.3 Legacy

Fox has likened the amino acid globules to cells, and proposed it bridged the macromolecule to cell transition. However, his hypothesis was later dismissed as proteinoids are not proteins, they feature mostly non-peptide bonds and amino acid cross-linkages not present in living organisms. Furthermore, they have no compartmentalization and there is no information content in the molecules.

His hypothesis, however, was a catalyst to further investigate other mechanisms that could have brought about abiogenesis on Earth, such as RNA world, PAH world hypothesis, Iron–sulfur world theory, protocells, etc.

## 12.4 See also

- Abiogenesis
- Jeewanu
- Protocell
- Proto-mitochondrion

## 12.5 References

[1] Fox, Sidney Walter; Dose, Klaus (1977). *Molecular Evolution and the Origin of Life*. W. H. Freeman & Co Ltd. ISBN 978-0716701637.

[2] Hayakawa, T.; Fox, S. W.; Windsor, C.R. (1967). *Arch. Biochem. Biophys.* **118**: 265–272. doi:10.1016/0003-9861(67)90347-5. Missing or empty |title= (help)

[3] Experiments on origin of organic molecules. Retrieved 13 January 2008.

[4] Proteinoid microspheres and methods for preparation and use thereof. Patent US 5601846 A (1997)

[5] Woodward, Robert J., Photo editor (1969). *Our amazing world of Nature: its marvels and mysteries*. Reader's Digest Association. ISBN 0-340-13000-8.

## 12.6 Further reading

- Fox, Sidney W.; Kaoru Harada (14 November 1958). "Thermal Copolymerization of Amino Acids to a Product Resembling Protein". *Science*. New Series **128** (3333): 1214. doi:10.1126/science.128.3333.1214.

# Chapter 13

# Autocatalysis

A single chemical reaction is said to have undergone **autocatalysis**, or be **autocatalytic**, if one of the reaction products is also a reactant and therefore a catalyst in the same or a coupled reaction.[1] The reaction is called an **autocatalytic reaction**.

The rate equations for autocatalytic reactions are fundamentally nonlinear. This nonlinearity can lead to the spontaneous generation of order. A dramatic example of this order is that which is found in living systems. The spontaneous order creation corresponds to a decrease in the entropy of the system, which must be compensated by a larger increase in the entropy of the surroundings in order to satisfy the Second Law of Thermodynamics.

A *set* of chemical reactions can be said to be "collectively autocatalytic" if a number of those reactions produce, as reaction products, catalysts for enough of the other reactions that the entire set of chemical reactions is self-sustaining given an input of energy and food molecules (see autocatalytic set).

## 13.1 Chemical reactions

Main articles: Chemical reaction and Chemical kinetics

A chemical reaction of two reactants and two products can be written as

$$\alpha A + \beta B \rightleftharpoons \sigma S + \tau T$$

where the Greek letters are stoichiometric coefficients and the capital Latin letters represent chemical species. The chemical reaction proceeds in both the forward and reverse direction. This equation is easily generalized to any number of reactants, products, and reactions.

### 13.1.1 Chemical equilibrium

In chemical equilibrium the forward and reverse reaction rates are such that each chemical species is being created at the same rate it is being destroyed. In other words, the rate of the forward reaction is equal to the rate of the reverse reaction.

$$k_+[A]^\alpha[B]^\beta = k_-[S]^\sigma[T]^\tau$$

Here, the brackets indicate the concentration of the chemical species, in moles per liter, and $k_+$ and $k_-$ are rate constants.

### 13.1.2 Far from equilibrium

Far from equilibrium, the forward and reverse reaction rates no longer balance and the concentration of reactants and products is no longer constant. For every forward reaction $\alpha$ molecules of A are destroyed. For every reverse reaction $\alpha$ molecules of A are created. In the case of an elementary reaction step the reaction order in each direction equals the molecularity, so that the rate of change in number of moles of A is then

$$\frac{d}{dt}[A] = -\alpha k_+[A]^\alpha[B]^\beta + \alpha k_-[S]^\sigma[T]^\tau$$

$$\frac{d}{dt}[B] = -\beta k_+[A]^\alpha[B]^\beta + \beta k_-[S]^\sigma[T]^\tau$$

$$\frac{d}{dt}[S] = \sigma k_+[A]^\alpha[B]^\beta - \sigma k_-[S]^\sigma[T]^\tau$$

$$\frac{d}{dt}[T] = \tau k_+[A]^\alpha[B]^\beta - \tau k_-[S]^\sigma[T]^\tau$$

This system of equations has a single stable fixed point when the forward rates and the reverse rates are equal. This means that the system evolves to the equilibrium state, and this is the only state to which it evolves.

### 13.1.3 Autocatalytic reactions

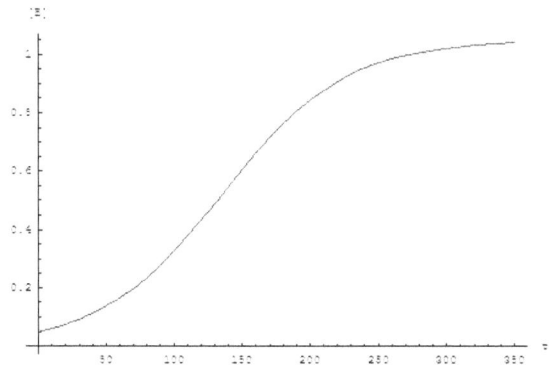

*Sigmoid variation of product concentration in autocatalytic reactions*

Autocatalytic reactions are those in which at least one of the products is a reactant. Perhaps the simplest autocatalytic reaction can be written[1]

$$A + B \rightleftharpoons 2B$$

with the rate equations (for an elementary reaction)

$$\frac{d}{dt}[A] = -k_+[A][B] + k_-[B]^2$$

$$\frac{d}{dt}[B] = +k_+[A][B] - k_-[B]^2$$

This reaction is one in which a molecule of species A interacts with a molecule of species B. The A molecule is converted into a B molecule. The final product consists of the original B molecule plus the B molecule created in the reaction.

The key feature of these rate equations is that they are nonlinear; the second term on the right varies as the square of the concentration of B. This feature can lead to multiple fixed points of the system, much like a quadratic equation can have multiple roots. Multiple fixed points allow for multiple states of the system. A system existing in multiple macroscopic states is more orderly (has lower entropy) than a system in a single state.

The concentrations of A and B vary in time according to[1][2]

$$[A] = \frac{[A]_0 + [B]_0}{1 + \frac{[B]_0}{[A]_0}e^{([A]_0+[B]_0)kt}}$$

and

$$[B] = \frac{[A]_0 + [B]_0}{1 + \frac{[A]_0}{[B]_0}e^{-([A]_0+[B]_0)kt}}$$

The graph for these equations is a sigmoid curve, which is typical for autocatalytic reactions: these chemical reactions proceed slowly at the start (the induction period) because there is little catalyst present, the rate of reaction increases progressively as the reaction proceeds as the amount of catalyst increases and then it again slows down as the reactant concentration decreases. If the concentration of a reactant or product in an experiment follows a sigmoid curve, the reaction may be autocatalytic.

These kinetic equations apply for example to the acid-catalyzed hydrolysis of some esters to carboxylic acids and alcohols.[2] There must be at least some acid present initially to start the catalyzed mechanism; if not the reaction must start by an alternate uncatalyzed path which is usually slower. The above equations for the catalyzed mechanism would imply that the concentration of acid product remains zero forever.[2]

## 13.2 Creation of order

### 13.2.1 Background

The Second Law of Thermodynamics states that the disorder (entropy) of a physical or chemical system and its surroundings (a closed system) must increase with time. Systems left to themselves become increasingly random, and orderly energy of a system like uniform motion degrades eventually to the random motion of particles in a heat bath.

There are, however, many instances in which physical systems spontaneously become emergent or orderly. For example, despite the destruction they cause, hurricanes have a very orderly vortex motion when compared to the random motion of the air molecules in a closed room. Even more spectacular is the order created by chemical systems; the most dramatic being the order associated with life.

This is consistent with the Second Law, which requires that the total disorder of a system *and its surroundings* must increase with time. Order can be created in a system by an even greater decrease in order of the systems surroundings.[3] In the hurricane example, hurricanes are formed from unequal heating within the atmosphere. The Earth's atmosphere is then far from thermal equilibrium. The order of the Earth's atmosphere increases, but at the expense of the order of the sun. The sun is becoming more disorderly as it ages and throws off light and material to the rest of the universe. The total disorder of the sun and the

## 13.2. CREATION OF ORDER

earth increases despite the fact that orderly hurricanes are generated on earth.

A similar example exists for living chemical systems. The sun provides energy to green plants. The green plants are food for other living chemical systems. The energy absorbed by plants and converted into chemical energy generates a system on earth that is orderly and far from chemical equilibrium. Here, the difference from chemical equilibrium is determined by an excess of reactants over the equilibrium amount. Once again, order on earth is generated at the expense of entropy increase of the sun. The total entropy of the earth and the rest of the universe increases, consistent with the Second Law.

Some autocatalytic reactions also generate order in a system at the expense of its surroundings. For example, (clock reactions) have intermediates whose concentrations oscillate in time, corresponding to temporal order. Other reactions generate spatial separation of chemical species corresponding to spatial order. More complex reactions are involved in metabolic pathways and metabolic networks in biological systems.

The transition to order as the distance from equilibrium increases is not usually continuous. Order typically appears abruptly. The threshold between the disorder of chemical equilibrium and order is known as a phase transition. The conditions for a phase transition can be determined with the mathematical machinery of non-equilibrium thermodynamics.

### 13.2.2 Temporal order

A chemical reaction cannot oscillate about a position of final equilibrium because the second law of thermodynamics requires that a thermodynamic system approach equilibrium and not recede from it. For a closed system at constant temperature and pressure, the Gibbs free energy must decrease continuously and not oscillate. However it is possible that the concentrations of some reaction intermediates oscillate, and also that the *rate* of formation of products oscillates.[4]

**Idealized example: Lotka-Volterra equation**

Consider a coupled set of two autocatalytic reactions in which the concentration of one of the reactants A is much larger than its equilibrium value. In this case the forward reaction rate is so much larger than the reverse rates that we can neglect the reverse rates.

$$A + X \to 2X$$

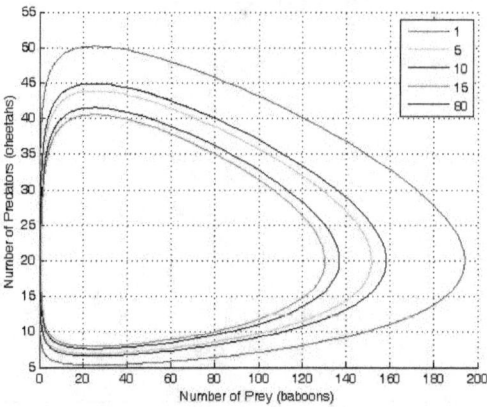

*The Lotka-Volterra equation is isomorphic with the predator prey model and the two reaction autocatalytic model. In this example baboons and cheetahs are equivalent to two different chemical species X and Y in autocatalytic reactions.*

$$X + Y \to 2Y$$

$$Y \to E$$

with the rate equations

$$\frac{d}{dt}[X] = k_1[A][X] - k_2[X][Y]$$

$$\frac{d}{dt}[Y] = k_2[X][Y] - k_3[Y]$$

Here, we have neglected the depletion of the reactant A, since its concentration is so large. The rate constants for the three reactions are $k_1$, $k_2$, and $k_3$, respectively.

This system of rate equations is known as the Lotka-Volterra equation and is most closely associated with population dynamics in predator-prey relationships. This system of equations can yield oscillating concentrations of the reaction intermediates X and Y. The amplitude of the oscillations depends on the concentration of A (which decreases without oscillation). Such oscillations are a form of emergent temporal order that is not present in equilibrium.

**Another idealized example: Brusselator**

Another example of a system that demonstrates temporal order is the Brusselator (see Prigogine reference). It is characterized by the reactions

$$A \to X$$

$$2X + Y \to 3X$$

$$B + X \to Y + D$$
$$X \to E$$

with the rate equations

$$\frac{d}{dt}[X] = [A] + [X]^2[Y] - [B][X] - [X]$$

$$\frac{d}{dt}[Y] = [B][X] - [X]^2[Y]$$

where, for convenience, the rate constants have been set to 1.

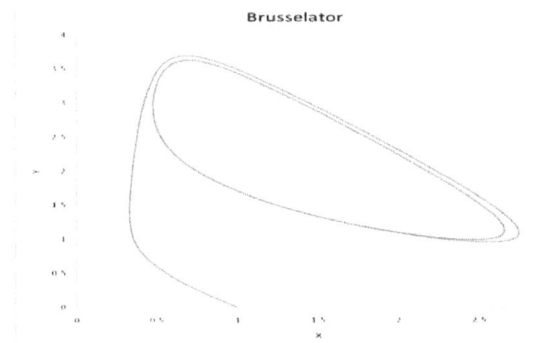

*The Brusselator in the unstable regime. A=1. B=2.5. X(0)=1. Y(0)=0. The system approaches a limit cycle. For B<1+A the system is stable and approaches a fixed point.*

The Brusselator has a fixed point at

$$[X] = A$$
$$[Y] = \frac{B}{A}$$

The fixed point becomes unstable when

$$B > 1 + A^2$$

leading to an oscillation of the system. Unlike the Lotka-Volterra equation, the oscillations of the Brusselator do not depend on the amount of reactant present initially. Instead, after sufficient time, the oscillations approach a limit cycle.[5]

**Real examples**

Real examples of clock reactions are the Belousov-Zhabotinsky reaction (BZ reaction), the Briggs-Rauscher reaction, the Bray-Liebhafsky reaction and the iodine clock reaction. These are oscillatory reactions, and the concentration of products and reactants can be approximated in terms of damped oscillations.

The best-known reaction, the BZ reaction, can be created with a mixture of potassium bromate, malonic acid, and manganese sulfate prepared in a heated solution of sulfuric acid.[6]

### 13.2.3 Spatial order

An idealized example of spatial spontaneous symmetry breaking is the case in which we have two boxes of material separated by a permeable membrane so that material can diffuse between the two boxes. It is assumed that identical Brusselators are in each box with nearly identical initial conditions. (see Prigogine reference)

$$\frac{d}{dt}[X_1] = [A] + [X_1]^2[Y_1] - [B][X_1] - [X_1] + D_x(X_2 - X_1)$$

$$\frac{d}{dt}[Y_1] = [B][X_1] - [X_1]^2[Y_1] + D_y(Y_2 - Y_1)$$

$$\frac{d}{dt}[X_2] = [A] + [X_2]^2[Y_2] - [B][X_2] - [X_2] + D_x(X_1 - X_2)$$

$$\frac{d}{dt}[Y_2] = [B][X_2] - [X_2]^2[Y_2] + D_y(Y_1 - Y_2)$$

Here, the numerical subscripts indicate which box the material is in. There are additional terms proportional to the diffusion coefficient D that account for the exchange of material between boxes.

If the system is initiated with the same conditions in each box, then a small fluctuation will lead to separation of materials between the two boxes. One box will have a predominance of X, and the other will have a predominance of Y.

## 13.3 Biological example

It is known that an important metabolic cycle, glycolysis, displays temporal order.[7] Glycolysis consists of the degradation of one molecule of glucose and the overall production of two molecules of ATP. The process is therefore of great importance to the energetics of living cells. The global glycolysis reaction involves glucose, ADP, NAD, pyruvate, ATP, and NADH.

glucose+2ADP+2P$_i$+2NAD → 2(pyruvate)+2ATP+2NADH

The details of the process are quite involved, however, a section of the process is autocatalyzed by phosphofructokinase (PFK). This portion of the process is responsible for oscillations in the pathway that lead to the process oscillating between an active and an inactive form. Thus, the autocatalytic reaction can modulate the process.

## 13.4 Phase transitions

The initial amounts of reactants determine the distance from chemical equilibrium of the system. The greater the initial concentrations the further the system is from equilibrium. As the initial concentration increases, an abrupt change in order occurs. This abrupt change is known as phase transition. At the phase transition, fluctuations in macroscopic quantities, such as chemical concentrations, increase as the system oscillates between the more ordered state (lower entropy, such as ice) and the more disordered state (higher entropy, such as liquid water). Also, at the phase transition, macroscopic equations, such as the rate equations, fail. Rate equations can be derived from microscopic considerations. The derivations typically rely on a mean field theory approximation to microscopic dynamical equations. Mean field theory breaks down in the presence of large fluctuations (see Mean field theory article for a discussion). Therefore, since large fluctuations occur in the neighborhood of a phase transition, macroscopic equations, such as rate equations, fail. As the initial concentration increases further, the system settles into an ordered state in which fluctuations are again small. (see Prigogine reference)

## 13.5 Asymmetric autocatalysis

Asymmetric autocatalysis occurs when the reaction product is chiral and thus acts as a chiral catalyst for its own production. Reactions of this type, such as the Soai reaction, have the property that they can amplify a very small enantiomeric excess into a large one. This has been proposed as an important step in the origin of biological homochirality.[8]

## 13.6 Role in origin of life

Main article: Abiogenesis

In 1995 Stuart Kauffman proposed that life initially arose as autocatalytic chemical networks.[9]

British ethologist Richard Dawkins wrote about autocatalysis as a potential explanation for abiogenesis in his 2004 book *The Ancestor's Tale*. He cites experiments performed by Julius Rebek and his colleagues at the Scripps Research Institute in California in which they combined amino adenosine and pentafluorophenyl ester with the autocatalyst amino adenosine triacid ester (AATE). One system from the experiment contained variants of AATE which catalysed the synthesis of themselves. This experiment demonstrated the possibility that autocatalysts could exhibit competition within a population of entities with heredity, which could be interpreted as a rudimentary form of natural selection, and that certain environmental changes (such as irradiation) could alter the chemical structure of some of these self-replicating molecules (an analogue for mutation) in such ways that could either boost or interfere with its ability to react, thus boosting or interfering with its ability to replicate and spread in the population.[10]

Autocatalysis plays a major role in the processes of life. Two researchers who have emphasised its role in the origins of life are Robert Ulanowicz[11] and Stuart Kauffman.[12]

Autocatalysis occurs in the initial transcripts of rRNA. The introns are capable of excising themselves by the process of two nucleophilic transesterification reactions. The RNA able to do this is sometimes referred to as a ribozyme. Additionally, the citric acid cycle is an autocatalytic cycle run in reverse.

Ultimately, biological metabolism itself can be seen as a vast autocatalytic set, in that all of the molecular constituents of a biological cell are produced by reactions involving this same set of molecules.

## 13.7 Examples of autocatalytic reactions

- DNA replication
- Haloform reaction
- Tin pest
- Reaction of Permanganate with Oxalic Acid
- The mechanism of the above reaction [13]
- Vinegar syndrome
- Binding of oxygen by hemoglobin
- The spontaneous degradation of aspirin into salicylic acid and acetic acid, causing very old aspirin in sealed containers to smell mildly of vinegar.
- The α-bromination of acetophenone with bromine.
- Liesegang rings

## 13.8 See also

- Autocatalytic reactions and order creation
- Catalytic cycle
- Reaction–diffusion system
- Abiogenesis
- Stuart Kauffman
- Morphogenesis

## 13.9 References

[1] Steinfeld J.I., Francisco J.S. and Hase W.L. *Chemical Kinetics and Dynamics* (2nd ed., Prentice-Hall 1999) p.151-2 ISBN 0-13-737123-3

[2] Moore J.W. and Pearson R.G. *Kinetics and Mechanism* (John Wiley 1981) p.26 ISBN 0-471-03558-0

[3] Ilya Prigogine (1980). *From Being to Becoming: Time and Complexity in the Physical Sciences*. San Francisco: W. H. Freeman. ISBN 0-7167-1107-9.

[4] Espenson, J.H. *Chemical Kinetics and Reaction Mechanisms* (2nd ed., McGraw-Hill 2002) p.190 ISBN 0-07-288362-6

[5] http://www.math.ohio-state.edu/~{}ault/Papers/Brusselator.pdf Dynamics of the Brusselator

[6] Peterson, Gabriel. "The Belousov-Zhabotinsky Reaction". Archived from the original on December 31, 2012.

[7] G. Nicolis and Ilya Prigogine (1977). *Self-Organization in Nonequilibrium Systems*. New York: John Wiley and Sons. ISBN 0-471-02401-5.

[8] Soai K, Sato I, Shibata T (2001). "Asymmetric autocatalysis and the origin of chiral homogeneity in organic compounds.". *Chem Rec* **1** (4): 321–32. doi:10.1002/tcr.1017. PMID 11893072.

[9] Stuart Kauffman (1995). *At Home in the Universe: The Search for the Laws of Self-Organization and Complexity*. Oxford University Press. ISBN 0-19-509599-5.

[10] Rebeck, Julius (July 1994). "Synthetic Self-Replicating Molecules". *Scientific American*: 48–55.

[11] Ecology, the Ascendent Perspective", Robert Ulanowicz, Columbia Univ. Press 1997

[12] Investigations, Stuart Kauffman.

[13] Kovacs KA, Grof P, Burai L, Riedel M (2004). "Revising the Mechanism of the Permanganate/Oxalate Reaction". *J. Phys. Chem. A* **108** (50): 11026. doi:10.1021/jp047061u.

## 13.10 External links

- Some Remarks on Autocatalysis and Autopoiesis (Barry McMullin)
- Jain, Sanjay; Krishna, Sandeep (21 December 1998). "Autocatalytic Sets and the Growth of Complexity in an Evolutionary Model". *Physical Review Letters* **81** (25): 5684–5687. doi:10.1103/PhysRevLett.81.5684.

# Chapter 14

# Homochirality

**Homochirality** describes a geometric property of some materials that are composed of chiral units. Chiral bodies are objects which are nonsuperimposable on their mirror images. For example, left and right hands are chiral. A substance is said to be *homochiral* if all the constituent units are molecules of the same chiral form (enantiomer).

In biology, homochirality is a common property of amino acids and sugars. The origin of this phenomenon is not clearly understood. It is unclear if homochirality has a purpose; however it appears to be a form of information storage. [1] One suggestion is that it reduces entropy barriers in the formation of large organized molecules.[2] It has been experimentally verified that amino acids form large aggregates in larger abundance from enantiopure substrates than from racemic ones.

Homochirality is said to evolve in three distinct steps: **mirror-symmetry breaking** creates a minute enantiomeric imbalance and is key to homochirality, **chiral amplification** is a process of enantiomeric enrichment and **chiral transmission** allows the transfer of chirality of one set of molecules to another.

It is also entirely possible that homochirality is simply a result of the natural autoamplification process of life —that either the formation of life as preferring one chirality or the other was a chance rare event which happened to occur with the chiralities we observe, or that all chiralities of life emerged rapidly but due to catastrophic events and strong competition, the other unobserved chiral preferences were wiped out by the preponderance and metabolic, enantiomeric enrichment from the 'winning' chirality choices . The emergence of chirality consensus as a natural autoamplification process has been associated with the 2nd law of thermodynamics.[3]

## 14.1 Mirror-symmetry breaking

Known mechanisms for the production of non-racemic mixtures from racemic starting materials include: asymmetric physical laws, such as the electroweak interaction; asymmetric environments, such as those caused by circularly polarized light, quartz crystals, or the Earth's rotation; and statistical fluctuations during racemic synthesis.[4] Once established, chirality would be selected for.[5] A small enantiomeric excess can be amplified into a large one by asymmetric autocatalysis, such as in the Soai reaction.[6] In asymmetric autocatalysis, the catalyst is a chiral molecule, which means that a chiral molecule is catalysing its own production. An initial enantiomeric excess, such as can be produced by polarized light, then allows the more abundant enantiomer to outcompete the other.[7]

One supposition is that the discovery of an enantiomeric imbalance in molecules in the Murchison meteorite supports an extraterrestrial origin of homochirality: there is evidence for the existence of circularly polarized light originating from Mie scattering on aligned interstellar dust particles which may trigger the formation of an enantiomeric excess within chiral material in space.[8] Another speculation (the Vester-Ulbricht hypothesis) suggests that fundamental chirality of physical processes such as that of the beta decay (see Parity violation) leads to slightly different half-lives of biologically relevant molecules. Homochirality may also result from spontaneous absolute asymmetric synthesis.[9][10]

## 14.2 Chiral amplification

Laboratory experiments exist demonstrating how in certain autocatalytic reaction systems the presence of a small amount of reaction product with enantiomeric excess at the start of the reaction can result in a much larger enantiomeric excess at the end of the reaction. In the Soai reaction,[11] pyrimidine-5-carbaldehyde (*Scheme 1*) is alkylated by diisopropylzinc to the corresponding pyrimidyl alcohol. Because the initial reaction product is also an effective catalyst the reaction is autocatalytic. The presence of just 0.2 equivalent of the alcohol S-enantiomer at the start of the reaction is sufficient to amplify the enantiomeric excess to

93%.

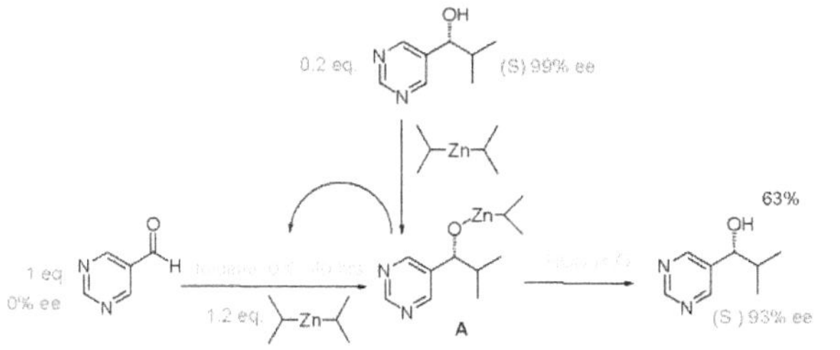

*Scheme 1. Soai autocatalysis*

Another study[12] concerns the proline catalyzed aminoxylation of propionaldehyde by nitrosobenzene (*scheme 2*). In this system too the presence of enantioenriched catalyst drives the reaction towards one of the two possible optical isomers.

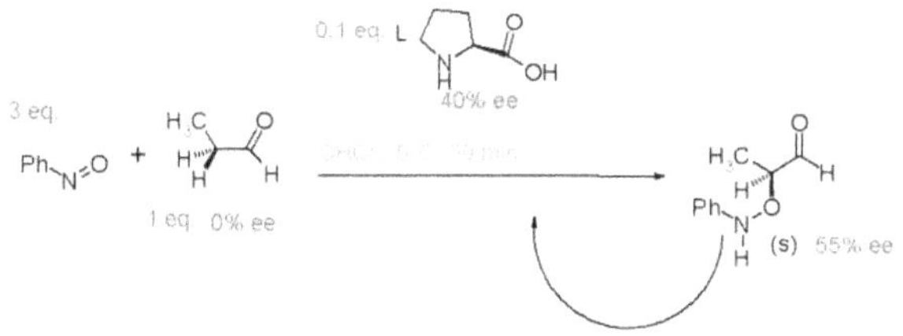

*Scheme 2. Blackmond autocatalysis*

Serine octamer clusters[13][14] are also contenders. These clusters of 8 serine molecules appear in mass spectrometry with an unusual homochiral preference, however there is no evidence that such clusters exist under non-ionizing conditions and amino acid phase behavior is far more prebiotically relevant.[15] The recent observation that partial sublimation of a 10% enantioenriched sample of leucine results in up to 82% enrichment in the sublimate shows that enantioenrichment of amino acids could occur in space.[16] Partial sublimation processes can take place on the surface of meteors where large variations in temperature exist. This finding may have consequences for the development of the Mars Organic Detector scheduled for launch in 2013 which aims to recover trace amounts of amino acids from the Mars surface exactly by a sublimation technique.

A high asymmetric amplification of the enantiomeric excess of sugars are also present in the amino acid catalyzed asymmetric formation of carbohydrates[17]

One classic study involves an experiment that takes place in the laboratory.[18] When sodium chlorate is allowed to crystallize from water and the collected crystals examined in a polarimeter, each crystal turns out to be chiral and either the L form or the D form. In an ordinary experiment the amount of L crystals collected equals the amount of D crystals (corrected for statistical effects). However, when the sodium chlorate solution is stirred during the crystallization process the crystals are either exclusively L or exclusively D. In 32 consecutive crystallization experiments 14 experiments deliver D-crystals and 18 others L-crystals. The explanation for this symmetry breaking is unclear but is related to autocatalysis taking place in the nucleation process.

In a related experiment, a crystal suspension of a racemic amino acid derivative continuously stirred, results in a 100% crystal phase of one of the enantiomers because the enantiomeric pair is able to equilibrate in solution (compare with dynamic kinetic resolution)[19]

## 14.3 Chiral transmission

Many strategies in asymmetric synthesis are built on chiral transmission. Especially important is the so-called organocatalysis of organic reactions by proline for example in Mannich reactions.

## 14.4 Optical resolution in racemic amino acids

There exists no theory elucidating correlations among L-amino acids. If one takes, for example, alanine, which has a small methyl group, and phenylalanine, which has a larger benzyl group, a simple question is in what aspect, L-alanine resembles L-phenylalanine more than D-phenylalanine, and what kind of mechanism causes the selection of all L-amino acids. Because it might be possible that alanine was L and phenylalanine was D.

It was reported[20] in 2004 that excess racemic D,L-asparagine (Asn), which spontaneously forms crystals of either isomer during recrystallization, induces asymmetric resolution of a co-existing racemic amino acid such as arginine (Arg), aspartic acid (Asp), glutamine (Gln), histidine (His), leucine (Leu), methionine (Met), phenylalanine (Phe), serine (Ser), valine (Val), tyrosine (Tyr), and tryptophan (Trp). The enantiomeric excess {ee=100x(L-D)/(L+D)} of these amino acids was correlated almost linearly with that of the inducer, i.e., Asn. When recrystallizations from a mixture of 12 D,L-amino acids (Ala, Asp, Arg, Glu, Gln, His, Leu, Met, Ser, Val, Phe, and Tyr) and excess D,L-Asn were made, all amino acids with the same configuration with Asn were preferentially co-crystallized.[20] It was incidental whether the enrichment took place in L- or D-Asn, however, once the selection was made, the co-existing amino acid with the same configuration at the α-carbon was preferentially involved because of thermodynamic stability in the crystal formation. The maximal ee was reported to be 100%. Based on these results, it is proposed that a mixture of racemic amino acids causes spontaneous and effective optical resolution, even if asymmetric synthesis of a single amino acid does not occur without an aid of an optically active molecule.

This is the first study elucidating reasonably the formation of chirality from racemic amino acids with experimental evidences.

## 14.5 History

This term was introduced by Kelvin in 1904, the year that he published his Baltimore Lecture of 1884. Kelvin used the term homochirality as a relationship between two molecules, i.e. two molecule are homochiral if they have the same chirality.[17][21] Recently, however, homochiral has been used in the same sense as enantiomerically pure. This is permitted in some journals (but not encouraged), its meaning changing into the preference of a process or system for a single optical isomer in a pair of isomers in these journals.

## 14.6 See also

- Chirality (biology)
- CIP system
- Stereochemistry
- Pfeiffer Effect
- Unsolved problems in chemistry

## 14.7 References

[1] (2009), *A New Definition of Life*, Carroll, J. D. Chirality, 21: 354–358, 2009. doi:10.1002/chir.20590

[2] *Do Homochiral Aggregates Have an Entropic Advantage?* Julian, R. R.; Myung, S.; Clemmer, D. E. J. Phys. Chem. B.; (Article); **2005**; 109(1); 440-444. doi:10.1021/jp046478x

[3] Jaakkola, S., Sharma, V. and Annila, A. (2008). "Cause of chirality consensus". *Curr. Chem. Biol.* **2** (2): 53–58. doi:10.2174/187231308784220536.

[4] Plasson, Raphaël; Kondepudi, Dilip K.; Bersini, Hugues; et al. (August 2007). "Emergence of homochirality in far-from-equilibrium systems: Mechanisms and role in prebiotic chemistry". *Chirality* (Hoboken, NJ: John Wiley & Sons) **19** (8): 589–600. doi:10.1002/chir.20440. ISSN 0899-0042. PMID 17559107. "Special Issue: Proceedings from the Eighteenth International Symposium on Chirality (ISCD-18), Busan, Korea, 2006"

[5] Clark, Stuart (July–August 1999). "Polarized Starlight and the Handedness of Life". *American Scientist* (Research Triangle Park, NC: Sigma Xi) **87** (4): 336. Bibcode:1999AmSci..87..336C. doi:10.1511/1999.4.336. ISSN 0003-0996.

[6] Shibata, Takanori; Morioka, Hiroshi; Hayase, Tadakatsu; et al. (17 January 1996). "Highly Enantioselective Catalytic Asymmetric Automultiplication of Chiral Pyrimidyl Alcohol". *Journal of the American Chemical Society* (Washington, D.C.: American Chemical Society) **118** (2): 471–472. doi:10.1021/ja953066g. ISSN 0002-7863.

[7] Soai, Kenso; Sato, Itaru; Shibata, Takanori (2001). "Asymmetric autocatalysis and the origin of chiral homogeneity in organic compounds". *The Chemical Record* (Hoboken, NJ: John Wiley & Sons on behalf of The Japan Chemical Journal Forum) **1** (4): 321–332. doi:10.1002/tcr.1017. ISSN 1528-0691. PMID 11893072.

[8] Uwe Meierhenrich. *Amino Acids and the Asymmetry of Life*; (Book) Springer-Verlag; **2008**. ISBN 978-3-540-76885-2

[9] Rajan, Aruna. "How did protein amino acids get left-handed while sugars got right-handed?" (PDF). Term Paper for Physics 569*. Retrieved June 18, 2014.

[10] "Interview: In the beginning...". *Highlights in Chemical Science* (5). 2008. Retrieved June 18, 2014.

[11] Takanori Shibata; Hiroshi Morioka; Tadakatsu Hayase; Kaori Choji; Kenso Soai (1996). "Highly Enantioselective Catalytic Asymmetric Automultiplication of Chiral Pyrimidyl Alcohol". *J. Am. Chem. Soc.* **118** (2): 471–472. doi:10.1021/ja953066g.

[12] Suju P. Mathew, Hiroshi Iwamura and Donna G. Blackmond (21 June 2004). "Amplification of Enantiomeric Excess in a Proline-Mediated Reaction". *Angewandte Chemie International Edition* **43** (25): 3317–3321. doi:10.1002/anie.200453997. PMID 15213963.

[13] Cooks, R. G., Zhang, D., Koch, K. J. (2001). "Chiroselective Self-Directed Octamerization of Serine: Implications for Homochirogenesis". *Anal. Chem.* **73** (15): 3646–3655. doi:10.1021/ac010284l. PMID 11510829.

[14] Nanita, S., Cooks, R. G. (2006). "Serine Octamers: Cluster Formation, Reactions, and Implications for Biomolecule Homochirality". *Angew. Chem. Int. Ed.* **45** (4): 554–569. doi:10.1002/anie.200501328. PMID 16404754.

[15] Donna G. Blackmond; Martin Klussmann (2007). "Spoilt for choice: assessing phase behaviour models for the evolution of homochirality". *Chem. Commun.* (39): 3990–3996. doi:10.1039/b709314b. PMID 17912393.

[16] Stephen P. Fletcher; Richard B. C. Jagt; Ben L. Feringa (2007). "An astrophysically relevant mechanism for amino acid enantiomer enrichment". *Chem. Commun.* **2007** (25): 2578–2580. doi:10.1039/b702882b. PMID 17579743.

[17] Armando Córdova; Magnus Engqvist; Ismail Ibrahem; Jesús Casas; Henrik Sundén (2005). "Plausible origins of homochirality in the amino acid catalyzed neogenesis of carbohydrates". *Chem. Commun.* **15** (15): 2047–2049. doi:10.1039/b500589b. PMID 15834501.

[18] Kondepudi, D. K., Kaufman, R. J. & Singh, N. (1990). "Chiral Symmetry Breaking in Sodium Chlorate Crystallization". *Science* **250** (4983): 975–976. Bibcode:1990Sci...250..975K. doi:10.1126/science.250.4983.975. PMID 17746924.

[19] *Emergence of a Single Solid Chiral State from a Nearly Racemic Amino Acid Derivative* Wim L. Noorduin, Toshiko Izumi, Alessia Millemaggi, Michel Leeman, Hugo Meekes, Willem J. P. Van Enckevort, Richard M. Kellogg, Bernard Kaptein, Elias Vlieg, and Donna G. Blackmond J. Am. Chem. Soc.; **2008**; 130(4) pp 1158 - 1159; (Communication) doi:10.1021/ja7106349

[20] S. Kojo; H. Uchino; M. Yoshimura; K. Tanaka (2004). "Racemic D,L-asparagine causes enantiomeric excess of other coexisting racemic D,L-amino acids during recrystallization: a hypothesis accounting for the origin of L-amino acids in the biosphere.". *Chem. Comm.* (19): 2146–2147. doi:10.1039/b409941a. PMID 15467844.

[21] *Stereochemistry* David G. Morris, Cambridge : Royal Society of Chemistry, **2001**, p30.

## 14.8 External links

- *On the Genesis of Homochirality* A. Maureen Rouhi Chemical & Engineering News June 17, **2004** Link

- *Observations Support Homochirality Theory* Photonics TechnologyWorld November **1998** Link

- Scienceweek digest **1998** Link

- *How left-handed amino acids got ahead: a demonstration of the evolution of biological homochirality in the lab* Press release Imperial College London **2004** Link

- *Origins of Homochirality* conference in Nordita Stockholm, **February 2008**, talks available online

- The Handedness of the Universe by Roger A Hegstrom and Dilip K Kondepudi http://quantummechanics.ucsd.edu/ph87/ScientificAmerican/Sciam/Hegstrom_The_Handedness_of_the_universe.pdf

# Chapter 15

# Protocell

Not to be confused with Proteobacteria.

A **protocell** (or **protobiont**) is a self-organized, endogenously ordered, spherical collection of lipids proposed as a stepping-stone to the origin of life.[1][2] A central question in evolution is how simple protocells first arose and how they could differ in reproductive output, thus enabling the accumulation of novel biological emergences over time, i.e. biological evolution. Although a functional protocell has not yet been achieved in a laboratory setting, the goal to understand the process appears well within reach.[3][4][5][6]

## 15.1 Overview

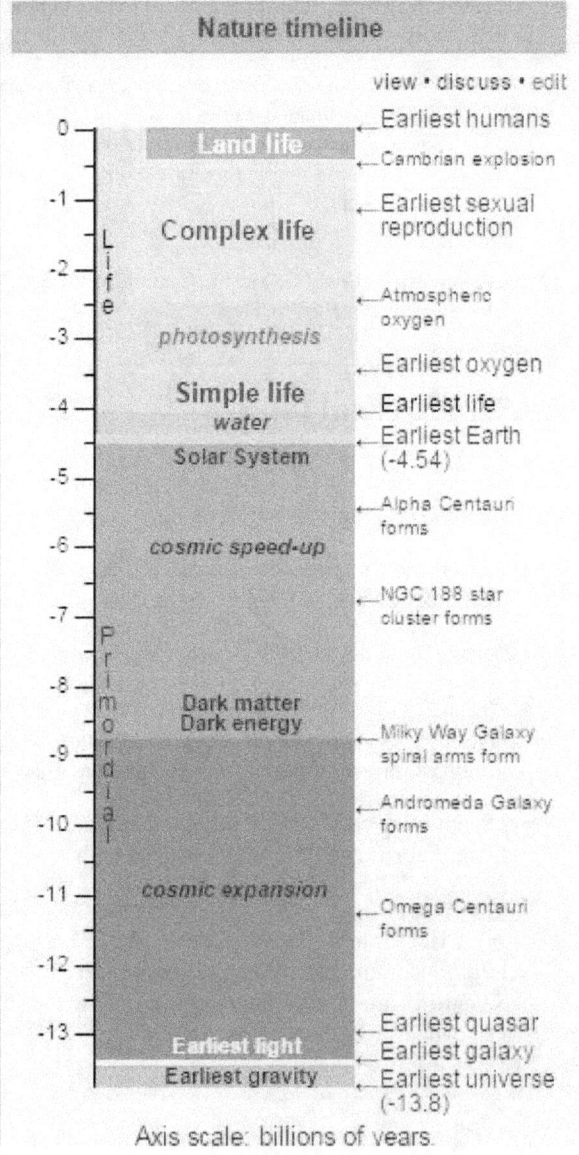

Compartmentalization was important in the origins of life. Membranes create enclosed compartments that are separate from the external environment, thus providing the cell with functionally specialized aqueous spaces. Because lipid bilayer of membranes is impermeable to most hydrophilic molecules (dissolved by water), the cell must have membrane transport systems that are in charge of import of nutritive molecules as well as export of waste.[7] It is very challenging to construct protocells from molecular assemblies. An important step in this challenge is the achievement of vesicle dynamics that are relevant to cellular functions, such as membrane trafficking and self-reproduction, using amphiphilic molecules. On the primitive Earth, numerous chemical reactions of organic compounds produced the ingredients of life. Of these substances, amphiphilic molecules might be the first player in the evolution from molecular assembly to cellular life.[8][9] A step from vesicle toward protocell might be to develop self-reproducing vesicles coupled with the metabolic system.[10]

## 15.2 Selectivity for compartmentalization

Self-assembled vesicles are essential components of primitive cells.[1] The second law of thermodynamics requires that the universe move in a direction in which disorder (or

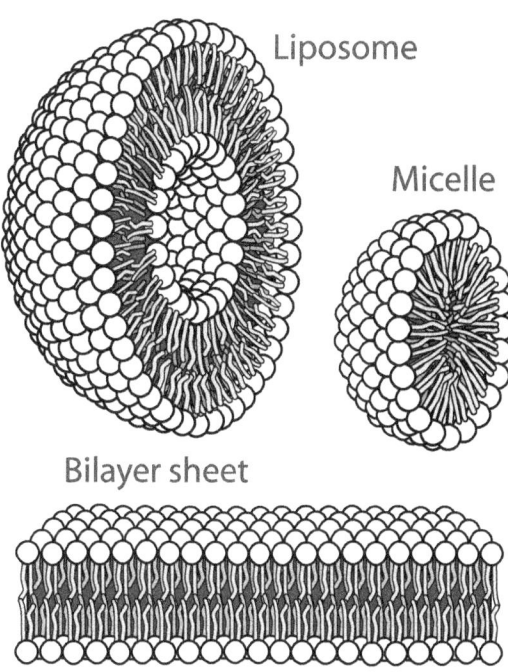

*The three main structures phospholipids form in solution; the liposome (a closed bilayer), the micelle and the bilayer.*

entropy) increases, yet life is distinguished by its great degree of organization. Therefore, a boundary is needed to separate life processes from non-living matter.[11] The cell membrane is the only cellular structure that is found in all of the cells of all of the organisms on Earth.[12]

Researchers Irene A. Chen and Jack W. Szostak (Nobel Prize in Physiology or Medicine 2009) amongst others, demonstrated that simple physicochemical properties of elementary protocells can give rise to essential cellular behaviors, including primitive forms of Darwinian competition and energy storage. Such cooperative interactions between the membrane and encapsulated contents could greatly simplify the transition from replicating molecules to true cells.[4] Furthermore, competition for membrane molecules would favor stabilized membranes, suggesting a selective advantage for the evolution of cross-linked fatty acids and even the phospholipids of today.[4] This micro-encapsulation allowed for metabolism within the membrane, exchange of small molecules and prevention of passage of large substances across it.[13] The main advantages of encapsulation include increased solubility of the cargo and creating energy in the form of chemical gradient. Energy is thus often said to be stored by cells in the structures of molecules of substances such as carbohydrates (including sugars), lipids, and proteins, which release energy when chemically combined with oxygen during cellular respiration.[14][15]

### 15.2.1 Energy gradient

A March 2014 study by NASA's Jet Propulsion Laboratory demonstrated a unique way to study the origins of life: fuel cells.[16] Fuel cells are similar to biological cells in that electrons are also transferred to and from molecules. In both cases, this results in electricity and power. The study states that one important factor was that the Earth provides electrical energy at the seafloor. "This energy could have kickstarted life and could have sustained life after it arose. Now, we have a way of testing different materials and environments that could have helped life arise not just on Earth, but possibly on Mars, Europa and other places in the Solar System."[16]

## 15.3 Vesicles and micelles

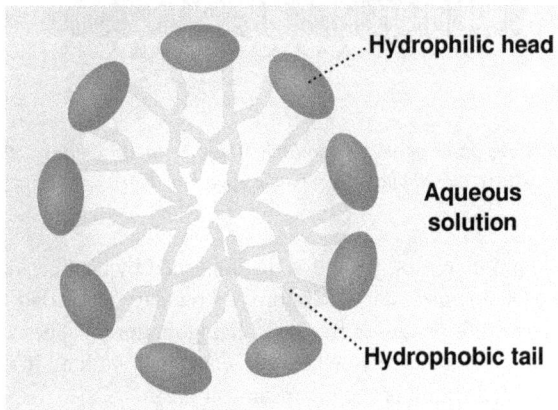

*Scheme of a micelle spontaneously formed by phospholipids in an aqueous solution*

When phospholipids are placed in water, the molecules spontaneously arrange such that the tails are shielded from the water, resulting in the formation of membrane structures such as bilayers, vesicles, and micelles.[2] In modern cells, vesicles are involved in metabolism, transport, buoyancy control,[17] and enzyme storage. They can also act as natural chemical reaction chambers. A typical vesicle or micelle in aqueous solution forms an aggregate with the hydrophilic "head" regions in contact with surrounding solvent, sequestering the hydrophobic single-tail regions in the micelle centre. This phase is caused by the packing behavior of single-tail lipids in a bilayer. Although the protocellular self-assembly process that spontaneously form lipid *monolayer* vesicles and micelles in nature resemble the kinds of primordial vesicles or protocells that might have existed at the beginning of evolution, they are not as sophisticated as the *bilayer* membranes of today's living organisms.[18]

Rather than being made up of phospholipids, however, early membranes may have formed from monolayers or bilayers of fatty acids, which may have formed more readily in a prebiotic environment.[19] Fatty acids have been synthesized in laboratories under a variety of prebiotic conditions and have been found on meteorites, suggesting their natural synthesis in nature.[4]

Oleic acid vesicles represent good models of membrane protocells that could have existed in prebiotic times.[20]

Electrostatic interactions induced by short, positively charged, hydrophobic peptides containing 7 amino acids in length or fewer, can attach RNA to a vesicle membrane, the basic cell membrane.[21][22]

### 15.3.1 Geothermal ponds and clay

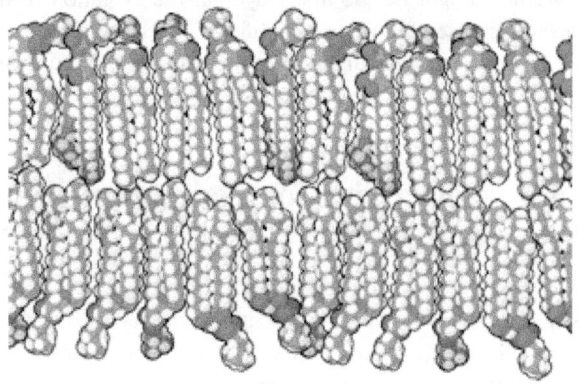

*This fluid lipid bilayer cross section is made up entirely of phosphatidylcholine.*

Scientists have come to conclude that life began in hydrothermal vents in the deep sea, but a 2012 study suggests that inland pools of condensed and cooled geothermal vapor have the ideal characteristics for the origin of life.[23] The conclusion is based mainly on the chemistry of modern cells, where the cytoplasm is rich in potassium, zinc, manganese, and phosphate ions, which are not widespread in marine environments. Such conditions, the researchers argue, are found only where hot hydrothermal fluid brings the ions to the surface — places such as geysers, mud pots, fumaroles and other geothermal features. Within these fuming and bubbling basins, water laden with zinc and manganese ions could have collected, cooled and condensed in shallow pools.[23]

Another study in the 1990s showed that montmorillonite clay can help create RNA chains of as many as 50 nucleotides joined together spontaneously into a single RNA molecule.[5] Later, in 2002, it was discovered that by adding montmorillonite to a solution of fatty acid micelles (lipid spheres), the clay sped up the rate of vesicle formation 100-

fold.[5]

Research has shown that some minerals can catalyze the stepwise formation of hydrocarbon tails of fatty acids from hydrogen and carbon monoxide gases - gases that may have been released from hydrothermal vents or geysers. Fatty acids of various lengths are eventually released into the surrounding water,[19] but vesicle formation requires a higher concentration of fatty acids, so it is suggested that protocell formation started at land-bound hydrothermal vents such as geysers, mud pots, fumaroles and other geothermal features where water evaporates and concentrates the solute.[5][24][25]

### 15.3.2 Montmorillonite bubbles

Another group suggests that primitive cells might have formed inside inorganic clay microcompartments, which can provide an ideal container for the synthesis and compartmentalization of complex organic molecules.[26] Clay-armored *bubbles* form naturally when particles of montmorillonite clay collect on the outer surface of air bubbles under water. This creates a semi permeable vesicle from materials that are readily available in the environment. The authors remark that montmorillonite is known to serve as a chemical catalyst, encouraging lipids to form membranes and single nucleotides to join into strands of RNA. Primitive reproduction can be envisioned when the clay bubbles burst, releasing the lipid membrane-bound product into the surrounding medium.[26]

### 15.3.3 Membrane transport

Instead of the more popular phospholipids of modern cells, the membrane of protocells in the RNA world would be composed of fatty acids,[27] and that such membranes have relatively high permeability to ions and small molecules,[1] such as nucleoside monophosphate (NMP), nucleoside diphosphate (NDP), and nucleoside triphosphate (NTP), and may withstand millimolar concentrations of $Mg^{2+}$.[28] Osmotic pressure also plays a significant role in protocell membrane transport.[1]

It has been proposed that electroporation resulting from lightning strikes could be a mechanism of natural horizontal gene transfer.[29] Electroporation is the rapid increase in bilayer permeability induced by the application of a large artificial electric field across the membrane. During electroporation in laboratory procedures, the lipid molecules are not chemically altered but simply shift position, opening up a pore (hole) that acts as the conductive pathway through the bilayer as it is filled with water. The mechanism is the creation of nanometer sized water-filled holes in the membrane. Experimentally, electroporation is used to introduce

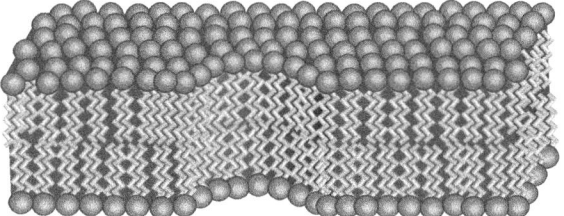

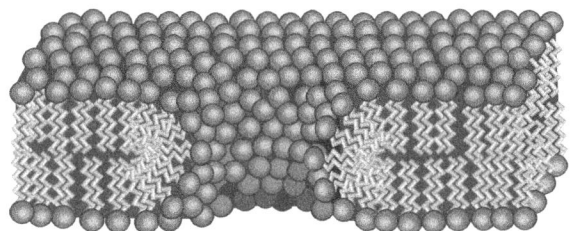

*Schematic showing two possible conformations of the lipids at the edge of a pore. In the top image the lipids have not rearranged, so the pore wall is hydrophobic. In the bottom image some of the lipid heads have bent over, so the pore wall is hydrophilic.*

hydrophilic molecules into cells. It is a particularly useful technique for large highly charged molecules such as DNA and RNA, which would never passively diffuse across the hydrophobic bilayer core.[30] Because of this, electroporation is one of the key methods of transfection as well as bacterial transformation.

**Fusion**

Some molecules or particles are too large or too hydrophilic to pass through a lipid bilayer, but can be moved across the cell membrane through fusion or budding of vesicles.[31] This may have eventually led to mechanisms that facilitate movement of molecules to the inside (endocytosis) or to release its contents into the extracellular space (exocytosis).

## 15.4 Artificial models

### 15.4.1 Langmuir-Blodgett deposition

Main article: Langmuir–Blodgett film

Starting with a technique commonly used to deposit molecules on a solid surface, Langmuir–Blodgett deposition, scientists are able to assemble phospholipid membranes of arbitrary complexity layer by layer.[32][33] These artificial phospholipid membranes support functional insertion both of purified and of *in situ* expressed membrane proteins.[33] The technique could help astrobiologists understand how the first living cells originated.[32]

### 15.4.2 Jeewanu

Main article: Jeewanu

Jeewanu protocells are synthetic chemical particles that

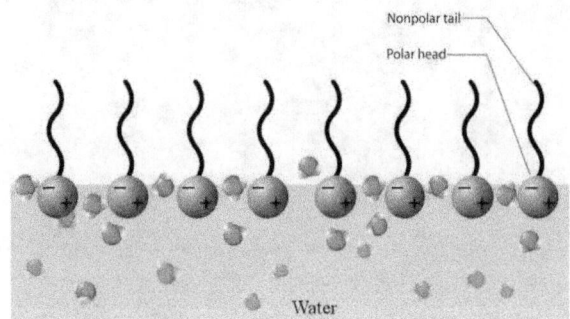

*Surfactant molecules arranged on an air – water interface*

possess cell-like structure and seem to have some functional living properties.[34] First synthesized in 1963 from simple minerals and basic organics while exposed to sunlight, it is still reported to have some metabolic capabilities, the presence of semipermeable membrane, amino acids, phospholipids, carbohydrates and RNA-like molecules.[34][35] However, the nature and properties of the Jeewanu remains to be clarified.[34][35][36]

In a similar synthesis experiment a frozen mixture of water, methanol, ammonia and carbon monoxide was exposed to ultraviolet (UV) radiation. This combination yielded large amounts of organic material that self-organised to form globules or vesicles when immersed in water.[37] The investigating scientist considered these globules to resemble cell membranes that enclose and concentrate the chemistry of life, separating their interior from the outside world. The globules were between 10 to 40 micrometres (0.00039 to 0.00157 in), or about the size of red blood cells. Remarkably, the globules fluoresced, or glowed, when exposed to UV light. Absorbing UV and converting it into visible light in this way was considered one possible way of providing energy to a primitive cell. If such globules played a role in the origin of life, the fluorescence could have been a precursor to primitive photosynthesis. Such fluorescence also provides the benefit of acting as a sunscreen, diffusing any damage that otherwise would be inflicted by UV radiation. Such a protective function would have been vital for life on the early Earth, since the ozone layer, which blocks out the sun's most destructive UV rays, did not form until after photosynthetic life began to produce oxygen.[38]

## 15.5 Ethics and controversy

Protocell research has created controversy and opposing opinions, including critics of the vague definition of "artificial life".[39] The creation of a basic unit of life is the most pressing ethical concern, although the most widespread worry about protocells is their potential threat to human health and the environment through uncontrolled replication.[40]

## 15.6 See also

- Abiogenesis
- Artificial cell
- Emergence
- Entropy and life
- Last universal ancestor
- Protocell Circus - film.
- Pseudo-panspermia
- RNA world hypothesis
- Synthetic biology

## 15.7 References

[1] Chen, Irene A.; Walde, Peter (July 2010). "From Self-Assembled Vesicles to Protocells" (PDF). *Cold Spring Harb Perspect Biol.* **2** (7): a002170. doi:10.1101/cshperspect.a002170. PMC 2890201. PMID 20519344.

[2] Garwood, Russell J. (2012). "Patterns In Palaeontology: The first 3 billion years of evolution". *Palaeontology Online* **2** (11): 1–14. Retrieved June 25, 2015.

[3] National Science Foundation (2013). "Exploring Life's Origins - Protocells". Retrieved 2014-03-18.

[4] Chen, Irene A. (8 December 2006). "The Emergence of Cells During the Origin of Life". *Science* **314** (5805): 1558–1559. doi:10.1126/science.1137541. PMID 17158315.

[5] Zimmer, Carl (26 June 2004). "What Came Before DNA?". *Discover Magazine*: 1–5.

[6] Rasmussen, Steen (2 July 2014). "Scientists Create Possible Precursor to Life". *A Letters Journal Exploring the Frontiers of Physics. Volume 107, Number 2, July 2014.* (Astrobiology Web). Retrieved 2014-10-24.

[7] Book: "Molecular Biology of the Cell." Alberts B., Johnson Lewis, et al. 2004.

[8] Deamer, D.W.; Dworkin, J.P. "Chemistry and Physics of Primitive Membranes." Top. Curr. Chem. 2005, 259, 1–27.

[9] Walde, P. "Surfactant Assemblies and their various possible roles for the origin(s) of life." Orig. Life Evol. Biosph. 2006, 36, 109–150.

[10] Sakuma, Yuka; Imai, Masayuki (2015). "From Vesicles to Protocells: The Roles of Amphiphilic Molecules". *Life* **5** (1): 651–675. doi:10.3390/life5010651. Retrieved 2015-04-13.

[11] Shapiro, Robert (12 February 2007). "A Simpler Origin for Life". *Scientific American*.

[12] Vodopich, Darrell S.; Moore., Randy (2002). "The Importance of Membranes". *Biology Laboratory Manual, 6/a*. McGraw-Hill. Retrieved 2014-03-17.

[13] Chang, Thomas Ming Swi (2007). *Artificial cells : biotechnology, nanomedicine, regenerative medicine, blood substitutes, bioencapsulation, cell/stem cell therapy*. Hackensack, N.J.: World Scientific. ISBN 981-270-576-7.

[14] Knowles, JR (1980). "Enzyme-catalyzed phosphoryl transfer reactions". *Annu. Rev. Biochem.* **49**: 877–919. doi:10.1146/annurev.bi.49.070180.004305. PMID 6250450.

[15] Campbell, Neil A.; Williamson, Brad; Heyden, Robin J. (2006). *Biology: Exploring Life*. Boston, Massachusetts: Pearson Prentice Hall. ISBN 0-13-250882-6.

[16] Clavin, Whitney (13 March 2014). "How Did Life Arise? Fuel Cells May Have Answers". *NASA*.

[17] Walsby, AE (1994). "Gas vesicles". *Microbiological reviews* **58** (1): 94–144. PMC 372955. PMID 8177173.

[18] Szostak, Jack W. (3 September 2004). "Battle of the Bubbles May Have Sparked Evolution". *Howard Hughes Medical Institute*.

[19] National Science Foundation (2013). "Membrane Lipids of Past and Present". *Exploring Life's Origins Project - A timeline of Life's Evolution*. Retrieved 2014-03-17.

[20] Douliez, Jean-Paul; Zhendre, Vanessa; Grélard, Axelle; Dufourc, Erick J. (24 November 2014). "Aminosilane/Oleic Acid Vesicles as Model Membranes of Protocells". *Langmuir* **30** (49): 14717–14724. doi:10.1021/la503908z. Retrieved 2015-04-13.

[21] "Peptide glue may have held first protocell components together".

[22] "Electrostatic Localization of RNA to Protocell Membranes by Cationic Hydrophobic Peptides".

[23] Switek, Brian (13 February 2012). "Debate bubbles over the origin of life". *Nature - News*.

[24] Szostak, Jack W. (4 June 2008). "Researchers Build Model Protocell Capable of Copying DNA". *HHMI News* (Howard Hughes Medical Institute).

[25] Cohen, Philip (23 October 2003). "Clay's matchmaking could have sparked life". *New Scientist*. Journal reference: *Science* (vol 302, p 618 )

[26] Stone, Howard A. (7 February 2011). "Clay-armored bubbles may have formed first protocells". *Harvard School of Engineering and Applied Sciences*.

[27] Müller, A. W. (June 2006). "Re-creating an RNA world". *Cell Mol Life Sci.* **63** (11): 1278–93. doi:10.1007/s00018-006-6047-1.

[28] Ma, Wentao; Yu, Chunwu; Zhang, Wentao; Hu., Jiming (Nov 2007). "Nucleotide synthetase ribozymes may have emerged first in the RNA world". *RNA* **13** (11): 2012–2019. doi:10.1261/rna.658507. PMC 2040096. PMID 17878321.

[29] Demanèche, S; Bertolla, F; Buret, F; et. al (August 2001). "Laboratory-scale evidence for lightning-mediated gene transfer in soil". *Appl. Environ. Microbiol.* **67** (8): 3440–4. doi:10.1128/AEM.67.8.3440-3444.2001. PMC 93040. PMID 11472916.

[30] Neumann, E; Schaefer-Ridder, M; Wang, Y; Hofschneider, PH (1982). "Gene transfer into mouse lyoma cells by electroporation in high electric fields". *EMBO J.* **1** (7): 841–5. PMC 553119. PMID 6329708.

[31] Norris, V.; Raine, D.J. (October 1998). "A fission-fusion origin for life". *Orig Life Evol Biosph* **28** (4-4): 523–537. PMID 9742727. Retrieved 2014-03-18.

[32] "Scientists Create Artificial Cell Membranes". *Astrobiology Magazine*. 4 October 2014. Retrieved 2014-05-07.

[33] Matosevic, Sandro; Paegel, Brian M. (29 September 2013). "Layer-by-layer cell membrane assembly". *Nature Chemistry* **5** (11): 958–963. doi:10.1038/nchem.1765. Retrieved 2014-05-07.

[34] Grote, M (September 2011). "Jeewanu, or the 'particles of life'" (PDF). *Journal of Biosciences* **36** (4): 563–570. doi:10.1007/s12038-011-9087-0. PMID 21857103. Archived (PDF) from the original on 2014-03-23.

[35] Gupta, V. K.; Rai, R. K. (2013). "Histochemical localisation of RNA-like material in photochemically formed self-sustaining, abiogenic supramolecular assemblies 'Jeewanu'". *Int. Res. J. of Science & Engineering* **1** (1): 1–4. ISSN 2322-0015.

[36] Caren, Linda D.; Ponnamperuma, Cyril (1967). "A review of some experiments on the synthesis of 'Jeewanu'" (PDF). *NASA Technical Memorandum X-1439* (Moffett Field, California: Ames Research Center).

[37] Dworkin, Jason P.; Deamer, David W.; Sandford, Scott A.; Allamandola, Louis J. (30 January 2001). "Self-assembling amphiphilic molecules: Synthesis in simulated interstellar/precometary ices". *Proceedings of the National Academy of Sciences of the United States of America* **98** (3): 815–9. Bibcode:2001PNAS...98..815D. doi:10.1073/pnas.98.3.815. PMC 14665. PMID 11158552.

[38] Mullen, L (5 September 2005). "Building Life from Star-Stuff". *Astrobiology Magazine*.

[39] "Life after the synthetic cell". *Nature* **465** (7297): 422–424. 27 May 2010. doi:10.1038/465422a. PMID 20495545.

[40] Beadau, Mark A.; Parke, Emily C. (2009). *The ethics of protocells moral and social implications of creating life in the laboratory* (Online ed.). Cambridge, Mass.: MIT Press. ISBN 978-0-262-51269-5.

## 15.8 External links

- "Protocells: Bridging Nonliving and Living Matter." Edited by Steen Rasmussen, Mark A. Bedau, Liaochai Chen, David Deamer, David Krakauer, Norman, H.Packard and Peter F. Stadler. MIT Press, Cambridge, Massachusetts. 2008.

- "Living Chemistry & A Natural History of Protocells." Synth-ethic: Art and Synthetic Biology Exhibition (2013) at the Natural History Museum, Vienna, Austria.

# Chapter 16

# Iron–sulfur world hypothesis

The **iron–sulfur world hypothesis** is a set of proposals for the origin of life and the early evolution of life advanced in a series of articles between 1988 and 1992 by Günter Wächtershäuser, a Munich patent lawyer with a degree in chemistry, who had been encouraged and supported by philosopher Karl R. Popper to publish his ideas. The hypothesis proposes that early life may have formed on the surface of iron sulfide minerals, hence the name.[1][2][3][4][5] It was developed by retrodiction from extant biochemistry in conjunction with chemical experiments.

## 16.1 Origin of life

### 16.1.1 Pioneer organism

Wächtershäuser proposes that the earliest form of life, termed "pioneer organism", originated in a volcanic hydrothermal flow at high pressure and high (100 °C) temperature. It had a composite structure of a mineral base with catalytic transition metal centers (predominantly iron and nickel, but also perhaps cobalt, manganese, tungsten and zinc). The catalytic centers catalyzed autotrophic carbon fixation pathways generating small molecule (non-polymer) organic compounds from inorganic gases (e.g. carbon monoxide, carbon dioxide, hydrogen cyanide and hydrogen sulfide). These organic compounds were retained on or in the mineral base as organic ligands of the transition metal centers with a flow retention time in correspondence with their mineral bonding strength thereby defining an autocatalytic "surface metabolism". The catalytic transition metal centers became autocatalytic by being accelerated by their organic products turned ligands. The carbon fixation metabolism became autocatalytic by forming a metabolic cycle in the form of a primitive sulfur-dependent version of the reductive citric acid cycle. Accelerated catalysts expanded the metabolism and new metabolic products further accelerated the catalysts. The idea is that once such a primitive autocatalytic metabolism was established, its intrinsically synthetic chemistry began to produce ever more complex organic compounds, ever more complex pathways and ever more complex catalytic centers.

The fundamental idea of abiogenesis, according to the iron–sulfur world hypothesis can be simplified in the following brief characterization: Pressurize and heat a water flow with dissolved volcanic gases (e.g. carbon monoxide, ammonia and hydrogen sulfide) to 100 °C. Pass the flow over catalytic transition metal solids (e.g. iron sulfide and nickel sulfide). Wait and locate the formation of catalytic metallo-peptides.

### 16.1.2 Nutrient conversions

The water gas shift reaction ($CO + H_2O \rightarrow CO_2 + H_2$) occurs in volcanic fluids with diverse catalysts or without catalysts.[6] The combination of ferrous sulfide and hydrogen sulfide as reducing agents in conjunction with pyrite formation – $FeS + H_2S \rightarrow FeS_2 + 2H^+ + 2e^-$ (or $H_2$ instead of $2H^+ + 2e^-$) – has been demonstrated under mild volcanic conditions.[7][8] This key result has been disputed.[9] Nitrogen fixation has been demonstrated for the isotope $^{15}N_2$ in conjunction with pyrite formation.[10] Ammonia forms from nitrate with $FeS/H_2S$ as reductant.[11] Methylmercaptan [$CH_3$-SH] and carbon oxysulfide [COS] form from $CO_2$ and $FeS/H_2S$,[12] or from CO and $H_2$ in the presence of NiS.[13]

### 16.1.3 Synthetic reactions

Reaction of carbon monoxide (CO), hydrogen sulfide ($H_2S$) and Methanethiol $CH_3SH$ in the presence of nickel sulfide and iron sulfide generates the methyl thioester of acetic acid [$CH_3$-CO-$SCH_3$] and presumably thioacetic acid ($CH_3$-CO-SH) as the simplest activated acetic acid analogues of acetyl-CoA. These activated acetic acid derivatives serve as starting materials for subsequent exergonic synthetic steps.[13] They also serve for energy coupling with endergonic reactions, notably the formation of (phospho)anhydride compounds.[14] However, Huber and Wächtershäuser reported a low 0.5% acetate yields based on input of $CH_3SH$ (Methanethiol) (8 mM) in the presence of 350 mM CO. This is about 500 times and 3700 times [15] the highest $CH_3SH$ and CO concentrations respectively measured to date in a natural hydrothermal vent fluid.[16]

Reaction of nickel hydroxide with hydrogen cyanide (HCN) (in the presence or absence of ferrous hydroxide, hydrogen sulfide or methyl mercaptan) generates nickel cyanide, which reacts with carbon monoxide (CO) to generate pairs of α-hydroxy and α-amino acids: e.g. glycolate/glycine, lactate/alanine, glycerate/serine; as well as pyruvic acid in significant quantities.[17] Pyruvic acid is also formed at high pressure and high temperature from CO, $H_2O$, FeS in the presence of nonyl mercaptan.[18] Reaction of pyruvic acid or other α-keto acids with ammonia in the presence of ferrous hydroxide or in the presence of ferrous sulfide and hydrogen sulfide generates alanine or other α-amino acids.[19] Reaction of α-amino acids in aqueous solution with COS or with CO and $H_2S$ generates a peptide cycle wherein dipeptides, tripeptides etc. are formed and subsequently degraded via N-terminal hydantoin moieties and N-terminal urea moieties and subsequent cleavage of the N-terminal amino acid unit.[20][21][22]

Proposed reaction mechanism for reduction of $CO_2$ on FeS: Ying et al. (2007) have proved that direct transformation of mackinawite (FeS) to pyrite ($FeS_2$) on reaction with $H_2S$ till 300 °C is not possible without the presence of critical amount of oxidant. In the absence of any oxidant, FeS reacts with $H_2S$ up to 300 °C to give pyrrhotite. Farid et al. have proved experimentally that mackinawite (FeS) has ability to reduce $CO_2$ to CO at temperature higher than 300 °C. They claimed surface of FeS is oxized which on reaction with $H_2S$ gives pyrite ($FeS_2$). It is expected that CO reacts with $H_2O$ in the Drobner experiment to give $H_2$.

## 16.2 Early evolution

Early evolution is defined as beginning with the origin of life and ending with the last universal common ancestor (LUCA). According to the iron–sulfur world theory it covers a coevolution of cellular organization (cellularization), the genetic machinery and enzymatization of the metabolism.

### 16.2.1 Cellularization

Cellularization occurs in several stages. It begins with the formation of primitive lipids (e.g. fatty acids or isoprenoid acids) in the surface metabolism. These lipids accumulate on or in the mineral base. This lipophilizes the outer or inner surfaces of the mineral base, which promotes condensation reactions over hydrolytic reactions by lowering the activity of water and protons.

In the next stage lipid membranes are formed. While still anchored to the mineral base they form a semi-cell bounded partly by the mineral base and partly by the membrane. Further lipid evolution leads to self-supporting lipid membranes and closed cells. The earliest closed cells are pre-cells (*sensu* Kandler) because they allow frequent exchange of genetic material (e.g. by fusions). According to Woese, this frequent exchange of genetic material is the cause for the existence of the common stem in the tree of life and for a very rapid early evolution.

### 16.2.2 Proto-ecological systems

William Martin and Michael Russell suggest that the first cellular life forms may have evolved inside alkaline hydrothermal vents at seafloor spreading zones in the deep sea.[23][24] These structures consist of microscale caverns that are coated by thin membraneous metal sulfide walls. Therefore, these structures would resolve several critical points germane to Wächtershäuser's suggestions at once:

1. the micro-caverns provide a means of concentrating newly synthesised molecules, thereby increasing the chance of forming oligomers;

2. the steep temperature gradients inside the hydrothermal vent allow for establishing "optimum zones" of partial reactions in different regions of the vent (e.g. monomer synthesis in the hotter, oligomerisation in the colder parts);

3. the flow of hydrothermal water through the structure provides a constant source of building blocks and energy (chemical disequilibrium between hydrothermal hydrogen and marine carbon dioxide);

4. the model allows for a succession of different steps of cellular evolution (prebiotic chemistry, monomer and oligomer synthesis, peptide and protein synthesis, RNA world, ribonucleoprotein assembly and DNA world) in a single structure, facilitating exchange between all developmental stages;

5. synthesis of lipids as a means of "closing" the cells against the environment is not necessary, until basically all cellular functions are developed.

This model locates the "last universal common ancestor" (LUCA) within the inorganically formed physical confines of an alkaline hydrothermal vent, rather than assuming the existence of a free-living form of LUCA. The last evolutionary step en route to bona fide free-living cells would be the synthesis of a lipid membrane that finally allows the organisms to leave the microcavern system of the vent. This postulated late acquisition of the biosynthesis of lipids as directed by genetically encoded peptides is consistent with the presence of completely different types of membrane lipids in archaea and bacteria (plus eukaryotes). The kind of vent at the foreground of their suggestion is chemically more similar to the warm (ca. 100 °C) off ridge vents such as Lost City than to the more familiar black smoker type vents (ca. 350 °C).

In an abiotic world, a thermocline of temperatures and a chemocline in concentration is associated with the prebiotic synthesis of organic molecules, hotter in proximity to the chemically rich vent, cooler but also less chemically rich at greater distances. The migration of synthesized compounds from areas of high concentration to areas of low concentration gives a directionality that provides both source and sink in a self-organizing fashion, enabling a proto-metabolic process by which acetic acid production and its eventual oxidization can be spatially organized.

In this way many of the individual reactions that are today found in central metabolism could initially have occurred independent of any developing cell membrane. Each vent microcompartment is functionally equivalent to a single cell. Chemical communities having greater structural integrity and resilience to wildly fluctuating conditions are then selected for; their success would lead to local zones of depletion for important precursor chemicals. Progressive incorporation of these precursor components within a cell membrane would gradually increase metabolic complexity within the cell membrane, whilst leading to greater environmental simplicity in the external environment. In principle, this could lead to the development of complex catalytic sets capable of self-maintenance.

Russell adds a significant factor to these ideas, by pointing out that semi-permeable mackinawite (an iron sulfide mineral) and silicate membranes could naturally develop under these conditions and electrochemically link reactions separated in space, if not in time.[25][26]

## 16.3 References

[1] Wächtershäuser, Günter (1988-12-01). "Before enzymes and templates: theory of surface metabolism". *Microbiol. Mol. Biol. Rev.* **52** (4): 452–84. PMC 373159. PMID 3070320. Retrieved 2009-05-02.

[2] Wächtershäuser, G (January 1990). "Evolution of the first metabolic cycles". *Proceedings of the National Academy of Sciences of the United States of America* **87** (1): 200–4. Bibcode:1990PNAS...87..200W. doi:10.1073/pnas.87.1.200. PMC 53229. PMID 2296579. Retrieved 2009-05-02.

[3] Günter Wächtershäuser, G (1992). "Groundworks for an evolutionary biochemistry: The iron-sulphur world". *Progress in Biophysics and Molecular Biology* **58** (2): 85–201. doi:10.1016/0079-6107(92)90022-X. PMID 1509092. Retrieved 2009-05-02.

[4] Günter Wächtershäuser, G (2006). "From volcanic origins of chemoautotrophic life to Bacteria, Archaea and Eukarya". *Philosophical Transactions of the Royal Society B: Biological Sciences* **361** (1474): 1787–806; discussion 1806–8. doi:10.1098/rstb.2006.1904. PMC 1664677. PMID 17008219.

[5] Wächtershäuser, Günter (2007). "On the Chemistry and Evolution of the Pioneer Organism". *Chemistry & Biodiversity* **4** (4): 584–602. doi:10.1002/cbdv.200790052. PMID 17443873.

[6] Seewald, Jeffrey S.; Mikhail Yu. Zolotov; Thomas McCollom (January 2006). "Experimental investigation of single carbon compounds under hydrothermal conditions". *Geochimica et Cosmochimica Acta* **70** (2): 446–460. Bibcode:2006GeCoA..70..446S. doi:10.1016/j.gca.2005.09.002. Retrieved 2009-05-02.

[7] Taylor, P.; T. E. Rummery; D. G. Owen (1979). "Reactions of iron monosulfide solids with aqueous hydrogen sulfide up to 160°C". *Journal of Inorganic and Nuclear Chemistry* **41** (12): 1683–1687. doi:10.1016/0022-1902(79)80106-2. Retrieved 2009-05-02.

[8] Drobner, E.; H. Huber; G. Wächtershäuser; D. Rose; K. O. Stetter (1990). "Pyrite formation linked with hydrogen evolution under anaerobic conditions". *Nature* **346** (6286): 742–744. Bibcode:1990Natur.346..742D. doi:10.1038/346742a0.

[9] Cahill, C. L.; L. G. Benning; H. L. Barnes; J. B. Parise (June 2000). "In situ time-resolved X-ray diffraction of iron sulfides during hydrothermal pyrite growth". *Chemical Geology* **167** (1–2): 53–63. doi:10.1016/S0009-2541(99)00199-0. Retrieved 2009-05-02.

[10] Mark Dorr, Mark; Johannes Käßbohrer; Renate Grunert; Günter Kreisel; Willi A. Brand; Roland A. Werner; Heike Geilmann; Christina Apfel; Christian Robl; Wolfgang Weigand (2003). "A Possible Prebiotic Formation of Ammonia from Dinitrogen on Iron Sulfide Surfaces". *Angewandte Chemie International Edition* **42** (13): 1540–3. doi:10.1002/anie.200250371. PMID 12698495.

[11] Blöchl, E; M Keller; G Wächtershäuser; K O Stetter (1992). "Reactions depending on iron sulfide and linking geochemistry with biochemistry". *Proceedings of the National Academy of Sciences of the United States of America* **89** (17): 8117–20. Bibcode:1992PNAS...89.8117B. doi:10.1073/pnas.89.17.8117. PMC 49867. PMID 11607321. Retrieved 2009-05-02.

[12] Heinen, Wolfgang; Anne Marie Lauwers (1996-04-01). "Organic sulfur compounds resulting from the interaction of iron sulfide, hydrogen sulfide and

carbon dioxide in an anaerobic aqueous environment". *Origins of Life and Evolution of Biospheres* **26** (2): 131–150. Bibcode:1996OLEB...26..131H. doi:10.1007/BF01809852.

[13] Huber, Claudia; Günter Wächtershäuser (1997-04-11). "Activated Acetic Acid by Carbon Fixation on (Fe,Ni)S Under Primordial Conditions". *Science* **276** (5310): 245–7. doi:10.1126/science.276.5310.245. PMID 9092471. Retrieved 2009-05-02.

[14] Günter Wächtershäuser; Michael W. W. Adams (1998). "The case for a hyperthermophilic, chemolithoautotrophic origin of life in an iron-sulfur world". In Juergen Wiegel (ed.). *Thermophiles: The Keys to Molecular Evolution and the Origin of Life*. pp. 47–57. ISBN 9780748407477.

[15] Chandru, Kuhan; Gilbert, Alexis; Butch, Christopher; Aono, Masashi; Cleaves, Henderson James II (21 July 2016). "The Abiotic Chemistry of Thiolated Acetate Derivatives and the Origin of Life". *Scientific Reports* **6** (29883). doi:10.1038/srep29883.

[16] Reeves, Eoghan P.; McDermott, Jill M.; Seewald, Jeffrey S. (April 15, 2014). "The origin of methanethiol in midocean ridge hydrothermal fluids". *Proceedings of the National Academy of Sciences of the United States of America PNAS, Proceedings of the National Academy of Sciences* **111** (15). doi:10.1073/pnas.1400643111.

[17] Huber, Claudia; Günter Wächtershäuser (2006-10-27). "α-Hydroxy and α-Amino Acids Under Possible Hadean, Volcanic Origin-of-Life Conditions". *Science* **314** (5799): 630–2. Bibcode:2006Sci...314..630H. doi:10.1126/science.1130895. PMID 17068257. Retrieved 2009-05-02.

[18] Cody, George D.; Nabil Z. Boctor; Timothy R. Filley; Robert M. Hazen; James H. Scott; Anurag Sharma; Hatten S. Yoder (2000-08-25). "Primordial Carbonylated Iron-Sulfur Compounds and the Synthesis of Pyruvate". *Science* **289** (5483): 1337–40. Bibcode:2000Sci...289.1337C. doi:10.1126/science.289.5483.1337. PMID 10958777. Retrieved 2009-05-02.

[19] Huber, Claudia; Günter Wächtershäuser (February 2003). "Primordial reductive amination revisited". *Tetrahedron Letters* **44** (8): 1695–1697. doi:10.1016/S0040-4039(02)02863-0. Retrieved 2009-05-02.

[20] Huber, Claudia; Günter Wächtershäuser (1998-07-31). "Peptides by Activation of Amino Acids with CO on (Ni,Fe)S Surfaces: Implications for the Origin of Life". *Science* **281** (5377): 670–2. Bibcode:1998Sci...281..670H. doi:10.1126/science.281.5377.670. PMID 9685253. Retrieved 2009-05-02.

[21] Huber, Claudia; Wolfgang Eisenreich; Stefan Hecht; Günter Wächtershäuser (2003-08-15). "A Possible Primordial Peptide Cycle". *Science* **301** (5635): 938–40. Bibcode:2003Sci...301..938H. doi:10.1126/science.1086501. PMID 12920291. Retrieved 2009-05-02.

[22] Wächtershäuser, Günter (2000-08-25). "ORIGIN OF LIFE: Life as We Don't Know It". *Science* **289** (5483): 1307–8. doi:10.1126/science.289.5483.1307. PMID 10979855. (requires nonfree AAAS member subscription)

[23] Martin, William; Michael J Russell (2003). "On the origins of cells: a hypothesis for the evolutionary transitions from abiotic geochemistry to chemoautotrophic prokaryotes, and from prokaryotes to nucleated cells". *Philosophical Transactions of the Royal Society of London. Series B: Biological Sciences* **358** (1429): 59–83; discussion 83–5. doi:10.1098/rstb.2002.1183. PMC 1693102. PMID 12594918.

[24] Martin, William; Michael J Russell (2007). "On the origin of biochemistry at an alkaline hydrothermal vent". *Philos Trans R Soc Lond B Biol Sci.* **362** (1486): 1887–925. doi:10.1098/rstb.2006.1881. PMC 2442388. PMID 17255002.

[25] Michael Russell, Michael (2006). "First Life". *American Scientist* **94** (1): 32. doi:10.1511/2006.1.32.

[26] Russell, Michael (Ed), (2010), "Origins, Abiogenesis and the Search for Life in the Universe" (Cosmology Science Publications)

# Chapter 17

# Panspermia

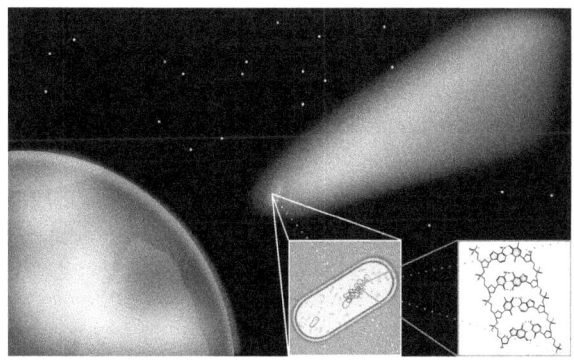

*Illustration of a comet (center) transporting a bacterial life form (inset) through space to the Earth (left)*

**Panspermia** (from Greek πᾶν *(pan)*, meaning "all", and σπέρμα *(sperma)*, meaning "seed") is the hypothesis that life exists throughout the Universe, distributed by meteoroids, asteroids, comets,[1] planetoids,[2] and, also, by spacecraft in the form of unintended contamination by microorganisms.[3][4]

Panspermia is a hypothesis proposing that microscopic life forms that can survive the effects of space, such as extremophiles, become trapped in debris that is ejected into space after collisions between planets and small Solar System bodies that harbor life. Some organisms may travel dormant for an extended amount of time before colliding randomly with other planets or intermingling with protoplanetary disks. If met with ideal conditions on a new planet's surfaces, the organisms become active and the process of evolution begins. Panspermia is not meant to address how life began, just the method that may cause its distribution in the Universe.[5][6][7]

Pseudo-panspermia (sometimes called *"soft panspermia"* or *"molecular panspermia"*) argues that the pre-biotic organic building blocks of life originated in space and were incorporated in the solar nebula from which the planets condensed and were further—and continuously—distributed to planetary surfaces where life then emerged (abiogenesis).[8][9] From the early 1970s it was becoming evident that interstellar dust consisted of a large component of organic molecules. Interstellar molecules are formed by chemical reactions within very sparse interstellar or circumstellar clouds of dust and gas.[10] The dust plays a critical role of shielding the molecules from the ionizing effect of ultraviolet radiation emitted by stars.[11]

The chemistry leading to life may have begun shortly after the Big Bang, 13.8 billion years ago, during a habitable epoch when the Universe was only 10–17 million years old. Though life is confirmed only on the Earth, some think that extraterrestrial life is not only plausible, but probable or inevitable. Other planets and moons in the Solar System and other planetary systems are being examined for evidence of having once supported simple life, and projects such as SETI are trying to detect radio transmissions from possible alien civilizations.

## 17.1 History

## 17.1. HISTORY

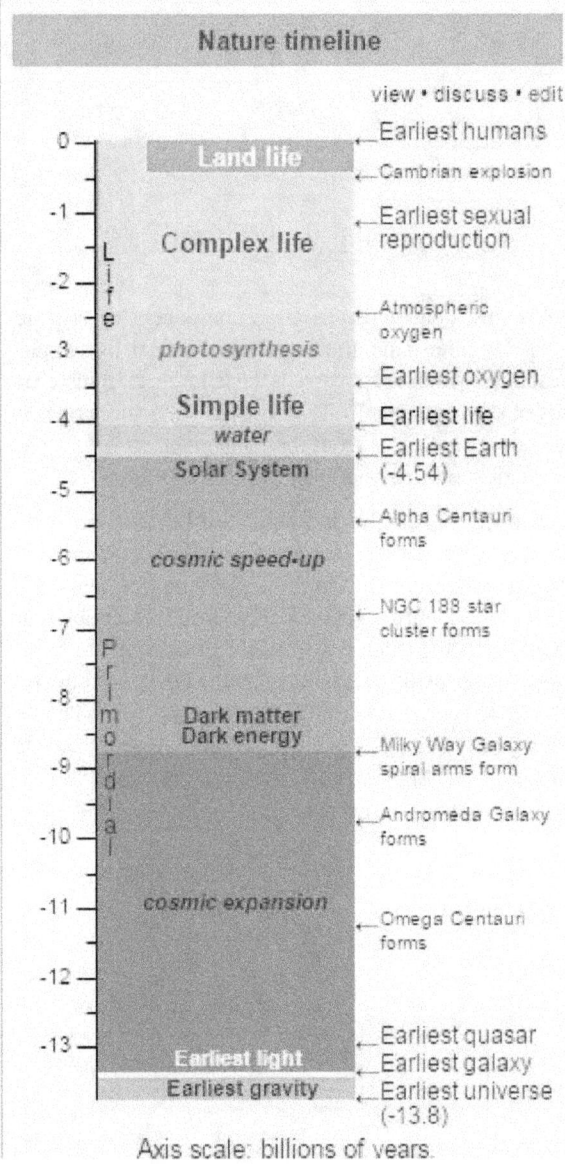

The first known mention of the term was in the writings of the 5th century BC Greek philosopher Anaxagoras.[12] Panspermia began to assume a more scientific form through the proposals of Jöns Jacob Berzelius (1834),[13] Hermann E. Richter (1865),[14] Kelvin (1871),[15] Hermann von Helmholtz (1879)[16][17] and finally reaching the level of a detailed hypothesis through the efforts of the Swedish chemist Svante Arrhenius (1903).[18]

Fred Hoyle (1915–2001) and Chandra Wickramasinghe (born 1939) were influential proponents of panspermia.[19][20] In 1974 they proposed the hypothesis that some dust in interstellar space was largely organic (containing carbon), which Wickramasinghe later proved to be correct.[21][22][23] Hoyle and Wickramasinghe further contended that life forms continue to enter the Earth's atmosphere, and may be responsible for epidemic outbreaks, new diseases, and the genetic novelty necessary for macroevolution.[24]

In an Origins Symposium presentation on April 7, 2009, physicist Stephen Hawking stated his opinion about what humans may find when venturing into space, such as the possibility of alien life through the theory of panspermia: "Life could spread from planet to planet or from stellar system to stellar system, carried on meteors."[25]

Three series of astrobiology experiments have been conducted outside the International Space Station between 2008 and 2015 (EXPOSE) where a wide variety of biomolecules, microorganisms, and their spores were exposed to the solar flux and vacuum of space for about 1.5 years. Some organisms survived in an inactive state for considerable lengths of time,[26][27] and those samples sheltered by simulated meteorite material provide experimental evidence for the likelihood of the hypothetical scenario of lithopanspermia.[28]

Several simulations in laboratories and in low Earth orbit suggest that ejection, entry and impact is survivable for some simple organisms. In 2015, "remains of biotic life" were found in 4.1 billion-year-old rocks in Western Australia, when the young Earth was about 400 million years old.[29][30] According to one of the researchers, "If life arose relatively quickly on Earth ... then it could be common in the universe."[29]

## 17.2 Proposed mechanisms

Panspermia can be said to be either interstellar (between star systems) or interplanetary (between planets in the same star system);[31][32] its transport mechanisms may include comets,[33][34] radiation pressure and lithopanspermia (microorganisms embedded in rocks).[35][36][37] Interplanetary transfer of nonliving material is well documented, as evidenced by meteorites of Martian origin found on Earth.[37] Space probes may also be a viable transport mechanism for interplanetary cross-pollination in the Solar System or even beyond. However, space agencies have implemented planetary protection procedures to reduce the risk of planetary contamination,[38][39] although, as recently discovered, some microorganisms, such as Tersicoccus phoenicis, may be resistant to procedures used in spacecraft assembly clean room facilities.[3][4] In 2012, mathematician Edward Belbruno and astronomers Amaya Moro-Martín and Renu Malhotra proposed that gravitational low energy transfer of rocks among the young planets of stars in their birth cluster is commonplace, and not rare in the general galactic stellar population.[40][41] Deliberate directed panspermia from space to seed Earth[42] or sent from Earth to seed other planetary systems have also been proposed.[43][44][45][46] One twist to the hypothesis by engineer Thomas Dehel (2006), proposes that plasmoid magnetic fields ejected from the magnetosphere may move the few spores lifted from the Earth's atmosphere with sufficient speed to cross interstellar space to other systems before the spores can be destroyed.[47][48]

### 17.2.1 Radiopanspermia

In 1903, Svante Arrhenius published in his article *The Distribution of Life in Space*,[49] the hypothesis now called radiopanspermia, that microscopic forms of life can be propagated in space, driven by the radiation pressure from stars.[50] Arrhenius argued that particles at a critical size below 1.5 μm would be propagated at high speed by radiation pressure of the Sun. However, because its effectiveness decreases with increasing size of the particle, this mechanism holds for very tiny particles only, such as single bacterial spores.[51] The main criticism of radiopanspermia hypothesis came from Shklovskii and Sagan, who pointed out the proofs of the lethal action of space radiations (UV and X-rays) in the cosmos.[52] Regardless of the evidence, Wallis and Wickramasinghe argued in 2004 that the transport of individual bacteria or clumps of bacteria, is overwhelmingly more important than lithopanspermia in terms of numbers of microbes transferred, even accounting for the death rate of unprotected bacteria in transit.[53]

Then, data gathered by the orbital experiments ERA, BIOPAN, EXOSTACK and EXPOSE, determined that isolated spores, including those of *B. subtilis*, were killed by several orders of magnitude if exposed to the full space environment for a mere few seconds, but if shielded against solar UV, the spores were capable of surviving in space for up to 6 years while embedded in clay or meteorite powder (artificial meteorites).[51][54] Though minimal protection is required to shelter a spore against UV radiation, exposure to solar UV and cosmic ionizing radiation of unprotected DNA, break it up into its bases.[55][56][57] Also, exposing DNA to the ultrahigh vacuum of space alone is sufficient to cause DNA damage, so the transport of unprotected DNA or RNA during interplanetary flights powered solely by light pressure is extremely unlikely.[57] The feasibility of other means of transport for the more massive shielded spores into the outer Solar System – for example, through gravitational capture by comets – is at this time unknown.

Based on experimental data on radiation effects and DNA stability, it has been concluded that for such long travel times, boulder sized rocks which are greater than or equal to 1 meter in diameter are required to effectively shield resistant microorganisms, such as bacterial spores against galactic cosmic radiation.[58][59] These results clearly negate the radiopanspermia hypothesis, which requires single spores accelerated by the radiation pressure of the Sun, requiring many years to travel between the planets, and support the likelihood of interplanetary transfer of microorganisms within asteroids or comets, the so-called **lithopanspermia** hypothesis.[51][54]

### 17.2.2 Lithopanspermia

Lithopanspermia, the transfer of organisms in rocks from one planet to another either through interplanetary or interstellar space, remains speculative. Although there is no evidence that lithopanspermia has occurred in the Solar System, the various stages have become amenable to experimental testing.[60]

- **Planetary ejection** — For lithopanspermia to occur, microorganisms must survive ejection from a planetary surface which involves extreme forces of acceleration and shock with associated temperature excursions. Hypothetical values of shock pressures experienced by ejected rocks are obtained with Martian meteorites, which suggest the shock pressures of approximately 5 to 55 GPa, acceleration of $3\times10^6$ m/s$^2$ and jerk of $6\times10^9$ m/s$^3$ and post-shock temperature increases of about 1 K to 1000 K.[61][62] To determine the effect of acceleration during ejection on microorganisms, rifle and ultracentrifuge methods were successfully used under simulated outer space conditions.[60]

- **Survival in transit** — The survival of microorganisms has been studied extensively using both simulated facilities and in low Earth orbit. A large number of microorganisms have been selected for exposure experiments. It is possible to separate these microorganisms into two groups, the human-borne, and the extremophiles. Studying the human-borne microorganisms is significant for human welfare and future manned missions; whilst the extremophiles are vital for studying the physiological requirements of survival in space.[60]

- **Atmospheric entry** — An important aspect of the lithopanspermia hypothesis to test is that microbes situated on or within rocks could survive hypervelocity entry from space through Earth's atmosphere (Cockell, 2008). As with planetary ejection, this is experimentally tractable, with sounding rockets and orbital vehicles being used for microbiological experiments.[60][61] *B. subtilis* spores inoculated onto granite domes were subjected to hypervelocity atmospheric transit (twice) by launch to a ~120 km altitude on an Orion two-stage rocket. The spores were shown to have survived on the sides of the rock, but they did not survive on the forward-facing surface that was subjected to a maximum temperature of 145 °C.[63] In separate experiments, as part of the ESA STONE experiment, numerous organisms were embedded in different types or rocks and were mounted in the heat shield of six Foton re-entry capsules. During reentry, the rock samples were subjected to temperatures and pressure loads comparable to those experienced in meteorites.[64] The exogenous arrival of photosynthetic microorganisms could have quite profound consequences for the course of biological evolution on the inoculated planet. As photosynthetic organisms must be close to the surface of a rock to obtain sufficient light energy, atmospheric transit might act as a filter against them by ablating the surface layers of the rock. Although cyanobacteria have been shown to survive the desiccating, freezing conditions of space in orbital experiments, this would be of no benefit as the STONE experiment showed that they cannot survive atmospheric entry.[65] Thus, non-photosynthetic organisms deep within rocks have a chance to survive the exit and entry process. (See also: Impact survival.) Research presented at the European Planetary Science Congress in 2015 suggests that ejection, entry and impact is survivable for some simple organisms.[66]

### 17.2.3 Accidental panspermia

Thomas Gold, a professor of astronomy, suggested in 1960 the hypothesis of "Cosmic Garbage", that life on Earth might have originated accidentally from a pile of waste products dumped on Earth long ago by extraterrestrial beings.[67]

### 17.2.4 Directed panspermia

Main article: Directed panspermia

Directed panspermia concerns the deliberate transport of microorganisms in space, sent to Earth to start life here, or sent from Earth to seed new planetary systems with life by introduced species of microorganisms on lifeless planets. The Nobel prize winner Francis Crick, along with Leslie Orgel proposed that life may have been purposely spread by an advanced extraterrestrial civilization,[42] but considering an early "RNA world" Crick noted later that life may have originated on Earth.[68] It has been suggested that 'directed' panspermia was proposed in order to counteract various objections, including the argument that microbes would be inactivated by the space environment and cosmic radiation before they could make a chance encounter with Earth.[69]

Conversely, active directed panspermia has been proposed to secure and expand life in space.[45] This may be motivated by biotic ethics that values, and seeks to propagate, the basic patterns of our organic gene/protein lifeform.[70] The panbiotic program would seed new planetary systems nearby, and clusters of new stars in interstellar clouds. These young targets, where local life would not have formed yet, avoid any interference with local life.

For example, microbial payloads launched by solar sails at speeds up to $0.0001\ c$ (30,000 m/s) would reach targets at 10 to 100 light-years in 0.1 million to 1 million years. Fleets of microbial capsules can be aimed at clusters of new stars in star-forming clouds, where they may land on planets or captured by asteroids and comets and later delivered to planets. Payloads may contain extremophiles for diverse environments and cyanobacteria similar to early microorganisms. Hardy multicellular organisms (rotifer cysts) may be included to induce higher evolution.[71]

The probability of hitting the target zone can be calculated from $P(target) = \frac{A(target)}{\pi(dy)^2} = \frac{ar(target)^2 v^2}{(tp)^2 d^4}$ where

A(target) is the cross-section of the target area, $dy$ is the positional uncertainty at arrival; $a$ – constant (depending on units), $r$(target) is the radius of the target area; $v$ the velocity of the probe; (tp) the targeting precision (arcsec/yr); and $d$ the distance to the target, guided by high-resolution astrometry of $1 \times 10^{-5}$ arcsec/yr (all units in SIU). These calculations show that relatively near target stars(Alpha PsA, Beta Pictoris) can be seeded by milligrams of launched microbes; while seeding the Rho Ophiochus star-forming cloud requires hundreds of kilograms of dispersed capsules.[45]

Directed panspermia to secure and expand life in space is becoming possible because of developments in solar sails, precise astrometry, extrasolar planets, extremophiles and microbial genetic engineering. After determining the composition of chosen meteorites, astroecologists performed laboratory experiments that suggest that many colonizing microorganisms and some plants could obtain many of their chemical nutrients from asteroid and cometary materials.[72] However, the scientists noted that phosphate ($PO_4$) and nitrate ($NO_3$–N) critically limit nutrition to many terrestrial lifeforms.[72] With such materials, and energy from long-lived stars, microscopic life planted by directed panspermia could find an immense future in the galaxy.[73]

A number of publications since 1979 have proposed the idea that directed panspermia could be demonstrated to be the origin of all life on Earth if a distinctive 'signature' message were found, deliberately implanted into either the genome or the genetic code of the first microorganisms by our hypothetical progenitor.[74][75][76][77] In 2013 a team of physicists claimed that they had found mathematical and semiotic patterns in the genetic code which, they believe, is evidence for such a signature.[78][79][80] Further investigations are needed.

A microscopic ball made of titanium and vanadium was found in Earth's upper atmosphere in early 2015. Milton Wainwright, a UK researcher and astrobiologist at the University of Buckingham claimed in a tabloid that the metal ball "could contain DNA." He speculates that it could be an alien device sent to Earth by extraterrestrials in order to continue seeding the planet with life.[81]

### 17.2.5 Pseudo-panspermia

Further information: List of interstellar and circumstellar molecules and Abiogenesis § Extraterrestrial organic molecules

Pseudo-panspermia (sometimes called soft panspermia, molecular panspermia or quasi-panspermia) proposes that the organic molecules used for life originated in space and were incorporated in the solar nebula, from which the planets condensed and were further —and continuously— distributed to planetary surfaces where life then emerged (abiogenesis).[8][9] From the early 1970s it was becoming evident that interstellar dust consisted of a large component of organic molecules. The first suggestion came from Chandra Wickramasinghe, who proposed a polymeric composition based on the molecule formaldehyde ($CH_2O$).[82] Interstellar molecules are formed by chemical reactions within very sparse interstellar or circumstellar clouds of dust and gas. Usually this occurs when a molecule becomes ionized, often as the result of an interaction with cosmic rays. This positively charged molecule then draws in a nearby reactant by electrostatic attraction of the neutral molecule's electrons. Molecules can also be generated by reactions between neutral atoms and molecules, although this process is generally slower.[10] The dust plays a critical role of shielding the molecules from the ionizing effect of ultraviolet radiation emitted by stars.[11]

A 2008 analysis of $^{12}C/^{13}C$ isotopic ratios of organic compounds found in the Murchison meteorite indicates a non-terrestrial origin for these molecules rather than terrestrial contamination. Biologically relevant molecules identified so far include uracil, an RNA nucleobase, and xanthine.[83][84] These results demonstrate that many organic compounds which are components of life on Earth were already present in the early Solar System and may have played a key role in life's origin.[85]

In August 2009, NASA scientists identified one of the fundamental chemical building-blocks of life (the amino acid glycine) in a comet for the first time.[86]

In August 2011, a report, based on NASA studies with meteorites found on Earth, was published suggesting building blocks of DNA (adenine, guanine and related organic molecules) may have been formed extraterrestrially in outer space.[87][88][89] In October 2011, scientists reported that cosmic dust contains complex organic matter ("amorphous organic solids with a mixed aromatic-aliphatic structure") that could be created naturally, and rapidly, by stars.[90][91][92] One of the scientists suggested that these complex organic compounds may have been related to the development of life on Earth and said that, "If this is the case, life on Earth may have had an easier time getting started as these organics can serve as basic ingredients for life."[90]

In August 2012, and in a world first, astronomers at Copenhagen University reported the detection of a specific sugar molecule, glycolaldehyde, in a distant star system. The molecule was found around the protostellar binary *IRAS 16293-2422*, which is located 400 light years from Earth.[93][94] Glycolaldehyde is needed to form ribonucleic acid, or RNA, which is similar in function to DNA. This finding suggests that complex organic molecules may form in stellar systems prior to the formation of planets, eventually arriving on young planets early in their formation.[95]

In September 2012, NASA scientists reported that polycyclic aromatic hydrocarbons (PAHs), subjected to interstellar medium (ISM) conditions, are transformed, through hydrogenation, oxygenation and hydroxylation, to more complex organics - "a step along the path toward amino acids and nucleotides, the raw materials of proteins and DNA, respectively".[96][97] Further, as a result of these transformations, the PAHs lose their spectroscopic signature which could be one of the reasons "for the lack of PAH detection in interstellar ice grains, particularly the outer regions of cold, dense clouds or the upper molecular layers of protoplanetary disks."[96][97]

In 2013, the Atacama Large Millimeter Array (ALMA Project) confirmed that researchers have discovered an important pair of prebiotic molecules in the icy particles in interstellar space (ISM). The chemicals, found in a giant cloud of gas about 25,000 light-years from Earth in ISM, may be a precursor to a key component of DNA and the other may have a role in the formation of an important amino acid. Researchers found a molecule called cyanomethanimine, which produces adenine, one of the four nucleobases that form the "rungs" in the ladder-like structure of DNA. The other molecule, called ethanamine, is thought to play a role in forming alanine, one of the twenty amino acids in the genetic code. Previously, scientists thought such processes took place in the very tenuous gas between the stars. The new discoveries, however, suggest that the chemical formation sequences for these molecules occurred not in gas, but on the surfaces of ice grains in interstellar space.[98] NASA ALMA scientist Anthony Remijan stated that finding these molecules in an interstellar gas cloud means that important building blocks for DNA and amino acids can 'seed' newly formed planets with the chemical precursors for life.[99]

In March 2013, a simulation experiment indicate that dipeptides (pairs of amino acids) that can be building blocks of proteins, can be created in interstellar dust.[100]

In February 2014, NASA announced a greatly upgraded database for tracking polycyclic aromatic hydrocarbons (PAHs) in the universe. According to scientists, more than 20% of the carbon in the universe may be associated with PAHs, possible starting materials for the formation of life. PAHs seem to have been formed shortly after the Big Bang, are widespread throughout the universe, and are associated with new stars and exoplanets.[101]

In March 2015, NASA scientists reported that, for the first time, complex DNA and RNA organic compounds of life, including uracil, cytosine and thymine, have been formed in the laboratory under outer space conditions, using starting chemicals, such as pyrimidine, found in meteorites. Pyrimidine, like polycyclic aromatic hydrocarbons (PAHs), the most carbon-rich chemical found in the Universe, may have been formed in red giants or in interstellar dust and gas clouds, according to the scientists.[102]

In May 2016, the Rosetta Mission team reported the presence of glycine, methylamine and ethylamine in the coma of 67P/Churyumov-Gerasimenko. This, plus the detection of phosphorus, is consistent with the hypothesis that comets played a crucial role in the emergence of life on Earth.

## 17.3 Extraterrestrial life

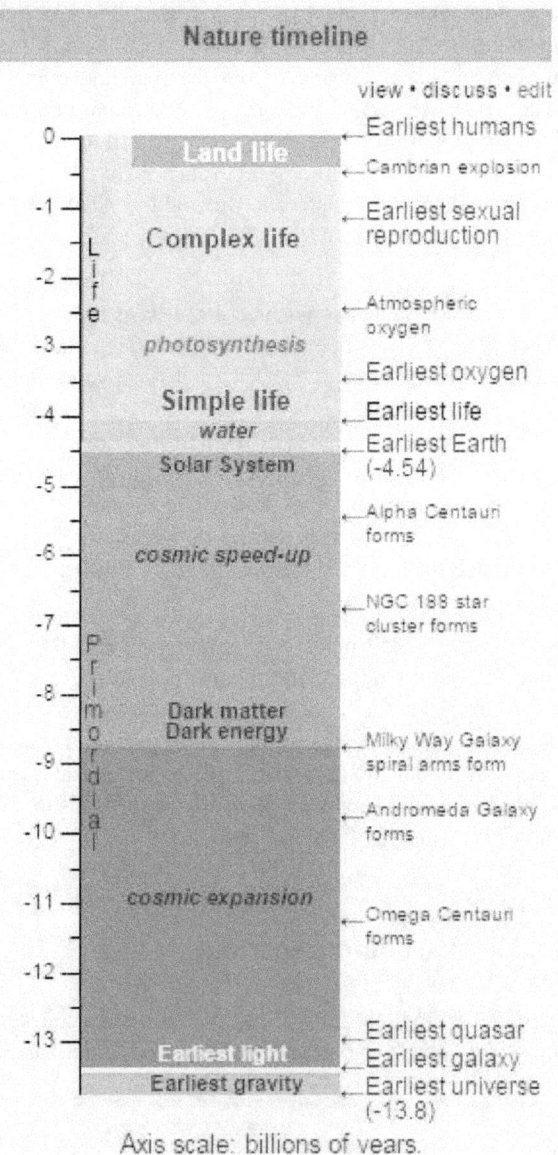

The chemistry of life may have begun shortly after the Big Bang, 13.8 billion years ago, during a habitable epoch when the Universe was only 10–17 million years old.[103][104][105] According to the panspermia hypothesis, microscopic life—distributed by meteoroids, asteroids and other small Solar System bodies—may exist throughout the universe.[106] Nonetheless, Earth is the only place in the universe known to harbor life.[107][108] The sheer number of planets in the Milky Way galaxy, however, may make it probable that life has arisen somewhere else in the galaxy and the universe. It is generally agreed that the conditions required for the evolution of intelligent life as we know it are probably exceedingly rare in the universe, while simultaneously noting that simple single-celled microorganisms may be more likely.[109]

The extrasolar planet results from the Kepler mission estimate 100–400 billion exoplanets, with over 3,500 as candidates or confirmed exoplanets.[110] On 4 November 2013, astronomers reported, based on Kepler space mission data, that there could be as many as 40 billion Earth-sized planets orbiting in the habitable zones of sun-like stars and red dwarf stars within the Milky Way Galaxy.[111][112] 11 billion of these estimated planets may be orbiting sun-like stars.[113] The nearest such planet may be 12 light-years away, according to the scientists.[111][112]

It is estimated that space travel over cosmic distances would take an incredibly long time to an outside observer, and with vast amounts of energy required. However, there are reasons to hypothesize that faster-than-light interstellar space travel might be feasible. This has been explored by NASA scientists since at least 1995.[114]

### 17.3.1 Hypotheses on extraterrestrial sources of illnesses

Hoyle and Wickramasinghe have speculated that several outbreaks of illnesses on Earth are of extraterrestrial origins, including the 1918 flu pandemic, and certain outbreaks of polio and mad cow disease. For the 1918 flu pandemic they hypothesized that cometary dust brought the virus to Earth simultaneously at multiple locations—a view almost universally dismissed by experts on this pandemic. Hoyle also speculated that HIV came from outer space.[115] After Hoyle's death, *The Lancet* published a letter to the editor from Wickramasinghe and two of his colleagues,[116] in which they hypothesized that the virus that causes severe acute respiratory syndrome (SARS) could be extraterrestrial in origin and not originated from chickens. *The Lancet* subsequently published three responses to this letter, showing that the hypothesis was not evidence-based, and casting doubts on the quality of the experiments referenced by

Wickramasinghe in his letter.[117][118][119] A 2008 encyclopedia notes that "Like other claims linking terrestrial disease to extraterrestrial pathogens, this proposal was rejected by the greater research community."[115]

In April 2016, Jiangwen Qu of the Department of Infectious Disease Control in China presented a statistical study suggesting that "extremes of sunspot activity to within plus or minus 1 year may precipitate influenza pandemics." He discussed possible mechanisms of epidemic initiation and early spread, including speculation on primary causation by externally derived viral variants from space via cometary dust.[120]

### 17.3.2 Case studies

- A meteorite originating from Mars known as ALH84001 was shown in 1996 to contain microscopic structures resembling small terrestrial nanobacteria. When the discovery was announced, many immediately conjectured that these were fossils and were the first evidence of extraterrestrial life — making headlines around the world. Public interest soon started to dwindle as most experts started to agree that these structures were not indicative of life, but could instead be formed abiotically from organic molecules. However, in November 2009, a team of scientists at Johnson Space Center, including David McKay, reasserted that there was "strong evidence that life may have existed on ancient Mars", after having reexamined the meteorite and finding magnetite crystals.[121][122]

- On May 11, 2001, two researchers from the University of Naples claimed to have found viable extraterrestrial bacteria inside a meteorite. Geologist Bruno D'Argenio and molecular biologist Giuseppe Geraci claim the bacteria were wedged inside the crystal structure of minerals, but were resurrected when a sample of the rock was placed in a culture medium.[123][124][125]

- An Indian and British team of researchers led by Chandra Wickramasinghe reported on 2001 that air samples over Hyderabad, India, gathered from the stratosphere by the Indian Space Research Organisation, contained clumps of living cells. Wickramasinghe calls this "unambiguous evidence for the presence of clumps of living cells in air samples from as high as 41 km, above which no air from lower down would normally be transported".[126][127] Two bacterial and one fungal species were later independently isolated from these filters which were identified as *Bacillus simplex*, *Staphylococcus pasteuri* and *Engyodontium album* respectively.[128][129] Similar isolation experiments at separate laboratories were unsuccessful.

Pushkar Ganesh Vaidya from the Indian Astrobiology Research Centre reported in 2009 that "the three microorganisms captured during the balloon experiment do not exhibit any distinct adaptations expected to be seen in microorganisms occupying a cometary niche".[130][131]

- In 2005 an improved experiment was conducted by ISRO. On April 10, 2005, air samples were collected from the upper atmosphere at altitudes ranging from 20 km to more than 40 km. The samples were tested at two labs in India. The labs found 12 bacterial and 6 different fungal species in these samples. The fungi were *Penicillium decumbens*, *Cladosporium cladosporioides*, *Alternaria sp.* and *Tilletiopsis albescens*. Out of the 12 bacterial samples, three were identified as new species and named *Janibacter hoyeli.sp.nov* (after Fred Hoyle), *Bacillus isronensis.sp.nov* (named after ISRO) and *Bacillus aryabhati* (named after the ancient Indian mathematician, Aryabhata). These three new species showed that they were more resistant to UV radiation than similar bacteria.[132][133]

Atmospheric sampling by NASA in 2010 before and after hurricanes, collected 314 different types of bacteria; the study suggests that large-scale convection during tropical storms and hurricanes can then carry this material from the surface higher up into the atmosphere.[134][135]

- On January 10, 2013, Chandra Wickramasinghe found fossil diatom frustules in what he thinks is a new kind of carbonaceous meteorite called Polonnaruwa that landed in the North Central Province of Sri Lanka on 29 December 2012.[136] Early on, there was criticism that that Wickramasinghe's report was not an examination of an actual meteorite but of some terrestrial rock passed off as a meteorite.[137]

Wickramasinghe's team remark that they are aware that a large number of unrelated stones have been submitted for analysis, and have no knowledge regarding the nature, source or origin of the stones their critics have examined, so Wickramasinghe clarifies that he is using the stones submitted by the Medical Research Institute in Sri Lanka.[138] In response to the criticism from other scientists, Wickramasinghe performed X-ray diffraction[139] and isotope[138] analyses to verify its meteoritic origin. His analysis revealed a 95% silica and

3% quartz content,[139] and interpreted this result as a "carbonaceous meteorite of unknown type".[139] In addition, Wickramasinghe's team remarked that the temperature at which sand must be heated by lightning to melt and form a fulgurite (1770 °C) would have vaporized and burned all carbon-rich organisms and melted and thus destroyed the delicately marked silica frustules of the diatoms,[138] and that the oxygen isotope data confirms its meteoric origin.[138] Wickramasinghe's team also argues that since living diatoms require nitrogen fixation to synthetize amino acids, proteins, DNA, RNA and other life-critical biomolecules, a population of extraterrestrial cyanobacteria must have been a required component of the comet (Polonnaruwa meteorite) "ecosystem".[138]

- In 2013, Dale Warren Griffin, a microbiologist working at the United States Geological Survey noted that viruses are the most numerous entities on Earth. Griffin speculates that viruses evolved in comets and on other planets and moons may be pathogenic to humans, so he proposed to also look for viruses on moons and planets of the Solar System.[140]

### 17.3.3 Hoaxes

A separate fragment of the Orgueil meteorite (kept in a sealed glass jar since its discovery) was found in 1965 to have a seed capsule embedded in it, whilst the original glassy layer on the outside remained undisturbed. Despite great initial excitement, the seed was found to be that of a European Juncaceae or Rush plant that had been glued into the fragment and camouflaged using coal dust. The outer "fusion layer" was in fact glue. Whilst the perpetrator of this hoax is unknown, it is thought that they sought to influence the 19th century debate on spontaneous generation — rather than panspermia — by demonstrating the transformation of inorganic to biological matter.[141]

## 17.4 Extremophiles

See also: Extremophile
Until the 1970s, life was believed to depend on its access to sunlight. Even life in the ocean depths, where sunlight cannot reach, was believed to obtain its nourishment either from consuming organic detritus rained down from the surface waters or from eating animals that did.[142] However, in 1977, during an exploratory dive to the Galapagos Rift in the deep-sea exploration submersible *Alvin*, scientists discovered colonies of assorted creatures clustered around undersea volcanic features known as black smokers.[142] It was soon determined that the basis for this food chain is a form of bacterium that derives its energy from oxidation of reactive chemicals, such as hydrogen or hydrogen sulfide, that bubble up from the Earth's interior. This chemosynthesis revolutionized the study of biology by revealing that terrestrial life need not be Sun-dependent; it only requires water and an energy gradient in order to exist.

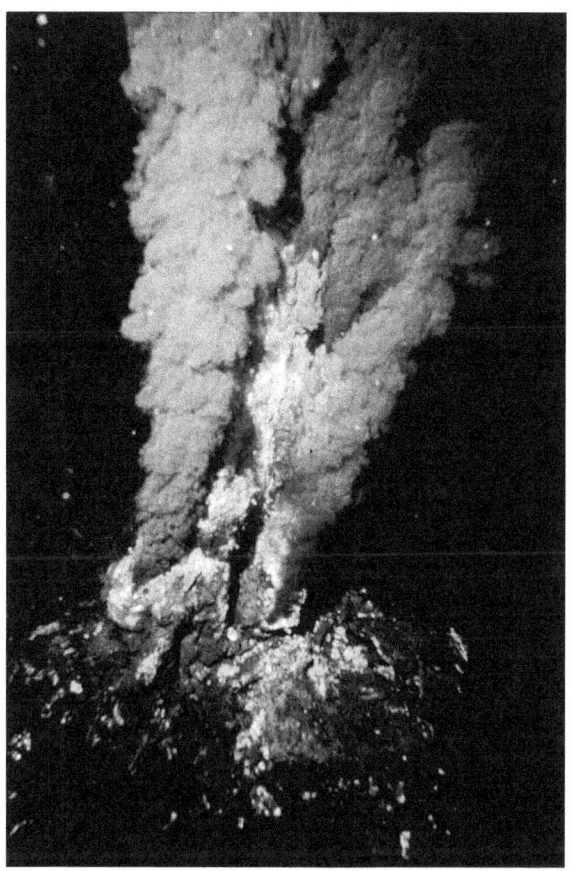

*Hydrothermal vents are able to support extremophile bacteria on Earth and may also support life in other parts of the cosmos.*

It is now known that extremophiles, microorganisms with extraordinary capability to thrive in the harshest environments on Earth, can specialize to thrive in the deep-sea,[143][144][145] ice, boiling water, acid, the water core of nuclear reactors, salt crystals, toxic waste and in a range of other extreme habitats that were previously thought to be inhospitable for life.[146][147][148][149] Living bacteria found in ice core samples retrieved from 3,700 metres (12,100 ft) deep at Lake Vostok in Antarctica, have provided data for extrapolations to the likelihood of microorganisms surviving frozen in extraterrestrial habitats or during interplanetary transport.[150] Also, bacteria have been discovered living within warm rock deep in the Earth's crust.[151]

In order to test some these organisms' potential re-

## 17.4. EXTREMOPHILES

silence in outer space, plant seeds and spores of bacteria, fungi and ferns have been exposed to the harsh space environment.[148][149][152] Spores are produced as part of the normal life cycle of many plants, algae, fungi and some protozoans, and some bacteria produce endospores or cysts during times of stress. These structures may be highly resilient to ultraviolet and gamma radiation, desiccation, lysozyme, temperature, starvation and chemical disinfectants, while metabolically inactive. Spores germinate when favourable conditions are restored after exposure to conditions fatal to the parent organism.

Although computer models suggest that a captured meteoroid would typically take some tens of millions of years before collision with a planet,[40] there are documented viable Earthly bacterial spores that are 40 million years old that are very resistant to radiation,[40][46] and others able to resume life after being dormant for 25 million years,[153] suggesting that lithopanspermia life-transfers are possible via meteorites exceeding 1 m in size.[40]

The discovery of deep-sea ecosystems, along with advancements in the fields of astrobiology, observational astronomy and discovery of large varieties of extremophiles, opened up a new avenue in astrobiology by massively expanding the number of possible extraterrestrial habitats and possible transport of hardy microbial life through vast distances.[60]

### 17.4.1 Research in outer space

See also: List of microorganisms tested in outer space

The question of whether certain microorganisms can survive in the harsh environment of outer space has intrigued biologists since the beginning of spaceflight, and opportunities were provided to expose samples to space. The first American tests were made in 1966, during the Gemini IX and XII missions, when samples of bacteriophage T1 and spores of *Penicillium roqueforti* were exposed to outer space for 16.8 h and 6.5 h, respectively.[51][60] Other basic life sciences research in low Earth orbit started in 1966 with the Soviet biosatellite program Bion and the U.S. Biosatellite program. Thus, the plausibility of panspermia can be evaluated by examining life forms on Earth for their capacity to survive in space.[154] The following experiments carried on low Earth orbit specifically tested some aspects of panspermia or lithopanspermia:

**ERA**

The Exobiology Radiation Assembly (ERA) was a 1992 experiment on board the European Retrievable Carrier (EURECA) on the biological effects of space radiation. EU-

*EURECA facility deployment in 1992*

RECA was an unmanned 4.5 tonne satellite with a payload of 15 experiments.[155] It was an astrobiology mission developed by the European Space Agency (ESA). Spores of different strains of *Bacillus subtilis* and the *Escherichia coli* plasmid pUC19 were exposed to selected conditions of space (space vacuum and/or defined wavebands and intensities of solar ultraviolet radiation). After the approximately 11-month mission, their responses were studied in terms of survival, mutagenesis in the *his* (*B. subtilis*) or *lac* locus (pUC19), induction of DNA strand breaks, efficiency of DNA repair systems, and the role of external protective agents. The data were compared with those of a simultaneously running ground control experiment:[156][157]

- The survival of spores treated with the vacuum of space, however shielded against solar radiation, is substantially increased, if they are exposed in multilayers and/or in the presence of glucose as protective.

- All spores in "artificial meteorites", i.e. embedded in clays or simulated Martian soil, are killed.

- Vacuum treatment leads to an increase of mutation frequency in spores, but not in plasmid DNA.

- Extraterrestrial solar ultraviolet radiation is mutagenic, induces strand breaks in the DNA and reduces survival substantially.

- Action spectroscopy confirms results of previous space experiments of a synergistic action of space vacuum and solar UV radiation with DNA being the critical target.

- The decrease in viability of the microorganisms could be correlated with the increase in DNA damage.

- The purple membranes, amino acids and urea were not measurably affected by the dehydrating condition of open space, if sheltered from solar radiation. Plasmid

DNA, however, suffered a significant amount of strand breaks under these conditions.[156]

## BIOPAN

BIOPAN is a multi-user experimental facility installed on the external surface of the Russian Foton descent capsule. Experiments developed for BIOPAN are designed to investigate the effect of the space environment on biological material after exposure between 13 and 17 days.[158] The experiments in BIOPAN are exposed to solar and cosmic radiation, the space vacuum and weightlessness, or a selection thereof. Of the 6 missions flown so far on BIOPAN between 1992 and 2007, dozens of experiments were conducted, and some analyzed the likelihood of panspermia. Some bacteria, lichens (*Xanthoria elegans*, *Rhizocarpon geographicum* and their mycobiont cultures, the black Antarctic microfungi *Cryomyces minteri* and *Cryomyces antarcticus*), spores, and even one animal (tardigrades) were found to have survived the harsh outer space environment and cosmic radiation.[159][160][161][162]

## EXOSTACK

*EXOSTACK on the Long Duration Exposure Facility satellite.*

The German EXOSTACK experiment was deployed on 7 April 1984 on board the Long Duration Exposure Facility statelite. 30% of *Bacillus subtilis* spores survived the nearly 6 years exposure when embedded in salt crystals, whereas 80% survived in the presence of glucose, which stabilize the structure of the cellular macromolecules, especially during vacuum-induced dehydration.[51][163]

If shielded against solar UV, spores of *B. subtilis* were capable of surviving in space for up to 6 years, especially if embedded in clay or meteorite powder (artificial meteorites). The data support the likelihood of interplanetary transfer of microorganisms within meteorites, the so-called lithopanspermia hypothesis.[51]

## EXPOSE

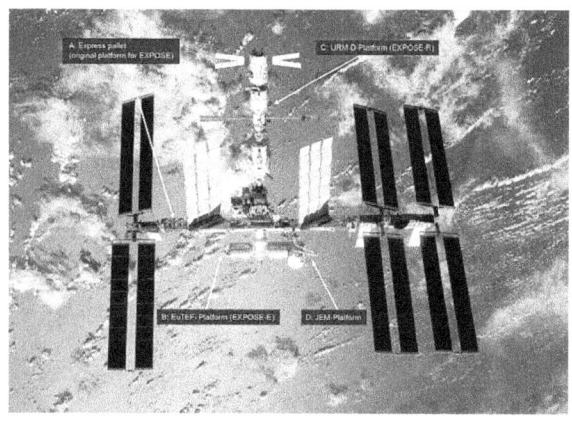

*Location of the astrobiology EXPOSE-E and EXPOSE-R facilities on the International Space Station*

EXPOSE is a multi-user facility mounted outside the International Space Station dedicated to astrobiology experiments.[152] There have been three EXPOSE experiments flown between 2008 and 2015: EXPOSE-E, EXPOSE-R and EXPOSE-R2.

Results from the orbital missions, especially the experiments *SEEDS*[164] and *LiFE*,[165] concluded that after an 18-month exposure, some seeds and lichens (*Stichococcus sp.* and *Acarospora sp.*, a lichenized fungal genus) may be capable to survive interplanetary travel if sheltered inside comets or rocks from cosmic radiation and UV radiation.[152][166] The *LIFE*, *SPORES*, and *SEEDS* parts of the experiments provided information about the likelihood of lithopanspermia.[167][168][169]

### Rosetta

In 2014, the *Rosetta* spacecraft arrived at comet 67P/Churyumov–Gerasimenko. A few months after arriving at the comet, *Rosetta* released a small lander, named *Philae*, onto its surface. The plan was to investigate Churyumov-Gerasimenko up close for two years. *Philae's* battery has since died. Scientists hoped that as the comet was travelling toward the sun greater solar energy would recharge *Philae* (via its solar panels) and *Philae* would resume operation, but eventually they released information that the chances of contact are close to zero.[170] Rosetta's

Project Scientist, Gerhard Schwehm, stated that sterilization is generally not crucial since comets are usually regarded as objects where prebiotic molecules can be found, but not living microorganisms.[171] Notwithstanding, other scientists think it would be an opportunity to gather evidence for one of panspermia's hypotheses: the possibility of both active and dormant microbes inside comets.[6][7]

In July 2015, scientists reported that upon the first touchdown of the *Philae* lander on comet 67/P's surface, measurements by the COSAC and Ptolemy instruments revealed sixteen organic compounds, four of which were seen for the first time on a comet, including acetamide, acetone, methyl isocyanate and propionaldehyde.[172][173][174]

In 2016 a paper[175] was published reporting that ROSINA mass spectrometer - one of the scientific instruments on-board Rosetta - discovered volatile glycine accompanied by methylamine and ethylamine in the coma of 67P/Churyumov-Gerasimenko.

**Tanpopo**

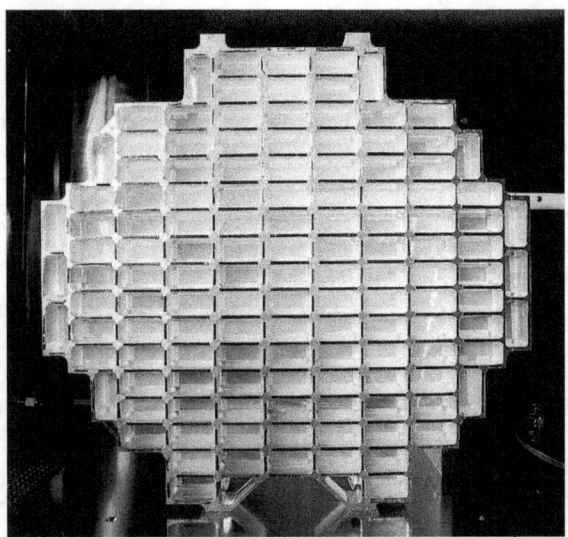

*Dust collector with aerogel blocks*

The Tanpopo mission is an orbital astrobiology experiment by Japan that is currently investigating the possible interplanetary transfer of life, organic compounds, and possible terrestrial particles in low Earth orbit. The Tanpopo experiment is taking place at the Exposed Facility located on the exterior of Kibo module of the International Space Station. The mission will collect cosmic dusts and other particles for three years by using an ultra-low density silica gel called aerogel. The purpose is to assess the panspermia hypothesis and the possibility of natural interplanetary transport of life and its precursors.[176][177] Some of these aerogels will be replaced every one or two years through 2018.[178] Sample collection began in May 2015, and the first samples will be returned to Earth in mid-2016.[179]

## 17.5 Criticism

Panspermia is often criticized because it does not answer the question of the origin of life but merely places it on another celestial body. It was also criticized because it was thought it could not be tested experimentally.[60]

Wallis and Wickramasinghe argued in 2004 that the transport of individual bacteria or clumps of bacteria, is overwhelmingly more important than lithopanspermia in terms of numbers of microbes transferred, even accounting for the death rate of unprotected bacteria in transit.[180] Then it was found that isolated spores of *B. subtilis* were killed by several orders of magnitude if exposed to the full space environment for a mere few seconds. These results clearly negate the original panspermia hypothesis, which requires single spores as space travelers accelerated by the radiation pressure of the Sun, requiring many years to travel between the planets. However, if shielded against solar UV, spores of *Bacillus subtilis* were capable of surviving in space for up to 6 years, especially if embedded in clay or meteorite powder (artificial meteorites). The data support the likelihood of interplanetary transfer of microorganisms within meteorites, the so-called **lithopanspermia** hypothesis.[51]

## 17.6 Science fiction

- Fred Hoyle's *The Black Cloud* is a science fiction novel published in 1957. The story details the arrival of an enormous cloud of gas that enters the Solar System and appears about to destroy most of the life on Earth by blocking the sunlight. The cloud is later revealed as a sentient alien gaseous entity.

- Jack Finney's novel *The Body Snatchers* (1955) and the subsequent film adaptations describe spores drifting through space to arrive on the surface of Earth, though the premise is most fully discussed in the second version *Invasion of the Body Snatchers (1978 film)*.

- Michael Crichton's 1969 novel, *The Andromeda Strain*, is based on the panspermiatic premise of a meteor bringing an crystalline alien bacterium to Earth. The phrase "Andromeda Strain" has become a shorthand for mysterious infectious diseases.

- Stephen King's short story "Weeds" (1976), later adapted into the Creepshow vignette "The Lonesome Death of Jordy Verrill" (1982; starring King,) involves a meteor crashing to Earth which carries with it a virulent plant/fungus which spreads rapidly.

- The opening sequence of Ridley Scott's 2012 *Alien* prequel, *Prometheus* depicts a humanoid species, referred to as 'the Engineers', seeding what is presumably the early Earth by disintegrating the body of one of their members and spilling his DNA into the water of the planet.

## 17.7 See also

- Abiogenesis
- Anthropic principle
- Astrobiology
- Cryptobiosis
- Drake equation
- Fermi paradox
- Fine-tuned Universe
- Interplanetary contamination
- Last universal ancestor
- List of microorganisms tested in outer space
- Planetary protection
- Rare Earth hypothesis
- Red rain in Kerala

## 17.8 References

[1] Wickramasinghe, Chandra (2011). "Bacterial morphologies supporting cometary panspermia: a reappraisal". *International Journal of Astrobiology* **10** (1): 25–30. Bibcode:2011IJAsB..10...25W. doi:10.1017/S1473550410000157.

[2] Rampelotto, P. H. (2010). Panspermia: A promising field of research. In: Astrobiology Science Conference. Abs 5224.

[3] Forward planetary contamination like *Tersicoccus phoenicis*, that has shown resistance to methods usually used in spacecraft assembly clean rooms: Madhusoodanan, Jyoti (May 19, 2014). "Microbial stowaways to Mars identified". *Nature*. doi:10.1038/nature.2014.15249. Retrieved May 23, 2014.

[4] Webster, Guy (November 6, 2013). "Rare New Microbe Found in Two Distant Clean Rooms". *NASA.gov*. Retrieved November 6, 2013.

[5] A variation of the panspermia hypothesis is **necropanspermia** which is described by astronomer Paul Wesson as follows: "The vast majority of organisms reach a new home in the Milky Way in a technically dead state ... Resurrection may, however, be possible." Grossman, Lisa (2010-11-10). "All Life on Earth Could Have Come From Alien Zombies". *Wired*. Retrieved 10 November 2010.

[6] Hoyle, F. and Wickramasinghe, N.C., 1981. Evolution from Space (Simon & Schuster Inc., NY, 1981 and J.M. Dent and Son, Lond, 1981), ch3 pp. 35-49.

[7] Wickramasinghe, J., Wickramasinghe, C. and Napier, W., 2010. Comets and the Origin of Life (World Scientific, Singapore. 1981), ch6 pp. 137-154.

[8] Klyce, Brig (2001). "Panspermia Asks New Questions". Retrieved 25 July 2013.

[9] Klyce, Brig (2001). Kingsley, Stuart A; Bhathal, Ragbir, eds. "The Search for Extraterrestrial Intelligence (SETI) in the Optical Spectrum III". *Proc. SPIE Vol. 4273*. The Search for Extraterrestrial Intelligence (SETI) in the Optical Spectrum III **4273**: 11. Bibcode:2001SPIE.4273...11K. doi:10.1117/12.435366. |chapter= ignored (help)

[10] Dalgarno, A. (2006). "The galactic cosmic ray ionization rate". *Proceedings of the National Academy of Sciences* **103** (33): 12269–73. Bibcode:2006PNAS..10312269D. doi:10.1073/pnas.0602117103. PMC 1567869. PMID 16894166.

[11] Brown, Laurie M.; Pais, Abraham; Pippard, A. B. (1995). "The physics of the interstellar medium". *Twentieth Century Physics* (2nd ed.). CRC Press. p. 1765. ISBN 0-7503-0310-7.

[12] Margaret O'Leary (2008) Anaxagoras and the Origin of Panspermia Theory, iUniverse publishing Group, ISBN 978-0-595-49596-2

[13] Berzelius (1799–1848), J. J. "Analysis of the Alais meteorite and implications about life in other worlds".

[14] Lynn J. Rothschild; Adrian M. Lister (June 2003). *Evolution on Planet Earth – The Impact of the Physical Environment*. Academic Press. pp. 109–127. ISBN 978-0-12-598655-7.

[15] Thomson (Lord Kelvin), W. (1871). "Inaugural Address to the British Association Edinburgh. "We must regard it as probably to the highest degree that there are countless seed-bearing meteoritic stones moving through space."". *Nature* **4** (92): 261–278 [262]. Bibcode:1871Natur...4..261.. doi:10.1038/004261a0.

[16] "The word: Panspermia". *New Scientist* (2541). 7 March 2006. Retrieved 25 July 2013.

[17] "History of Panspermia". Retrieved 25 July 2013.

[18] Arrhenius, S., *Worlds in the Making: The Evolution of the Universe*. New York, Harper & Row, 1908.

[19] Napier, W.M. (2007). "Pollination of exoplanets by nebulae". *Int.J.Astrobiol* **6** (3): 223–228. Bibcode:2007IJAsB...6..223N. doi:10.1017/S1473550407003710.

[20] Line, M.A. (2007). "Panspermia in the context of the timing of the origin of life and microbial phylogeny". *Int. J. Astrobiol.* 3 **6** (3): 249–254. Bibcode:2007IJAsB...6..249L. doi:10.1017/S1473550407003813.

[21] Wickramasinghe, D. T.; Allen, D. A. (1980). "The 3.4-μm interstellar absorption feature". *Nature* **287** (5782): 518–519. Bibcode:1980Natur.287..518W. doi:10.1038/287518a0.

[22] Allen, D. A.; Wickramasinghe, D. T. (1981). "Diffuse interstellar absorption bands between 2.9 and 4.0 μm". *Nature* **294** (5838): 239–240. Bibcode:1981Natur.294..239A. doi:10.1038/294239a0.

[23] Wickramasinghe, D. T.; Allen, D. A. (1983). "Three components of 3?4 ?m absorption bands". *Astrophysics and Space Science* **97** (2): 369–378. Bibcode:1983Ap&SS..97..369W. doi:10.1007/BF00653492.

[24] Fred Hoyle; Chandra Wickramasinghe & John Watson (1986). *Viruses from Space and Related Matters*. University College Cardiff Press.

[25] Weaver, Rheyanne (April 7, 2009). "Ruminations on other worlds". *statepress.com*. Retrieved 25 July 2013.

[26] Cockell, Charles S. (19 May 2011). "Exposure of phototrophs to 548 days in low Earth orbit: microbial selection pressures in outer space and on early earth". *The ISME Journal* **5** (10): 1671–1682. doi:10.1038/ismej.2011.46. PMC 3176519. PMID 21593797. Retrieved 11 February 2016.

[27] Amos, Jonathan (23 Aug 2010). "Beer microbes live 553 days outside ISS". BBC News. BBC. Retrieved 11 February 2016.

[28] Panitz, Corinna; Horneck, Gerda; Rabbow, Elke; Petra Rettberg, Petra; Moeller, Ralf (January 2015). "The SPORES experiment of the EXPOSE-R mission: *Bacillus subtilis* spores in artificial meteorites" (PDF). *International Journal of Astrobiology* **14** (Special Issue 1): 105–114. Bibcode:2015IJAsB..14..105P. doi:10.1017/S1473550414000251. Retrieved 2015-05-08.

[29] Borenstein, Seth (19 October 2015). "Hints of life on what was thought to be desolate early Earth". *Excite* (Yonkers, NY: Mindspark Interactive Network). Associated Press. Retrieved 2015-10-20.

[30] Bell, Elizabeth A.; Boehnike, Patrick; Harrison, T. Mark; et al. (19 October 2015). "Potentially biogenic carbon preserved in a 4.1 billion-year-old zircon" (PDF). *Proc. Natl. Acad. Sci. U.S.A* (Washington, D.C.: National Academy of Sciences) **112** (47): 214518–14521. Bibcode:2015PNAS..11214518B. doi:10.1073/pnas.1517557112. ISSN 1091-6490. PMC 4664351. PMID 26483481. Retrieved 2015-10-20. Early edition, published online before print.

[31] Khan, Amina (7 March 2014). "Did two planets around nearby star collide? Toxic gas holds hints". *LA Times*. Retrieved 9 March 2014.

[32] Dent, W. R. F.; Wyatt, M. C.; Roberge, A.; Augereau, J.-C.; et al. (6 March 2014). "Molecular Gas Clumps from the Destruction of Icy Bodies in the β Pictoris Debris Disk". *Science* **343** (6178): 1490–1492. Bibcode:2014Sci...343.1490D. doi:10.1126/science.1248726. Retrieved 9 March 2014.

[33] Wickramasinghe, Chandra; Wickramasinghe, Chandra; Napier, William (2009). *Comets and the Origin of Life*. World Scientific Press. doi:10.1142/6008. ISBN 978-981-256-635-5.

[34] Wall, Mike. "Comet Impacts May Have Jump-Started Life on Earth". space.com. Retrieved 1 August 2013.

[35] Weber, P; Greenberg, J. M. (1985). "Can spores survive in interstellar space?". *Nature* **316** (6027): 403–407. Bibcode:1985Natur.316..403W. doi:10.1038/316403a0.

[36] Melosh, H. J. (1988). "The rocky road to panspermia". *Nature* **332** (6166): 687–688. Bibcode:1988Natur.332..687M. doi:10.1038/332687a0. PMID 11536601.

[37] C. Mileikowsky; F. A. Cucinotta; J. W. Wilson; B. Gladman; et al. (2000). "Risks threatening viable transfer of microbes between bodies in our solar system". *Planetary and Space Science* **48** (11): 1107–1115. Bibcode:2000P&SS...48.1107M. doi:10.1016/S0032-0633(00)00085-4.

[38] Studies Focus On Spacecraft Sterilization

[39] European Space Agency: Dry heat sterilisation process to high temperatures

[40] Edward Belbruno; Amaya Moro-Martı́n; Malhotra, Renu & Savransky, Dmitry (2012). "Chaotic Exchange of Solid Material between Planetary". *Astrobiology* **12** (8): 754–74. arXiv:1205.1059. Bibcode:2012AsBio..12..754B. doi:10.1089/ast.2012.0825. PMC 3440031. PMID 22897115.

[41] Slow-moving rocks better odds that life crashed to Earth from space News at Princeton, September 24, 2012.

[42] Crick, F. H.; Orgel, L. E. (1973). "Directed Panspermia". *Icarus* **19** (3): 341–348. Bibcode:1979JBIS...32..419M. doi:10.1016/0019-1035(73)90110-3.

[43] Mautner, Michael N. (2000). *Seeding the Universe with Life: Securing Our Cosmological Future* (PDF). Washington D. C.: Legacy Books (www.amazon.com). ISBN 0-476-00330-X.

[44] Mautner, M; Matloff, G. (1979). "Directed panspermia: A technical evaluation of seeding nearby planetary systems" (PDF). *J. British Interplanetary Soc.* **32**: 419.

[45] Mautner, M. N. (1997). "Directed panspermia. 3. Strategies and motivation for seeding star-forming clouds" (PDF). *J. British Interplanetary Soc.* **50**: 93–102. Bibcode:1997JBIS...50...93M.

[46] BBC Staff (23 August 2011). "Impacts 'more likely' to have spread life from Earth". BBC. Retrieved 24 August 2011.

[47] "Electromagnetic space travel for bugs? - space – 21 July 2006 – New Scientist Space". Space.newscientist.com. Archived from the original on January 11, 2009. Retrieved December 8, 2014.

[48] Dehel, T. (2006-07-23). "Uplift and Outflow of Bacterial Spores via Electric Field". *36th COSPAR Scientific Assembly. Held 16–23 July 2006* (Adsabs.harvard.edu) **36**: 1. arXiv:hep-ph/0612311. Bibcode:2006cosp..36....1D.

[49] "Die Verbreitung des Lebens im Weltenraum" (the "Distribution of Life in Space"). Published in Die Umschau. 1903.

[50] Nicholson, Wayne L. (2009). "Ancient micronauts: interplanetary transport of microbes by cosmic impacts". *Trends in Microbiology* **17** (6): 243–250. doi:10.1016/j.tim.2009.03.004. PMID 19464895.

[51] Horneck, G.; Klaus, D. M.; Mancinelli, R. L. (2010). "Space Microbiology". *Microbiology and Molecular Biology Reviews* **74** (1): 121–56. doi:10.1128/MMBR.00016-09. PMC 2832349. PMID 20197502.

[52] I. S. Shklovskii; Carl Sagan (1966). *Intelligent Life in the Universe*. Emerson-Adams Press, Incorporated. ISBN 978-1-892803-02-3.

[53] Wickramasinghe, M.K.; Wickramasinghe, C. (2004). "Interstellar transfer of planetary microbiota". *Mon. Not.R. Astr. Soc.* **348**: 52–57. Bibcode:2004MNRAS.348...52W. doi:10.1111/j.1365-2966.2004.07355.x.

[54] *Protection of Bacterial Spores in Space, a Contribution to the Discussion on Panspermia.* Gerda Horneck, Petra Rettberg, Günther Reitz, Jörg Wehner, Ute Eschweiler, Karsten Strauch, Corinna Panitz, Verena Starke, Christa Baumstark-Khan. Origins of life and evolution of the biosphere. December 2001, Volume 31, Issue 6, pp. 527-547.

[55] Rahn, R.O.; Hosszu, J.L. (1969). "Influence of relative humidity on the photochemistry of DNA films". *Biochim Biophys Acta* **190** (1): 126–131. doi:10.1016/0005-2787(69)90161-0. PMID 4898489.

[56] Patrick, M.H.; Gray, D.M. (1976). "Independence of photproduct formation on DNA conformation". *Photochem. Photobiol* **24** (6): 507–513. doi:10.1111/j.1751-1097.1976.tb06867.x. PMID 1019243.

[57] Wayne L. Nicholson; Andrew C. Schuerger; Peter Setlow (21 January 2005). "The solar UV environment and bacterial spore UV resistance: considerations for Earth-to-Mars transport by natural processes and human spaceflight" (PDF). *Mutation Research* **571** (1–2): 249–264. doi:10.1016/j.mrfmmm.2004.10.012. PMID 15748651. Retrieved 2 August 2013.

[58] Clark BC., Planetary interchange of bioactive material: probability factors and implications Origins Life Evol Biosphere 2001; 31: 185-97

[59] Mileikowsky C. et al. Natural Transfer of Microbes in space, part I: from Mars to Earth and Earth to Mars Icarus 2000; 145; 391-427

[60] Olsson-Francis, Karen; Cockell, Charles S. (2010). "Experimental methods for studying microbial survival in extraterrestrial environments". *Journal of Microbiological Methods* **80** (1): 1–13. doi:10.1016/j.mimet.2009.10.004. PMID 19854226.

[61] Cockell, Charles S. (2007). "The Interplanetary Exchange of Photosynthesis". *Origins of Life and Evolution of Biospheres* **38**: 87–104. Bibcode:2008OLEB...38...87C. doi:10.1007/s11084-007-9112-3.

[62] Horneck, Gerda; Stöffler, Dieter; Ott, Sieglinde; Hornemann, Ulrich; et al. (2008). "Microbial Rock Inhabitants Survive Hypervelocity Impacts on Mars-Like Host Planets: First Phase of Lithopanspermia Experimentally Tested". *Astrobiology* **8** (1): 17–44. Bibcode:2008AsBio...8...17H. doi:10.1089/ast.2007.0134. PMID 18237257.

[63] Fajardo-Cavazos, Patricia; Link, Lindsey; Melosh, H. Jay; Nicholson, Wayne L. (2005). "Bacillus subtilisSpores on Artificial Meteorites Survive Hypervelocity Atmospheric Entry: Implications for Lithopanspermia". *Astrobiology* **5** (6): 726–36. Bibcode:2005AsBio...5..726F. doi:10.1089/ast.2005.5.726. PMID 16379527.

[64] Brack, A.; Baglioni, P.; Borruat, G.; Brandstätter, F.; et al. (2002). "Do meteoroids of sedimentary origin survive terrestrial atmospheric entry? The ESA artificial meteorite experiment STONE". *Planetary and Space Science* **50** (7–8): 763–772. Bibcode:2002P&SS...50..763B. doi:10.1016/S0032-0633(02)00018-1.

[65] Cockell, Charles S.; Brack, André; Wynn-Williams, David D.; Baglioni, Pietro; et al. (2007). "Interplanetary Transfer of Photosynthesis: An Experimental Demonstration of a Selective Dispersal Filter in Planetary Island Biogeography". *Astrobiology* **7** (1): 1–9. Bibcode:2007AsBio...7....1C. doi:10.1089/ast.2006.0038. PMID 17407400.

[66] "Could Life Have Survived a Fall to Earth?". *EPSC*. 12 September 2013. Retrieved 2015-04-21.

[67] Gold, T. "Cosmic Garbage", Air Force and Space Digest, 65 (May 1960).

[68] "Anticipating an RNA world. Some past speculations on the origin of life: where are they today?" by L. E. Orgel and F. H. C. Crick in *FASEB J.* (1993) Volume 7 pages 238-239.

[69] Clark, Benton C. Clark (February 2001). "Planetary Interchange of Bioactive Material: Probability Factors and Implications". *Origins of life and evolution of the biosphere* **31** (1–2): 185–197. Bibcode:2001OLEB...31..185C. doi:10.1023/A:1006757011007. PMID 11296521.

[70] Mautner, Michael N. (2009). "Life-centered ethics, and the human future in space" (PDF). *Bioethics* **23** (8): 433–440. doi:10.1111/j.1467-8519.2008.00688.x. PMID 19077128.

[71] Mautner, Michael Noah Ph.D. (2000). *Seeding the Universe with Life: Securing our Cosmological Future* (PDF). Legacy Books (www.amazon.com). ISBN 0-476-00330-X.

[72] Mautner, Michael N. (2002). "Planetary bioresources and astroecology. 1. Planetary microcosm bioessays of Martian and meteorite materials: soluble electrolytes, nutrients, and algal and plant responses" (PDF). *Icarus* **158**: 72–86. Bibcode:2002Icar..158...72M. doi:10.1006/icar.2002.6841.

[73] Mautner, Michael N. (2005). "Life in the cosmological future: Resources, biomass and populations" (PDF). *Journal of the British Interplanetary Society* **58**: 167–180. Bibcode:2005JBIS...58..167M.

[74] G. Marx (1979). "Message through time". *Acta Astronautica* **6** (1–2): 221–225. Bibcode:1979AcAau...6..221M. doi:10.1016/0094-5765(79)90158-9.

[75] H. Yokoo; T. Oshima (1979). "Is bacteriophage φX174 DNA a message from an extraterrestrial intelligence?". *Icarus* **38** (1): 148–153. Bibcode:1979Icar...38..148Y. doi:10.1016/0019-1035(79)90094-0.

[76] Overbye, Dennis (26 June 2007). "Human DNA, the Ultimate Spot for Secret Messages (Are Some There Now?)". Retrieved 2014-10-09.

[77] Davies, Paul C.W. (2010). *The Eerie Silence: Renewing Our Search for Alien Intelligence*. Boston, Massachusetts: Houghton Mifflin Harcourt. ISBN 978-0-547-13324-9.

[78] V. I. shCherbak; M. A. Makukov (2013). "The "Wow! signal" of the terrestrial genetic code". *Icarus* **224** (1): 228–242. Bibcode:2013Icar..224..228S. doi:10.1016/j.icarus.2013.02.017.

[79] Makukov, Maxim (4 October 2014). "Claim to have identified extraterrestrial signal in the universal genetic code thereby confirming directed panspermia.". *Maxim Makukov*. The New Reddit Journal of Science. Retrieved 2014-10-09.

[80] M. A. Makukov; V. I. shCherbak (2014). "Space ethics to test directed panspermia". *Life Sciences in Space Research* **3**: 10–17. Bibcode:2014LSSR....3...10M. doi:10.1016/j.lssr.2014.07.003.

[81] "'Seed of Life' From Outer Space Suggests Aliens Created Life On Earth, U.K. Scientists Say". *The Inquisitr*. February 13, 2015. Retrieved 2015-03-11.

[82] Wickramasinghe, N.C. (1974). "Formaldehyde Polymers in Interstellar Space". *Nature* **252** (5483): 462–463. Bibcode:1974Natur.252..462W. doi:10.1038/252462a0.

[83] Martins, Zita; Botta, Oliver; Fogel, Marilyn L.; Sephton, Mark A.; Glavin, Daniel P.; Watson, Jonathan S.; Dworkin, Jason P.; Schwartz, Alan W.; Ehrenfreund, Pascale (2008). "Extraterrestrial nucleobases in the Murchison meteorite". *Earth and Planetary Science Letters* **270**: 130–136. Bibcode:2008E&PSL.270..130M. doi:10.1016/j.epsl.2008.03.026.

[84] AFP Staff (20 August 2009). "We may all be space aliens: study". AFP. Archived from the original on June 17, 2008. Retrieved 8 November 2014.

[85] Martins, Zita; Botta, Oliver; Fogel, Marilyn L.; Sephton, Mark A.; Glavin, Daniel P.; Watson, Jonathan S.; Dworkin, Jason P.; Schwartz, Alan W.; Ehrenfreund, Pascale (2008). "Extraterrestrial nucleobases in the Murchison meteorite". *Earth and Planetary Science Letters* **270**: 130–136. Bibcode:2008E&PSL.270..130M. doi:10.1016/j.epsl.2008.03.026.

[86] "'Life chemical' detected in comet". *NASA* (BBC News). 18 August 2009. Retrieved 6 March 2010.

[87] Callahan, M. P.; Smith, K. E.; Cleaves, H. J.; Ruzicka, J.; et al. (2011). "Carbonaceous meteorites contain a wide range of extraterrestrial nucleobases". *Proceedings of the National Academy of Sciences* **108** (34): 13995–8. Bibcode:2011PNAS..10813995C. doi:10.1073/pnas.1106493108. PMC 3161613. PMID 21836052.

[88] Steigerwald, John (8 August 2011). "NASA Researchers: DNA Building Blocks Can Be Made in Space". NASA. Retrieved 10 August 2011.

[89] ScienceDaily Staff (9 August 2011). "DNA Building Blocks Can Be Made in Space, NASA Evidence Suggests". ScienceDaily. Retrieved 9 August 2011.

[90] Chow, Denise (26 October 2011). "Discovery: Cosmic Dust Contains Organic Matter from Stars". Space.com. Retrieved 26 October 2011.

[91] ScienceDaily Staff (26 October 2011). "Astronomers Discover Complex Organic Matter Exists Throughout the Universe". ScienceDaily. Retrieved 27 October 2011.

[92] Kwok, Sun; Zhang, Yong (2011). "Mixed aromatic–aliphatic organic nanoparticles as carriers of unidentified infrared emission features". *Nature* **479** (7371): 80–3. Bibcode:2011Natur.479...80K. doi:10.1038/nature10542. PMID 22031328.

[93] Than, Ker (August 29, 2012). "Sugar Found In Space". *National Geographic*. Retrieved August 31, 2012.

[94] Staff (August 29, 2012). "Sweet! Astronomers spot sugar molecule near star". AP News. Retrieved August 31, 2012.

[95] Jørgensen, Jes K.; Favre, Cécile; Bisschop, Suzanne E.; Bourke, Tyler L.; et al. (2012). "Detection of the Simplest Sugar, Glycolaldehyde, in a Solar-Type Protostar with Alma". *The Astrophysical Journal* **757**: L4. Bibcode:2012ApJ...757L...4J. doi:10.1088/2041-8205/757/1/L4.

[96] Staff (September 20, 2012). "NASA Cooks Up Icy Organics to Mimic Life's Origins". Space.com. Retrieved September 22, 2012.

[97] Gudipati, Murthy S.; Yang, Rui (2012). "In-Situ Probing of Radiation-Induced Processing of Organics in Astrophysical Ice Analogs—Novel Laser Desorption Laser Ionization Time-Of-Flight Mass Spectroscopic Studies". *The Astrophysical Journal* **756**: L24. Bibcode:2012ApJ...756L..24G. doi:10.1088/2041-8205/756/1/L24.

[98] Loomis, Ryan A.; Zaleski, Daniel P.; Steber, Amanda L.; Neill, Justin L.; et al. (2013). "The Detection of Interstellar Ethanimine (Ch3Chnh) from Observations Taken During the Gbt Primos Survey". *The Astrophysical Journal* **765**: L9. Bibcode:2013ApJ...765L...9L. doi:10.1088/2041-8205/765/1/L9.

[99] Finley, Dave,*Discoveries Suggest Icy Cosmic Start for Amino Acids and DNA Ingredients,* The National Radio Astronomy Observatory, Feb. 28, 2013

[100] Kaiser, R. I.; Stockton, A. M.; Kim, Y. S.; Jensen, E. C.; et al. (March 5, 2013). "On the Formation of Dipeptides in Interstellar Model Ices". *The Astrophysical Journal* **765** (2): 111. Bibcode:2013ApJ...765..111K. doi:10.1088/0004-637X/765/2/111. Lay summary – Phys.org.

[101] Hoover, Rachel (February 21, 2014). "Need to Track Organic Nano-Particles Across the Universe? NASA's Got an App for That". *NASA*. Retrieved 22 February 2014.

[102] Marlaire, Ruth (3 March 2015). "NASA Ames Reproduces the Building Blocks of Life in Laboratory". *NASA*. Retrieved 5 March 2015.

[103] Loeb, Abraham (October 2014). "The Habitable Epoch of the Early Universe". *International Journal of Astrobiology* (Cambridge) **13** (4): 337–39. Bibcode:2014IJAsB..13..337L. doi:10.1017/S1473550414000196. Retrieved 15 December 2014.

[104] Loeb, Abraham (2 December 2013). "The Habitable Epoch of the Early Universe". *International Journal of Astrobiology* **13** (4): 337–339. arXiv:1312.0613v3. Bibcode:2014IJAsB..13..337L. doi:10.1017/S1473550414000196.

[105] Dreifus, Claudia (2 December 2014). "Much-Discussed Views That Go Way Back – Avi Loeb Ponders the Early Universe, Nature and Life". *The New York Times*. Retrieved 3 December 2014.

[106] Rampelotto, P.H. (2010). "Panspermia: A Promising Field Of Research" (PDF). *Astrobiology Science Conference*. Harvard: USRA. Retrieved 3 December 2014. External link in |work= (help)

[107] Graham, Robert W (February 1990). "Extraterrestrial Life in the Universe" (PDF). *Technical Memorandum* (Lewis Research Center, OH: NASA). 102363. Retrieved 7 July 2014.

[108] Altermann, Wladyslaw (2008). "From Fossils to Astrobiology – A Roadmap to Fata Morgana?". In Seckbach, Joseph; Walsh, Maud. *From Fossils to Astrobiology: Records of Life on Earth and the Search for Extraterrestrial Biosignatures* **12**. p. xvii. ISBN 1-4020-8836-1.

[109] Webb, Stephen (2002), *If the universe is teeming with aliens, where is everybody? Fifty solutions to the Fermi paradox and the problem of extraterrestrial life*, Copernicus, Springer.

[110] Steffen, Jason H.; Batalha, Natalie M.; Borucki, William J; Buchhave, Lars A.; et al. (9 November 2010). "Five Kepler target stars that show multiple transiting exoplanet candidates". *Astrophysical Journal* **725**: 1226–41. arXiv:1006.2763. Bibcode:2010ApJ...725.1226S. doi:10.1088/0004-637X/725/1/1226.

[111] Overbye, Dennis (November 4, 2013). "Far-Off Planets Like the Earth Dot the Galaxy". *The New York Times*. Retrieved 5 November 2013.

[112] Petigura, Eric A.; Howard, Andrew W.; Marcy, Geoffrey W (October 31, 2013). "Prevalence of Earth-size planets orbiting Sun-like stars". *Proceedings of the National Academy of Sciences of the United States of America* **110** (48): 19273–78. Bibcode:2013PNAS..11019273P. doi:10.1073/pnas.1319909110. Retrieved 5 November 2013.

[113] Khan, Amina (November 4, 2013). "Milky Way may host billions of Earth-size planets". *The Los Angeles Times*. Retrieved 5 November 2013.

[114] Crawford, I.A. (Sep 1995). "Some Thoughts on the Implications of Faster-Than-Light Interstellar Space Travel". *Quarterly Journal of the Royal Astronomical Society* **36** (3): 205. Bibcode:1995QJRAS..36..205C.

[115] Byrne, Joseph Patrick (2008). "Panspermia". *Encyclopedia of Pestilence, Pandemics, and Plagues* (entry). ABC-CLIO. pp. 454–55. ISBN 978-0-313-34102-1.

[116] Wickramasinghe, C; Wainwright, M; Narlikar, J (May 24, 2003). "SARS—a clue to its origins?". *Lancet* **361** (9371): 1832. doi:10.1016/S0140-6736(03)13440-X. PMID 12781581.

[117] Willerslev, E; Hansen, AJ; Rønn, R; Nielsen, OJ (Aug 2, 2003). "Panspermia – true or false?". *Lancet* **362** (9381): 406; author reply 407–8. doi:10.1016/S0140-6736(03)14039-1. PMID 12907025.

[118] Bhargava, PM (Aug 2, 2003). "Panspermia – true or false?". *Lancet* **362** (9381): 407; author reply 407–8. doi:10.1016/S0140-6736(03)14041-X. PMID 12907028.

[119] Ponce de Leon, S; Lazcano, A (Aug 2, 2003). "Panspermia – true or false?". *Lancet* **362** (9381): 406–7; author reply 407–8. doi:10.1016/s0140-6736(03)14040-8. PMID 12907026.

## 17.8. REFERENCES

[120] Qu, Jiangwen (2016). "Is sunspot activity a factor in influenza pandemics?". *Reviews in Medical Virology*. doi:10.1002/rmv.1887.

[121] "New Study Adds to Finding of Ancient Life Signs in Mars Meteorite". NASA. 2009-11-30. Retrieved 1 December 2009.

[122] Thomas-Keprta, K.; Clemett, S; McKay, D; Gibson, E & Wentworth, S (2009). "Origin of Magnetite Nanocrystals in Martian Meteorite ALH84001". *Geochimica et Cosmochimica Acta* **73** (21): 6631–6677. Bibcode:2009GeCoA..73.6631T. doi:10.1016/j.gca.2009.05.064.

[123] "Alien visitors". *New Scientist Space*. 11 May 2001. Retrieved 20 August 2009.

[124] D'Argenio, Bruno; Geraci, Giuseppe & del Gaudio, Rosanna (March 2001). "Microbes in rocks and meteorites: a new form of life unaffected by time, temperature, pressure". *Rendiconti Lincei* **12** (1): 51–68. doi:10.1007/BF02904521. Retrieved 13 October 2009.

[125] Geraci, Giuseppe; del Gaudio, Rosanna; D'Argenio, Bruno (2001), "Microbes in rocks and meteorites: a new form of life unaffected by time, temperature, pressure" (PDF), *Rend. Fis. Acc. Linceis* **9**: 51–68.

[126] "Scientists Say They Have Found Extraterrestrial Life in the Stratosphere But Peers Are Skeptical: Scientific American". Sciam. 2001-07-31. Retrieved 20 August 2009.

[127] Narlikar, JV; Lloyd, D; Wickramasinghe, NC; Turner; Al-Mufti; Wallis; Wainwright; Rajaratnam; Shivaji; Reddy; Ramadurai; Hoyle (2003). "Balloon experiment to detect micro-organisms in the outer space". *Astrophys Space Sci* **285** (2): 555–62. Bibcode:2003Ap&SS.285..555N. doi:10.1023/A:1025442021619.

[128] Wainwright, M; Wickramasinghe, N.C; Narlikar, J.V; Rajaratnam, P. "Microorganisms cultured from stratospheric air samples obtained at 41 km". Retrieved 11 May 2007.

[129] Wainwright, M (2003). "A microbiologist looks at panspermia". *Astrophys Space Sci* **285** (2): 563–70. Bibcode:2003Ap&SS.285..563W. doi:10.1023/A:1025494005689.

[130] Vaidya, Pushkar Ganesh (July 2009). "Critique on Vindication of Panspermia" (PDF). *Apeiron* **16** (3). Retrieved 28 November 2009.

[131] *Mumbai scientist challenges theory that bacteria came from space*, IN: AOL.

[132] *Janibacter hoylei sp. nov., Bacillus isronensis sp. nov. and Bacillus aryabhattai sp. nov.*, isolated from cryotubes used for collecting air from upper atmosphere. *International Journal of Systematic and Evolutionary Microbiology* 2009. http://ijs.sgmjournals.org/cgi/content/abstract/ijs.0.002527-0v1

[133] Discovery of New Microorganisms in the Stratosphere.

[134] Timothy Oleson (May 5, 2013). "Lofted by hurricanes, bacteria live the high life". *NASA* (Earth Magazine). Retrieved 21 September 2013.

[135] Helen Shen (28 January 2013). "High-flying bacteria spark interest in possible climate effects". *Nature News*. doi:10.1038/nature.2013.12310. Retrieved 21 September 2013.

[136] Wickramasinghe, N. C.; Wallis, J.; Wallis, D. H.; Samaranayake, Anil (January 10, 2013). "Fossil Diatoms in a New Carbonaceous Meteorite" (PDF). *Journal of Cosmology* **21** (37): 1–14. arXiv:1303.2398. Bibcode:2013JCos...21.9560W. Retrieved January 14, 2013.

[137] Phil Plait (15 January 2013). "No, Diatoms Have Not Been Found in a Meteorite". *Slate.com – Astronomy*. Retrieved 16 January 2013.

[138] Wallis, Jamie; Miyake, Nori; Hoover, Richard B.; Oldroyd, Andrew; et al. (5 March 2013). "The Polonnaruwa meteorite: oxygen isotope, crystalline and biological composition" (PDF). *Journal of Cosmology* **22** (2): 1845. arXiv:1303.1845. Bibcode:2013JCos...2210004W. Retrieved 7 March 2013.

[139] Wickramasinghe, N.C.; J. Wallis; N. Miyake; Anthony Oldroyd; et al. (4 February 2013). "Authenticity of the life-bearing Polonnaruwa meteorite" (PDF). *Journal of Cosmology*. Retrieved 4 February 2013.

[140] Griffin, Dale Warren (14 August 2013). "The Quest for Extraterrestrial Life: What About the Viruses?". *Astrobiology* **13** (8): 774–783. Bibcode:2013AsBio..13..774G. doi:10.1089/ast.2012.0959.

[141] Edward Anders, Eugene R. DuFresne, Ryoichi Hayatsu, Albert Cavaille, Ann DuFresne, and Frank W. Fitch. "Contaminated Meteorite", *Science, New Series*, Volume 146, Issue 3648 (Nov.27, 1964), 1157–1161.

[142] Chamberlin, Sean (1999). "Black Smokers and Giant Worms". *Fullerton College*. Retrieved 11 February 2011.

[143] Choi, Charles Q. (17 March 2013). "Microbes Thrive in Deepest Spot on Earth". LiveScience. Retrieved 17 March 2013.

[144] Oskin, Becky (14 March 2013). "Intraterrestrials: Life Thrives in Ocean Floor". LiveScience. Retrieved 17 March 2013.

[145] Glud, Ronnie; Wenzhöfer, Frank; Middelboe, Mathias; Oguri, Kazumasa; et al. (17 March 2013). "High rates of microbial carbon turnover in sediments in the deepest oceanic trench on Earth". *Nature Geoscience* **6** (4): 284–288. Bibcode:2013NatGe...6..284G. doi:10.1038/ngeo1773. Retrieved 17 March 2013.

[146] Carey, Bjorn (7 February 2005). "Wild Things: The Most Extreme Creatures". *Live Science*. Retrieved 20 October 2008.

[147] Cavicchioli, R. (Fall 2002). "Extremophiles and the search for extraterrestrial life". *Astrobiology* **2** (3): 281–92. Bibcode:2002AsBio...2..281C. doi:10.1089/153110702762027862. PMID 12530238.

[148] The BIOPAN experiment MARSTOX II of the FOTON M-3 mission July 2008.

[149] Surviving the Final Frontier. 25 November 2002.

[150] Christner, Brent C. (2002). "Detection, recovery, isolation, and characterization of bacteria in glacial ice and Lake Vostok accretion ice". *Ohio State University*. Retrieved 4 February 2011.

[151] Nanjundiah, V. (2000). "The smallest form of life yet?" (PDF). *Journal of Biosciences* **25** (1): 9–10. doi:10.1007/BF02985175. PMID 10824192.

[152] Rabbow, Elke Rabbow; Gerda Horneck; Petra Rettberg; Jobst-Ulrich Schott; et al. (9 July 2009). "EXPOSE, an Astrobiological Exposure Facility on the International Space Station – from Proposal to Flight" (PDF). *Orig Life Evol Biosph* **39** (6): 581–98. doi:10.1007/s11084-009-9173-6. PMID 19629743. Retrieved 8 July 2013.

[153] Bacterium revived from 25 million year sleep Digital Center for Microbial Ecology

[154] Tepfer, David Tepfer (December 2008). "The origin of life, panspermia and a proposal to seed the Universe". *Plant Science* **175** (6): 756–760. doi:10.1016/j.plantsci.2008.08.007.

[155] "Exobiology and Radiation Assembly (ERA)". *ESA*. NASA. 1992. Retrieved 22 July 2013.

[156] Zhang, K. Dose; A. Bieger-Dose; R. Dillmann; M. Gill; et al. (1995). "ERA-experiment "space biochemistry"". *Advances in Space Research* **16** (8): 119–129. doi:10.1016/0273-1177(95)00280-R. PMID 11542696.

[157] Vaisberg, Horneck G; Eschweiler U; Reitz G; Wehner J; et al. (1995). "Biological responses to space: results of the experiment "Exobiological Unit" of ERA on EURECA I". *Adv Space Res.* **16** (8): 105–18. Bibcode:1995AdSpR..16..105V. doi:10.1016/0273-1177(95)00279-N. PMID 11542695.

[158] "BIOPAN Pan for exposure to space environment". *Kayser Italia*. 2013. Retrieved 17 July 2013.

[159] De La Torre Noetzel, Rosa (2008). "Experiment lithopanspermia: Test of interplanetary transfer and re-entry process of epi- and endolithic microbial communities in the FOTON-M3 Mission". *37th COSPAR Scientific Assembly.* Held 13–20 July 2008 **37**: 660. Bibcode:2008cosp...37..660D.

[160] "Life in Space for Life ion Earth – Biosatelite Foton M3". June 26, 2008. Retrieved 13 October 2009.

[161] Jönsson, K. Ingemar Jönsson; Elke Rabbow; Ralph O. Schill; Mats Harms-Ringdahl; et al. (9 September 2008). "Tardigrades survive exposure to space in low Earth orbit". *Current Biology* **18** (17): R729–R731. doi:10.1016/j.cub.2008.06.048. PMID 18786368.

[162] de Vera; J.P.P.; et al. (2010). "COSPAR 2010 Conference". Research Gate. Retrieved 17 July 2013

[163] Paul Clancy (Jun 23, 2005). *Looking for Life, Searching the Solar System*. Cambridge University Press. Retrieved 26 March 2014.

[164] Tepfer, David Tepfer; Andreja Zalar & Sydney Leach. (May 2012). "Survival of Plant Seeds, Their UV Screens, and nptII DNA for 18 Months Outside the International Space Station". *Astrobiology* **12** (5): 517–528. Bibcode:2012AsBio..12..517T. doi:10.1089/ast.2011.0744. PMID 22680697.

[165] Scalzi, Giuliano Scalzi; Laura Selbmann; Laura Zucconi; Elke Rabbow; et al. (1 June 2012). "LIFE Experiment: Isolation of Cryptoendolithic Organisms from Antarctic Colonized Sandstone Exposed to Space and Simulated Mars Conditions on the International Space Station". *Origins of Life and Evolution of Biospheres* **42** (2 – 3): 253–262. doi:10.1007/s11084-012-9282-5.

[166] Onofri, Silvano Onofri; Rosa de la Torre; Jean-Pierre de Vera; Sieglinde Ott; et al. (May 2012). "Survival of Rock-Colonizing Organisms After 1.5 Years in Outer Space". *Astrobiology* **12** (5): 508–516. Bibcode:2012AsBio..12..508O. doi:10.1089/ast.2011.0736. PMID 22680696.

[167] Neuberger, Katja; Lux-Endrich, Astrid; Panitz, Corinna; Horneck, Gerda (January 2015). "Survival of Spores of Trichoderma longibrachiatum in Space: data from the Space Experiment SPORES on EXPOSE-R" (PDF). *International Journal of Astrobiology* **14** (Special Issue 1): 129–135. Bibcode:2015IJAsB..14..129N. doi:10.1017/S1473550414000408. Retrieved 2015-05-09.

[168] Schulze-Makuch, Dirk (3 September 2014). "New ISS Experiment Tests Organisms' Survival Skills in Space". *Air and Space Magazine*. Retrieved 2014-09-04.

[169] "Spacewalk Marks End of ESA's Exposed Space Chemistry Experiment". *ESA*. February 3, 2016. Retrieved 2016-02-09.

[170] esa. "Rosetta's lander faces eternal hibernation". *European Space Agency*. Retrieved 2016-06-08.

[171] "Nol bugs please, this is a clean planet!". European Space Agency (ESA). 30 July 2002. Retrieved 16 July 2013.

[172] Jordans, Frank (30 July 2015). "Philae probe finds evidence that comets can be cosmic labs". *The Washington Post*. Associated Press. Retrieved 30 July 2015.

[173] "Science on the Surface of a Comet". European Space Agency. 30 July 2015. Retrieved 30 July 2015.

[174] Bibring, J.-P.; Taylor, M.G.G.T.; Alexander, C.; Auster, U.; Biele, J.; Finzi, A. Ercoli; Goesmann, F.; Klinghoefer, G.; Kofman, W.; Mottola, S.; Seidenstiker, K.J.; Spohn, T.; Wright, I. (31 July 2015). "Philae's First Days on the Comet – Introduction to Special Issue". *Science* **349** (6247): 493. Bibcode:2015Sci...349..493B. doi:10.1126/science.aac5116. Retrieved 30 July 2015.

[175] Altwegg, Kathrin; Balsiger, Hans; Bar-Nun, Akiva; Berthelier, Jean-Jacques; Bieler, Andre; Bochsler, Peter; Briois, Christelle; Calmonte, Ursina; Combi, Michael R. (2016-05-01). "Prebiotic chemicals—amino acid and phosphorus—in the coma of comet 67P/Churyumov-Gerasimenko". *Science Advances* **2** (5): e1600285. doi:10.1126/sciadv.1600285. ISSN 2375-2548. PMID 27386550.

[176] Microbe space exposure experiment at International Space Station (ISS) proposed in "Tanpopo" mission. Research Gate, July 2010.

[177] "Tanpopo Experiment for Wastrobiology Exposure and Micrometeoroid Capture Onboard the ISS-JEM Exposed Facility." (PDF) H. Yano, A. Yamagishi, H. Hashimoto1, S. Yokobori, K. Kobayashi, H. Yabuta, H. Mita, M. Tabata H., Kawai, M. Higashide, K. Okudaira, S. Sasaki, E. Imai, Y. Kawaguchi, Y. Uchibori11, S. Kodaira and the Tanpopo Project Team. 45th Lunar and Planetary Science Conference (2014).

[178] Tanpopo mission to search space for origins of life. *The Japan News*, April 16, 2015.

[179] Yuko, Kawaguchi; et al. (13 May 2016). "Investigation of the Interplanetary Transfer of Microbes in the Tanpopo Mission at the Exposed Facility of the International Space Station". *Astrobiology* **16** (5): 363–376. Bibcode:2016AsBio..16..363K. doi:10.1089/ast.2015.1415. PMID 27176813. Retrieved 2016-05-29.

[180] Wickramasinghe, M.K.; Wickramasinghe, C. (2004). "Interstellar transfer of planetary microbiota". *Mon. Not.R. Astr. Soc.* **348**: 52–57. Bibcode:2004MNRAS.348...52W. doi:10.1111/j.1365-2966.2004.07355.x.

## 17.9 Further reading

- Crick, F (1981), *Life, Its Origin and Nature*, Simon & Schuster, ISBN 0-7088-2235-5.

- Hoyle, F (1983), *The Intelligent Universe*, London: Michael Joseph, ISBN 0-7181-2298-4.

## 17.10 External links

- A.E. Zlobin, 2013, Tunguska similar impacts and origin of life (mathematical theory of origin of life; incoming of pattern recognition algorithm due to comets)

- Francis Crick's notes for a lecture on directed panspermia, dated 5 November 1976.

- "Earth sows its seeds in space". *Nature News*. 23 February 2004. doi:10.1038/news040216-20 (inactive 2016-07-13).

- Warmflash, D.; Weiss, B. (24 October 2005). "Did Life Come from Another World?". *Scientific American* **293** (5): 64–71. doi:10.1038/scientificamerican1105-64.

# Chapter 18

# List of interstellar and circumstellar molecules

This is a list of molecules that have been detected in the interstellar medium and circumstellar envelopes, grouped by the number of component atoms. The chemical formula is listed for each detected compound, along with any ionized form that has also been observed.

## 18.1 Detection

The molecules listed below were detected by spectroscopy. Their spectral features are generated by transitions of component electrons between different energy levels, or by rotational or vibrational spectra. Detection usually occurs in radio, microwave, or infrared portions of the spectrum.[1]

Interstellar molecules are formed by chemical reactions within very sparse interstellar or circumstellar clouds of dust and gas. Usually this occurs when a molecule becomes ionized, often as the result of an interaction with a cosmic ray. This positively charged molecule then draws in a nearby reactant by electrostatic attraction of the neutral molecule's electrons. Molecules can also be generated by reactions between neutral atoms and molecules, although this process is generally slower.[2] The dust plays a critical role of shielding the molecules from the ionizing effect of ultraviolet radiation emitted by stars.[3]

### 18.1.1 History

The chemistry of life may have begun shortly after the Big Bang, 13.8 billion years ago, during a habitable epoch when the Universe was only 10–17 million years old.[4][5]

The first carbon-containing molecule detected in the interstellar medium was the methylidyne radical (CH) in 1937.[6] From the early 1970s it was becoming evident that interstellar dust consisted of a large component of more complex organic molecules (COMs),[7] probably polymers.

Chandra Wickramasinghe proposed the existence of polymeric composition based on the molecule formaldehyde ($H_2CO$).[8] Fred Hoyle and Chandra Wickramasinghe later proposed the identification of bicyclic aromatic compounds from an analysis of the ultraviolet extinction absorption at 2175 Å,[9] thus demonstrating the existence of polycyclic aromatic hydrocarbon molecules in space.

In 2004, scientists reported[10] detecting the spectral signatures of anthracene and pyrene in the ultraviolet light emitted by the Red Rectangle nebula (no other such complex molecules had ever been found before in outer space). This discovery was considered a confirmation of a hypothesis that as nebulae of the same type as the Red Rectangle approach the ends of their lives, convection currents cause carbon and hydrogen in the nebulae's core to get caught in stellar winds, and radiate outward.[11] As they cool, the atoms supposedly bond to each other in various ways and eventually form particles of a million or more atoms. The scientists inferred[10] that since they discovered polycyclic aromatic hydrocarbons (PAHs) — which may have been vital in the formation of early life on Earth — in a nebula, by necessity they must originate in nebulae.[11]

In 2010, fullerenes (or "buckyballs") were detected in nebulae.[12] Fullerenes have been implicated in the origin of life; according to astronomer Letizia Stanghellini, "It's possible that buckyballs from outer space provided seeds for life on Earth."[13]

In October 2011, scientists found using spectroscopy that cosmic dust contains complex organic compounds ("amorphous organic solids with a mixed aromatic-aliphatic structure") that could be created naturally, and rapidly, by stars.[14][15][16] The compounds are so complex that their chemical structures resemble the makeup of coal and petroleum; such chemical complexity was previously thought to arise only from living organisms.[14] These observations suggest that organic compounds introduced on Earth by interstellar dust particles could serve as basic ingredients for life due to their surface-catalytic activities.[17][18] One of the scientists suggested that these compounds may have been related to the development of life on Earth and said that, "If this is the case, life on Earth may have had an easier time getting started as these organics can serve as basic ingredients for life."[14]

In August 2012, astronomers at Copenhagen University reported the detection of a specific sugar molecule, glycolaldehyde, in a distant star system. The molecule was found around the protostellar binary *IRAS 16293-2422*, which is located 400 light years from Earth.[19][20] Glycolaldehyde is needed to form ribonucleic acid, or RNA, which is similar in function to DNA. This finding suggests that complex organic molecules may form in stellar systems prior to the formation of planets, eventually arriving on young planets early in their formation.[21]

In September 2012, NASA scientists reported that PAHs, subjected to interstellar medium (ISM) conditions, are transformed, through hydrogenation, oxygenation, and hydroxylation, to more complex organics — "a step along the path toward amino acids and nucleotides, the raw materials of proteins and DNA, respectively".[22][23] Further, as a result of these transformations, the PAHs lose their spectroscopic signature which could be one of the reasons "for the lack of PAH detection in interstellar ice grains, particularly the outer regions of cold, dense clouds or the upper molecular layers of protoplanetary disks."[22][23]

PAHs are found everywhere in deep space[24] and, in June 2013, PAHs were detected in the upper atmosphere of Titan, the largest moon of the planet Saturn.[25]

In 2013, Dwayne Heard at the University of Leeds suggested[26] that quantum mechanical tunneling could explain a reaction his group observed taking place, at a significantly higher than expected rate, between cold (around 63 Kelvin) hydroxyl and methanol molecules, apparently bypassing intramolecular energy barriers which would have to be overcome by thermal energy or ionization events for the same rate to exist at warmer temperatures. The proposed tunneling mechanism may help explain the common observation of fairly complex molecules (up to tens of atoms) in interstellar space.

A particularly large and rich region for detecting interstellar molecules is Sagittarius B2 (Sgr B2). This giant molecular cloud lies near the center of the Milky Way galaxy and is a frequent target for new searches. About half of the molecules listed below were first found near Sgr B2, and nearly every other molecule has since been detected in this feature.[27] A rich source of investigation for circumstellar molecules is the relatively nearby star CW Leonis (IRC +10216), where about 50 compounds have been identified.[28]

In March 2015, NASA scientists reported that, for the first time, complex DNA and RNA organic compounds of life, including uracil, cytosine and thymine, have been formed in the laboratory under outer space conditions, using starting chemicals, such as pyrimidine, found in meteorites. Pyrimidine, like polycyclic aromatic hydrocarbons (PAHs), the most carbon-rich chemical found in the Universe, may have been formed in red giants or in interstellar dust and gas clouds, according to the scientists.[29]

## 18.2 Molecules

The following tables list molecules that have been detected in the interstellar medium, grouped by the number of component atoms. If there is no entry in the molecule column, only the ionized form has been detected. For molecules where no designation was given in the scientific literature, that field is left empty. Mass is given in atomic mass units. The total number of unique species, including distinct ionization states, is listed in parentheses in each section header.

Most of the molecules detected so far are organic. Only one inorganic species has been observed in molecules which contain at least five atoms, $SiH_4$.[30] Larger molecules have so far all had at least one carbon atom, with no N–N or O–O bonds.[30]

*Carbon monoxide is frequently used to trace the distribution of mass in molecular clouds.*[31]

The H$_3^+$ cation is one of the most abundant ions in the universe. It was first detected in 1993.[69][70]

Methane, the primary component of natural gas, has also been detected on comets and in the atmosphere of several planets in the Solar System.[121]

### 18.2.1 Diatomic (43)

### 18.2.2 Triatomic (43)

### 18.2.3 Four atoms (27)

### 18.2.4 Five atoms (19)

### 18.2.5 Six atoms (16)

### 18.2.6 Seven atoms (10)

### 18.2.7 Eight atoms (11)

### 18.2.8 Nine atoms (10)

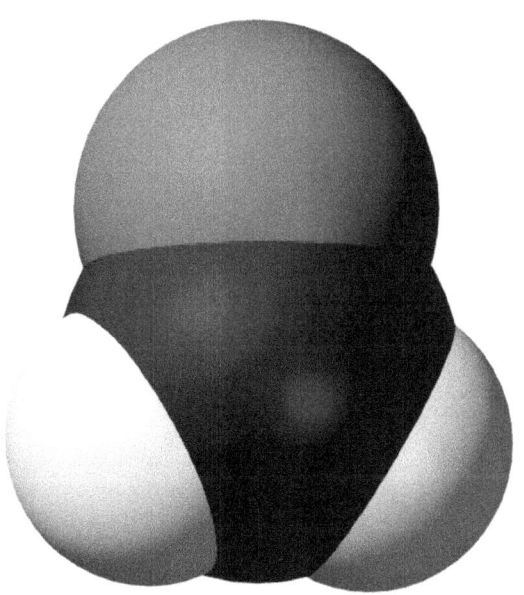

Formaldehyde is an organic molecule that is widely distributed in the interstellar medium.[100]

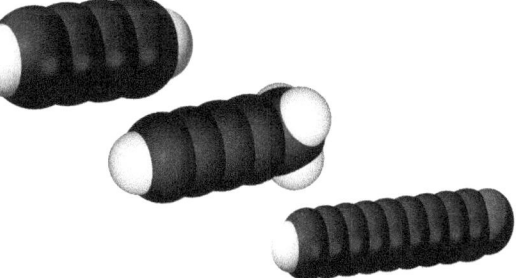

number of polyyne-derived chemicals are among the heaviest molecules found in the interstellar medium.

## 18.4 Unconfirmed (13)

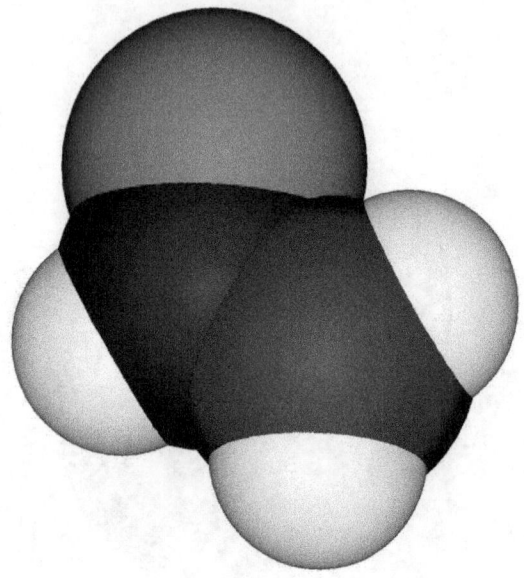

*In the ISM, formamide (above) can combine with methylene to form acetamide.*[140]

*The radio signature of acetic acid, a compound found in vinegar, was confirmed in 1997.*[157]

Evidence for the existence of the following molecules has been reported in scientific literature, but the detections are either described as tentative by the authors, or have been challenged by other researchers. They await independent confirmation.

## 18.5 See also

- Abiogenesis
- Astrobiology
- Astrochemistry
- Atomic and molecular astrophysics
- Cosmic dust
- Cosmic ray
- Cosmochemistry
- Diffuse interstellar band
- Extraterrestrial liquid water
- Forbidden mechanism
- Intergalactic dust
- Interplanetary medium
- Interstellar medium
- Organic compound

*Acetaldehyde (above) and its isomers vinyl alcohol and ethylene oxide have all been detected in interstellar space.*[152]

### 18.2.9 Ten or more atoms (15)

## 18.3 Deuterated molecules (20)

These molecules all contain one or more deuterium atoms, a heavier isotope of hydrogen.

- Outer space
- Panspermia
- Polycyclic aromatic hydrocarbon (PAH)
- Spectroscopy

## 18.6 References

[1] Shu, Frank H. (1982), *The Physical Universe: An Introduction to Astronomy*, University Science Books, ISBN 0-935702-05-9

[2] Dalgarno, A. (2006), "Interstellar Chemistry Special Feature: The galactic cosmic ray ionization rate", *Proceedings of the National Academy of Sciences* **103** (33): 12269–12273, Bibcode:2006PNAS..10312269D, doi:10.1073/pnas.0602117103, PMC 1567869, PMID 16894166

[3] Brown, Laurie M.; Pais, Abraham; Pippard, A. B. (1995), "The physics of the interstellar medium", *Twentieth Century Physics* (2nd ed.), CRC Press, p. 1765, ISBN 0-7503-0310-7

[4] Loeb, Abraham (October 2014). "The Habitable Epoch of the Early Universe". *International Journal of Astrobiology* **13** (04): 337–339. arXiv:1312.0613. Bibcode:2014IJAsB..13..337L. doi:10.1017/S1473550414000196. Retrieved 15 December 2014.

[5] Dreifus, Claudia (2 December 2014). "Much-Discussed Views That Go Way Back - Avi Loeb Ponders the Early Universe, Nature and Life". *New York Times*. Retrieved 3 December 2014.

[6] Woon, D. E. (May 2005), *Methylidyne radical*, The Astrochemist, retrieved 2007-02-13

[7] Ruaud, M.; Loison, J.C.; Hickson, K.M.; Gratier, P.; Hersant, F.; Wakelam, V. (2015). "Modeling Complex Organic Molecules in dense regions: Eley-Rideal and complex induced reaction". *Monthly Notices of the Royal Astronomical Society* **447** (4): 4004–4017. arXiv:1412.6256. Bibcode:2015MNRAS.447.4004R. doi:10.1093/mnras/stu2709.

[8] N.C. Wickramasinghe, Formaldehyde Polymers in Interstellar Space, Nature, 252, 462, 1974

[9] F. Hoyle and N.C. Wickramasinghe, Identification of the lambda 2200Å interstellar absorption feature, Nature, 270, 323, 1977

[10] Battersby, S. (2004). "Space molecules point to organic origins". New Scientist. Retrieved 11 December 2009.

[11] Mulas, G.; Malloci, G.; Joblin, C.; Toublanc, D. (2006). "Estimated IR and phosphorescence emission fluxes for specific polycyclic aromatic hydrocarbons in the Red Rectangle". *Astronomy and Astrophysics* **446** (2): 537–549. arXiv:astro-ph/0509586. Bibcode:2006A&A...446..537M. doi:10.1051/0004-6361:20053738.

[12] García-Hernández, D. A.; Manchado, A.; García-Lario, P.; Stanghellini, L.; Villaver, E.; Shaw, R. A.; Szczerba, R.; Perea-Calderón, J. V. (2010-10-28). "Formation Of Fullerenes In H-Containing Planetary Nebulae". *The Astrophysical Journal Letters* **724** (1): L39–L43. arXiv:1009.4357. Bibcode:2010ApJ...724L..39G. doi:10.1088/2041-8205/724/1/L39.

[13] Atkinson, Nancy (2010-10-27). "Buckyballs Could Be Plentiful in the Universe". Universe Today. Retrieved 2010-10-28.

[14] Chow, Denise (26 October 2011). "Discovery: Cosmic Dust Contains Organic Matter from Stars". Space.com. Retrieved 2011-10-26.

[15] ScienceDaily Staff (26 October 2011). "Astronomers Discover Complex Organic Matter Exists Throughout the Universe". ScienceDaily. Retrieved 2011-10-27.

[16] Kwok, Sun; Zhang, Yong (26 October 2011). "Mixed aromatic–aliphatic organic nanoparticles as carriers of unidentified infrared emission features". *Nature* **479** (7371): 80–3. Bibcode:2011Natur.479...80K. doi:10.1038/nature10542. PMID 22031328.

[17] Gallori, Enzo (November 2010). "Astrochemistry and the origin of genetic material". *Rendiconti Lincei* **22** (2): 113–118. doi:10.1007/s12210-011-0118-4. Retrieved 2011-08-11.

[18] Martins, Zita (February 2011). "Organic Chemistry of Carbonaceous Meteorites". *Elements* **7** (1): 35–40. doi:10.2113/gselements.7.1.35. Retrieved 2011-08-11.

[19] Than, Ker (August 29, 2012). "Sugar Found In Space". *National Geographic*. Retrieved August 31, 2012.

[20] Staff (August 29, 2012). "Sweet! Astronomers spot sugar molecule near star". AP News. Retrieved August 31, 2012.

[21] Jørgensen, J. K.; Favre, C.; Bisschop, S.; Bourke, T.; Dishoeck, E.; Schmalzl, M. (2012). "Detection of the simplest sugar, glycolaldehyde, in a solar-type protostar with ALMA" (PDF). *The Astrophysical Journal Letters*. eprint **757**: L4. arXiv:1208.5498. Bibcode:2012ApJ...757L...4J. doi:10.1088/2041-8205/757/1/L4.

[22] Staff (September 20, 2012). "NASA Cooks Up Icy Organics to Mimic Life's Origins". Space.com. Retrieved September 22, 2012.

[23] Gudipati, Murthy S.; Yang, Rui (September 1, 2012). "In-Situ Probing Of Radiation-Induced Processing Of Organics In Astrophysical Ice Analogs—Novel Laser Desorption Laser Ionization Time-Of-Flight Mass Spectroscopic Studies". *The Astrophysical Journal Letters* **756** (1):

## 18.6. REFERENCES

L24. Bibcode:2012ApJ...756L..24G. doi:10.1088/2041-8205/756/1/L24. Retrieved September 22, 2012.

[24] Clavin, Whitney (10 February 2015). "Why Comets Are Like Deep Fried Ice Cream". *NASA*. Retrieved 10 February 2015.

[25] López-Puertas, Manuel (June 6, 2013). "PAH's in Titan's Upper Atmosphere". *CSIC*. Retrieved June 6, 2013.

[26] http://www.sciencenews.org/view/generic/id/351444/description/Interstellar_chemistry_makes_use_of_quantum_shortcut#comment_351468

[27] Cummins, S. E.; Linke, R. A.; Thaddeus, P. (1986), "A survey of the millimeter-wave spectrum of Sagittarius B2", *Astrophysical Journal Supplement Series* **60**: 819–878, Bibcode:1986ApJS...60..819C, doi:10.1086/191102

[28] Kaler, James B. (2002), *The hundred greatest stars*, Copernicus Series, Springer, ISBN 0-387-95436-8, retrieved 2011-05-09

[29] Marlaire, Ruth (3 March 2015). "NASA Ames Reproduces the Building Blocks of Life in Laboratory". *NASA*. Retrieved 5 March 2015.

[30] Klemperer, William (2011), "Astronomical Chemistry", *Annual Review of Physical Chemistry* **62**: 173–184, doi:10.1146/annurev-physchem-032210-103332

[31] *The Structure of Molecular Cloud Cores*, Centre for Astrophysics and Planetary Science, University of Kent, retrieved 2007-02-16

[32] Ziurys, Lucy M. (2006), "The chemistry in circumstellar envelopes of evolved stars: Following the origin of the elements to the origin of life", *Proceedings of the National Academy of Sciences* **103** (33): 12274–12279, Bibcode:2006PNAS..10312274Z, doi:10.1073/pnas.0602277103, PMC 1567870, PMID 16894164

[33] Cernicharo, J.; Guelin, M. (1987), "Metals in IRC+10216 - Detection of NaCl, AlCl, and KCl, and tentative detection of AlF", *Astronomy and Astrophysics* **183** (1): L10–L12, Bibcode:1987A&A...183L..10C

[34] Ziurys, L. M.; Apponi, A. J.; Phillips, T. G. (1994), "Exotic fluoride molecules in IRC +10216: Confirmation of AlF and searches for MgF and CaF", *Astrophysical Journal* **433** (2): 729–732, Bibcode:1994ApJ...433..729Z, doi:10.1086/174682

[35] Tenenbaum, E. D.; Ziurys, L. M. (2009), "Millimeter Detection of AlO ($X^2\Sigma^+$): Metal Oxide Chemistry in the Envelope of VY Canis Majoris", *Astrophysical Journal* **694**: L59–L63, Bibcode:2009ApJ...694L..59T, doi:10.1088/0004-637X/694/1/L59

[36] Barlow, M. J.; Swinyard, B. M.; Owen, P. J.; Cernichoro, J.; Gomez, H. L.; Ivison, R. J.; Lim, T. L.; Matsuura, M.; Miller, S.; Olofsson, G.; Polehampton, E. T. (2013), "Detection of a Noble Gas Molecular Ion, $^{36}$ArH+, in the Crab Nebula", *Science* **342** (6164): 1343–1345, doi:10.1126/science.124358213

[37] Quenqua, Douglas (13 December 2013). "Noble Molecules Found in Space". *New York Times*. Retrieved 13 December 2013.

[38] Lambert, D. L.; Sheffer, Y.; Federman, S. R. (1995), "Hubble Space Telescope observations of $C_2$ molecules in diffuse interstellar clouds", *Astrophysical Journal* **438**: 740–749, Bibcode:1995ApJ...438..740L, doi:10.1086/175119

[39] Galazutdinov, G. A.; Musaev, F. A.; Krelowski, J. (2001), "On the detection of the linear $C_5$ molecule in the interstellar medium", *Monthly Notices of the Royal Astronomical Society* **325** (4): 1332–1334, Bibcode:2001MNRAS.325.1332G, doi:10.1046/j.1365-8711.2001.04388.x

[40] Neufeld, D. A.; et al. (2006), "Discovery of interstellar CF$^+$", *Astronomy and Astrophysics* **454** (2): L37–L40, arXiv:astro-ph/0603201, Bibcode:2006A&A...454L..37N, doi:10.1051/0004-6361:200600015

[41] Adams, Walter S. (1941), "Some Results with the COUDÉ Spectrograph of the Mount Wilson Observatory", *Astrophysical Journal* **93**: 11–23, Bibcode:1941ApJ....93...11A, doi:10.1086/144237

[42] Smith, D. (1988), "Formation and Destruction of Molecular Ions in Interstellar Clouds", *Philosophical Transactions of the Royal Society of London* **324** (1578): 257–273, Bibcode:1988RSPTA.324..257S, doi:10.1098/rsta.1988.0016

[43] Fuente, A.; et al. (2005), "Photon-dominated Chemistry in the Nucleus of M82: Widespread HOC$^+$ Emission in the Inner 650 Parsec Disk", *Astrophysical Journal* **619** (2): L155–L158, arXiv:astro-ph/0412361, Bibcode:2005ApJ...619L.155F, doi:10.1086/427990

[44] Guelin, M.; Cernicharo, J.; Paubert, G.; Turner, B. E. (1990), "Free CP in IRC + 10216", *Astronomy and Astrophysics* **230**: L9–L11, Bibcode:1990A&A...230L...9G

[45] Dopita, Michael A.; Sutherland, Ralph S. (2003), *Astrophysics of the diffuse universe*, Springer-Verlag, ISBN 3-540-43362-7

[46] Agúndez, M.; et al. (2010-07-30), "Astronomical identification of CN$^-$, the smallest observed molecular anion", *Astronomy & Astrophysics* **517**: L2, arXiv:1007.0662, Bibcode:2010A&A...517L...2A, doi:10.1051/0004-6361/201015186, retrieved 2010-09-03

[47] Khan, Amina. "Did two planets around nearby star collide? Toxic gas holds hints". *LA Times*. Retrieved March 9, 2014.

[48] Dent, W.R.F.; Wyatt, M.C.;Roberge, A.; Augereau,J.-C.; Casassus, S.;Corder, S.; Greaves, J.S.; de Gregorio-Monsalvo, I; Hales, A.; Jackson, A.P.; Hughes, A. Meredith; Lagrange, A.-M; Matthews, B.; Wilner, D. (March 6,

2014). "Molecular Gas Clumps from the Destruction of Icy Bodies in the β Pictoris Debris Disk". *Science* **343**: 1490–1492. arXiv:1404.1380. Bibcode:2014Sci...343.1490D. doi:10.1126/science.1248726. Retrieved March 9, 2014.

[49] Latter, W. B.; Walker, C. K.; Maloney, P. R. (1993), "Detection of the Carbon Monoxide Ion ($CO^+$) in the Interstellar Medium and a Planetary Nebula", *Astrophysical Journal Letters* **419**: L97, Bibcode:1993ApJ...419L..97L, doi:10.1086/187146

[50] Furuya, R. S.; et al. (2003), "Interferometric observations of FeO towards Sagittarius B2", *Astronomy and Astrophysics* **409** (2): L21–L24, Bibcode:2003A&A...409L..21F, doi:10.1051/0004-6361:20031304

[51] Adams, Walter S. (1970), "Rocket Observation of Interstellar Molecular Hydrogen", *Astrophysical Journal* **161**: L81–L85, Bibcode:1970ApJ...161L..81C, doi:10.1086/180575

[52] Blake, G. A.; Keene, J.; Phillips, T. G. (1985), "Chlorine in dense interstellar clouds - The abundance of HCl in OMC-1", *Astrophysical Journal, Part 1* **295**: 501–506, Bibcode:1985ApJ...295..501B, doi:10.1086/163394

[53] De Luca, M.; Gupta, H.; Neufeld, D.; Gerin, M.; Teyssier, D.; Drouin, B. J.; Pearson, J. C.; Lis, D. C.; et al. (2012), "Herschel/HIFI Discovery of HCl+ in the Interstellar Medium", *The Astrophysical Journal Letters* **751** (2): L37, Bibcode:2012ApJ...751L..37D, doi:10.1088/2041-8205/751/2/L37

[54] Neufeld, David A.; et al. (1997), "Discovery of Interstellar Hydrogen Fluoride", *Astrophysical Journal Letters* **488** (2): L141–L144, arXiv:astro-ph/9708013, Bibcode:1997ApJ...488L.141N, doi:10.1086/310942

[55] Wyrowski, F.; et al. (2009), "First interstellar detection of $OH^+$", *Astronomy & Astrophysics* **518**: A26, arXiv:1004.2627, Bibcode:2010A&A...518A..26W, doi:10.1051/0004-6361/201014364

[56] Meyer, D. M.; Roth, K. C. (1991), "Discovery of interstellar NH", *Astrophysical Journal Letters* **376**: L49–L52, Bibcode:1991ApJ...376L..49M, doi:10.1086/186100

[57] Wagenblast, R.; et al. (January 1993), "On the origin of NH in diffuse interstellar clouds", *Monthly Notices of the Royal Astronomical Society* **260** (2): 420–424, Bibcode:1993MNRAS.260..420W, doi:10.1093/mnras/260.2.420

[58] <Please add first missing authors to populate metadata.> (June 9, 2004), *Astronomers Detect Molecular Nitrogen Outside Solar System*, Space Daily, retrieved 2010-06-25

[59] Knauth, D. C; et al. (2004), "The interstellar $N_2$ abundance towards HD 124314 from far-ultraviolet observations", *Nature* **429** (6992): 636–638, Bibcode:2004Natur.429..636K, doi:10.1038/nature02614, PMID 15190346, retrieved 2010-06-25

[60] McGonagle, D.; et al. (1990), "Detection of nitric oxide in the dark cloud L134N", *Astrophysical Journal, Part 1* **359**: 121–124, Bibcode:1990ApJ...359..121M, doi:10.1086/169040

[61] Whiteoak, J. B.; Gardner, F. F. (1985), "Interstellar NaI absorption towards the stellar association ARA OB1", *Astronomical Society of Australia, Proceedings* (Sydney) **6** (2): 164–171, Bibcode:1985PASAu...6..164W

[62] Staff writers (March 27, 2007), *Elusive oxygen molecule finally discovered in interstellar space*, Physorg.com, retrieved 2007-04-02

[63] Ziurys, L. M. (1987), "Detection of interstellar PN - The first phosphorus-bearing species observed in molecular clouds", *Astrophysical Journal Letters* **321**: L81–L85, Bibcode:1987ApJ...321L..81Z, doi:10.1086/185010

[64] Tenenbaum, E. D.; Woolf, N. J.; Ziurys, L. M. (2007), "Identification of phosphorus monoxide ($X\ ^2\Pi_r$) in VY Canis Majoris: Detection of the first PO bond in space", *Astrophysical Journal Letters* **666**: L29–L32, Bibcode:2007ApJ...666L..29T, doi:10.1086/521361

[65] Yamamura, S. T.; Kawaguchi, K.; Ridgway, S. T. (2000), "Identification of SH v=1 Ro-vibrational Lines in R Andromedae", *The Astrophysical Journal* **528** (1): L33–L36, arXiv:astro-ph/9911080, Bibcode:2000ApJ...528L..33Y, doi:10.1086/312420, PMID 10587489

[66] Menten, K. M.; et al. (2011), "Submillimeter Absorption from $SH^+$, a New Widespread Interstellar Radical, $^{13}CH^+$ and HCl", *Astronomy & Astrophysics* **525**: A77, arXiv:1009.2825, Bibcode:2011A&A...525A..77M, doi:10.1051/0004-6361/201014363, retrieved 2010-12-03.

[67] Pascoli, G.; Comeau, M. (1995), "Silicon Carbide in Circumstellar Environment", *Astrophysics and Space Science* **226**: 149–163, Bibcode:1995Ap&SS.226..149P, doi:10.1007/BF00626907

[68] Kamiński, T.; et al. (2013), "Pure rotational spectra of TiO and $TiO_2$ in VY Canis Majoris", *Astronomy and Astrophysics* **551**: A113, arXiv:1301.4344, Bibcode:2013A&A...551A.113K, doi:10.1051/0004-6361/201220290

[69] Oka, Takeshi (2006), "Interstellar $H_3^+$", *Proceedings of the National Academy of Sciences* **103** (33): 12235–12242, Bibcode:2006PNAS..10312235O, doi:10.1073/pnas.0601242103, PMC 1567864, PMID 16894171, retrieved 2007-02-04

[70] Geballe, T. R.; Oka, T. (1996), "Detection of $H_3^+$ in Interstellar Space", *Nature* **384** (6607): 334–335, Bibcode:1996Natur.384..334G, doi:10.1038/384334a0, PMID 8934516

[71] Tenenbaum, E. D.; Ziurys, L. M. (2010), "Exotic Metal Molecules in Oxygen-rich Envelopes: Detection of AlOH (X1Σ+) in VY Canis Majoris", *Astrophysical*

*Journal* **712**: L93–L97, Bibcode:2010ApJ...712L..93T, doi:10.1088/2041-8205/712/1/L93

[72] Anderson, J. K.; et al. (2014), "Detection of CCN (X$^2\Pi_r$) in IRC+10216: Constraining Carbon-chain Chemistry", *Astrophysical Journal* **795**: L1, Bibcode:2014ApJ...795L...1A, doi:10.1088/2041-8205/795/1/L1

[73] Ohishi, Masatoshi, Masatoshi; et al. (1991), "Detection of a new carbon-chain molecule, CCO", *Astrophysical Journal Letters* **380**: L39–L42, Bibcode:1991ApJ...380L..39O, doi:10.1086/186168

[74] Irvine, William M.; et al. (1988), "Newly detected molecules in dense interstellar clouds", *Astrophysical Letters and Communications* **26**: 167–180, Bibcode:1988ApL&C..26..167I, PMID 11538461

[75] Halfen, D. T.; Clouthier, D. J.; Ziurys, L. M. (2008), "Detection of the CCP Radical (X $^2\Pi_r$) in IRC +10216: A New Interstellar Phosphorus-containing Species", *Astrophysical Journal* **677** (2): L101–L104, Bibcode:2008ApJ...677L.101H, doi:10.1086/588024

[76] Whittet, D. C. B.; Walker, H. J. (1991), "On the occurrence of carbon dioxide in interstellar grain mantles and ion-molecule chemistry", *Monthly Notices of the Royal Astronomical Society* **252**: 63–67, Bibcode:1991MNRAS.252...63W, doi:10.1093/mnras/252.1.63

[77] Zack, L. N.; Halfen, D. T.; Ziurys, L. M. (June 2011), "Detection of FeCN (X $^4\Delta_i$) in IRC+10216: A New Interstellar Molecule", *The Astrophysical Journal Letters* **733** (2): L36, Bibcode:2011ApJ...733L..36Z, doi:10.1088/2041-8205/733/2/L36

[78] Hollis, J. M.; Jewell, P. R.; Lovas, F. J. (1995), "Confirmation of interstellar methylene", *Astrophysical Journal, Part 1* **438**: 259–264, Bibcode:1995ApJ...438..259H, doi:10.1086/175070

[79] Lis, D. C.; et al. (2010-10-01), "Herschel/HIFI discovery of interstellar chloronium ($H_2Cl^+$)", *Astronomy & Astrophysics* **521**: L9, arXiv:1007.1461, Bibcode:2010A&A...521L...9L, doi:10.1051/0004-6361/201014959.

[80] *Europe's space telescope ISO finds water in distant places*, ESO, April 29, 1997, archived from the original on 2006-12-22, retrieved 2007-02-08

[81] Ossenkopf, V.; et al. (2010), "Detection of interstellar oxidaniumyl: Abundant $H2O^+$ towards the star-forming regions DR21, Sgr B2, and NGC6334", *Astronomy & Astrophysics* **518**: L111, arXiv:1005.2521, Bibcode:2010A&A...518L.111O, doi:10.1051/0004-6361/201014577.

[82] Parise, B.; Bergman, P.; Du, F. (2012), "Detection of the hydroperoxyl radical $HO_2$ toward ρ Ophiuchi A. Additional constraints on the water chemical network", *Astronomy & Astrophysics Letters* **541**: L11–L14, arXiv:1205.0361, Bibcode:2012A&A...541L..11P, doi:10.1051/0004-6361/201219379

[83] Snyder, L. E.; Buhl, D. (1971), "Observations of Radio Emission from Interstellar Hydrogen Cyanide", *Astrophysical Journal* **163**: L47–L52, Bibcode:1971ApJ...163L..47S, doi:10.1086/180664

[84] Schilke, P.; Benford, D. J.; Hunter, T. R.; Lis, D. C., Phillips, T. G.; Phillips, T. G. (2001), "A Line Survey of Orion-KL from 607 to 725 GHz", *Astrophysical Journal Supplement Series* **132** (2): 281–364, Bibcode:2001ApJS..132..281S, doi:10.1086/318951

[85] Schenewerk, M. S.; Snyder, L. E.; Hjalmarson, A. (1986), "Interstellar HCO - Detection of the missing 3 millimeter quartet", *Astrophysical Journal Letters* **303**: L71–L74, Bibcode:1986ApJ...303L..71S, doi:10.1086/184655

[86] Kawaguchi, Kentarou; et al. (1994), "Detection of a new molecular ion HC3NH(+) in TMC-1", *Astrophysical Journal* **420**: L95, Bibcode:1994ApJ...420L..95K, doi:10.1086/187171

[87] Agúndez, M.; Cernicharo, J.; Guélin, M. (2007), "Discovery of Phosphaethyne (HCP) in Space: Phosphorus Chemistry in Circumstellar Envelopes", *The Astrophysical Journal* **662** (2): L91, Bibcode:2007ApJ...662L..91A, doi:10.1086/519561, retrieved 2007-06-02

[88] Schilke, P.; Comito, C.; Thorwirth, S. (2003), "First Detection of Vibrationally Excited HNC in Space", *The Astrophysical Journal* **582** (2): L101–L104, Bibcode:2003ApJ...582L.101S, doi:10.1086/367628, retrieved 2008-09-14

[89] Hollis, J. M.; et al. (1991), "Interstellar HNO: Confirming the Identification - Atoms, ions and molecules: New results in spectral line astrophysics", *Atoms* (San Francisco: ASP) **16**: 407–412, Bibcode:1991ASPC...16..407H

[90] van Dishoeck, Ewine F.; et al. (1993), "Detection of the Interstellar NH 2 Radical", *Astrophysical Journal Letters* **416**: L83–L86, Bibcode:1993ApJ...416L..83V, doi:10.1086/187076

[91] Womack, M.; Ziurys, L. M.; Wyckoff, S. (1992), "A survey of $N_2H$(+) in dense clouds - Implications for interstellar nitrogen and ion-molecule chemistry", *Astrophysical Journal, Part 1* **387**: 417–429, Bibcode:1992ApJ...387..417W, doi:10.1086/171094

[92] Ziurys, L. M.; et al. (1994), "Detection of interstellar $N_2O$: A new molecule containing an N-O bond", *Astrophysical Journal Letters* **436**: L181–L184, Bibcode:1994ApJ...436L.181Z, doi:10.1086/187662

[93] Hollis, J. M.; Rhodes, P. J. (November 1, 1982), "Detection of interstellar sodium hydroxide in self-absorption toward the galactic center", *Astrophysical Journal Letters* **262**: L1–L5, Bibcode:1982ApJ...262L...1H, doi:10.1086/183900

[94] Goldsmith, P. F.; Linke, R. A. (1981), "A study of interstellar carbonyl sulfide", *Astrophysical Journal, Part 1* **245**: 482–494, Bibcode:1981ApJ...245..482G, doi:10.1086/158824

[95] Phillips, T. G.; Knapp, G. R. (1980), "Interstellar Ozone", *American Astronomical Society Bulletin* **12**: 440, Bibcode:1980BAAS...12..440P

[96] Johansson, L. E. B.; et al. (1984), "Spectral scan of Orion A and IRC+10216 from 72 to 91 GHz", *Astronomy and Astrophysics* **130** (2): 227–256, Bibcode:1984A&A...130..227J

[97] Cernicharo, José; et al. (2015), "Discovery of SiCSi in IRC+10216: a Missing Link Between Gas and Dust Carriers OF Si–C Bonds", *Astrophysical Journal Letters* **806**: L3, Bibcode:2015ApJ...806L...3C, doi:10.1088/2041-8205/806/1/L3

[98] Guélin, M.; et al. (2004), "Astronomical detection of the free radical SiCN", *Astronomy and Astrophysics* **363**: L9–L12, Bibcode:2000A&A...363L...9G

[99] Guélin, M.; et al. (2004), "Detection of the SiNC radical in IRC+10216", *Astronomy and Astrophysics* **426** (2): L49–L52, Bibcode:2004A&A...426L..49G, doi:10.1051/0004-6361:200400074

[100] Snyder, Lewis E.; et al. (1999), "Microwave Detection of Interstellar Formaldehyde", *Physical Review Letters* **61** (2): 77–115, Bibcode:1969PhRvL..22..679S, doi:10.1103/PhysRevLett.22.679

[101] Feuchtgruber, H.; et al. (June 2000), "Detection of Interstellar $CH_3$", *The Astrophysical Journal* **535** (2): L111–L114, arXiv:astro-ph/0005273, Bibcode:2000ApJ...535L.111F, doi:10.1086/312711, PMID 10835311

[102] Irvine, W. M.; et al. (1984), "Confirmation of the Existence of Two New Interstellar Molecules: $C_3H$ and $C_3O$", *Bulletin of the American Astronomical Society* **16**: 877, Bibcode:1984BAAS...16..877I

[103] Pety, J.; et al. (2012), "The IRAM-30 m line survey of the Horsehead PDR. II. First detection of the l-$C_3MH^+$ hydrocarbon cation", *Astronomy & Astrophysics* **548**: A68, arXiv:1210.8178, Bibcode:2012A&A...548A..68P, doi:10.1051/0004-6361/201220062

[104] Mangum, J. G.; Wootten, A. (1990), "Observations of the cyclic $C_3H$ radical in the interstellar medium", *Astronomy and Astrophysics* **239**: 319–325, Bibcode:1990A&A...239..319M

[105] Bell, M. B.; Matthews, H. E. (1995), "Detection of $C_3N$ in the spiral arm gas clouds in the direction of Cassiopeia A", *Astrophysical Journal, Part 1* **438**: 223–225, Bibcode:1995ApJ...438..223B, doi:10.1086/175066

[106] Thaddeus, P.; et al. (2008), "Laboratory and Astronomical Detection of the Negative Molecular Ion $C_3N^-$", *The Astrophysical Journal* **677** (2): 1132–1139, Bibcode:2008ApJ...677.1132T, doi:10.1086/528947

[107] Wootten, Alwyn; et al. (1991), "Detection of interstellar $H_3O(+)$ - A confirming line", *Astrophysical Journal Letters* **380**: L79–L83, Bibcode:1991ApJ...380L..79W, doi:10.1086/186178

[108] Ridgway, S. T.; et al. (1976), "Circumstellar acetylene in the infrared spectrum of IRC+10216", *Nature* **264**: 345, 346, Bibcode:1976Natur.264..345R, doi:10.1038/264345a0

[109] Ohishi, Masatoshi; et al. (1994), "Detection of a new interstellar molecule, $H_2CN$", *Astrophysical Journal Letters* **427**: L51–L54, Bibcode:1994ApJ...427L..51O, doi:10.1086/187362

[110] Minh, Y. C.; Irvine, W. M.; Brewer, M. K. (1991), "$H_2CS$ abundances and ortho-to-para ratios in interstellar clouds", *Astronomy and Astrophysics* **244**: 181–189, Bibcode:1991A&A...244..181M, PMID 11538284

[111] Guelin, M.; Cernicharo, J. (1991), "Astronomical detection of the HCCN radical - Toward a new family of carbon-chain molecules?", *Astronomy and Astrophysics* **244**: L21–L24, Bibcode:1991A&A...244L..21G

[112] Agúndez, M.; et al. (2015), "Discovery of interstellar ketenyl (HCCO), a surprisingly abundant radical", *Astronomy and Astrophysics* **577**: L5, Bibcode:2015A&A...577L...5A, doi:10.1051/0004-6361/201526317

[113] Minh, Y. C.; Irvine, W. M.; Ziurys, L. M. (1988), "Observations of interstellar HOCO(+) - Abundance enhancements toward the Galactic center", *Astrophysical Journal, Part 1* **334**: 175–181, Bibcode:1988ApJ...334..175M, doi:10.1086/166827

[114] Marcelino, Núria; et al. (2009), "Discovery of fulminic acid, HCNO, in dark clouds", *Astrophysical Journal* **690**: L27–L30, arXiv:0811.2679, Bibcode:2009ApJ...690L..27M, doi:10.1088/0004-637X/690/1/L27

[115] Brünken, S.; et al. (2010-07-22), "Interstellar HOCN in the Galactic center region", *Astronomy & Astrophysics* **516**: A109, arXiv:1005.2489, Bibcode:2010A&A...516A.109B, doi:10.1051/0004-6361/200912456

[116] Bergman; Parise; Liseau; Larsson; Olofsson; Menten; Güsten (2011), "Detection of interstellar hydrogen peroxide", *Astronomy & Astrophysics* **531**: L8, arXiv:1105.5799, Bibcode:2011A&A...531L...8B, doi:10.1051/0004-6361/201117170.

[117] Frerking, M. A.; Linke, R. A.; Thaddeus, P. (1979), "Interstellar isothiocyanic acid", *Astrophysical Journal Letters* **234**: L143–L145, Bibcode:1979ApJ...234L.143F, doi:10.1086/183126

[118] Nguyen-Q-Rieu; Graham, D.; Bujarrabal, V. (1984), "Ammonia and cyanotriacetylene in the envelopes of CRL 2688 and IRC + 10216", *Astronomy and Astrophysics* **138** (1): L5–L8, Bibcode:1984A&A...138L...5N

## 18.6. REFERENCES

[119] Halfen, D. T.; et al. (September 2009), "Detection of a New Interstellar Molecule: Thiocyanic Acid HSCN", *The Astrophysical Journal Letters* **702** (2): L124–L127, Bibcode:2009ApJ...702L.124H, doi:10.1088/0004-637X/702/2/L124

[120] Cabezas, C.; et al. (2013), "Laboratory and Astronomical Discovery of Hydromagnesium Isocyanide", *Astrophysical Journal* **775**: 133, arXiv:1309.0371, Bibcode:2013ApJ...775..133C, doi:10.1088/0004-637X/775/2/133

[121] Butterworth, Anna L.; et al. (2004), "Combined element (H and C) stable isotope ratios of methane in carbonaceous chondrites", *Monthly Notices of the Royal Astronomical Society* **347** (3): 807–812, Bibcode:2004MNRAS.347..807B, doi:10.1111/j.1365-2966.2004.07251.x

[122] http://www.astro.uni-koeln.de/site/vorhersagen/molecules/ism/Ammonium.html

[123] http://iopscience.iop.org/2041-8205/771/1/L10/

[124] Lacy, J. H.; et al. (1991), "Discovery of interstellar methane - Observations of gaseous and solid $CH_4$ absorption toward young stars in molecular clouds", *Astrophysical Journal* **376**: 556–560, Bibcode:1991ApJ...376..556L, doi:10.1086/170304

[125] Cernicharo, J.; Marcelino, N.; Roueff, E.; Gerin, M.; Jiménez-Escobar, A.; Muñoz Caro, G. M. (2012), "Discovery of the Methoxy Radical, $CH_3O$, toward B1: Dust Grain and Gas-phase Chemistry in Cold Dark Clouds", *The Astrophysical Journal Letters* **759** (2): L43–L46, Bibcode:2012ApJ...759L..43C, doi:10.1088/2041-8205/759/2/L43

[126] Finley, Dave (August 7, 2006), *Researchers Use NRAO Telescope to Study Formation Of Chemical Precursors to Life*, National Radio Astronomy Observatory, retrieved 2006-08-10

[127] Fossé, David; et al. (2001), "Molecular Carbon Chains and Rings in TMC-1", *Astrophysical Journal* **552** (1): 168–174, arXiv:astro-ph/0012405, Bibcode:2001ApJ...552..168F, doi:10.1086/320471, retrieved 2008-09-14

[128] Irvine, W. M.; et al. (1988), "Identification of the interstellar cyanomethyl radical ($CH_2CN$) in the molecular clouds TMC-1 and Sagittarius B2", *Astrophysical Journal Letters* **334**: L107–L111, Bibcode:1988ApJ...334L.107I, doi:10.1086/185323

[129] Dickens, J. E.; et al. (1997), "Hydrogenation of Interstellar Molecules: A Survey for Methylenimine ($CH_2NH$)", *Astrophysical Journal* **479** (1 Pt 1): 307–12, Bibcode:1997ApJ...479..307D, doi:10.1086/303884, PMID 11541227

[130] McGuire, B.A.; et al. (2012), "Interstellar Carbodiimide (HNCNH): A New Astronomical Detection from the GBT PRIMOS Survey via Maser Emission Features", *The Astrophysical Journal Letters* **758** (2): L33–L38, arXiv:1209.1590, Bibcode:2012ApJ...758L..33M, doi:10.1088/2041-8205/758/2/L33

[131] Ohishi, Masatoshi; et al. (1996), "Detection of a New Interstellar Molecular Ion, $H_2COH^+$ (Protonated Formaldehyde)", *Astrophysical Journal* **471** (1): L61–4, Bibcode:1996ApJ...471L..61O, doi:10.1086/310325, PMID 11541244

[132] Cernicharo, J.; et al. (2007), "Astronomical detection of $C_4H^-$, the second interstellar anion", *Astronomy and Astrophysics* **61** (2): L37–L40, Bibcode:2007A&A...467L..37C, doi:10.1051/0004-6361:20077415

[133] Liu, S.-Y.; Mehringer, D. M.; Snyder, L. E. (2001), "Observations of Formic Acid in Hot Molecular Cores", *Astrophysical Journal* **552** (2): 654–663, Bibcode:2001ApJ...552..654L, doi:10.1086/320563

[134] Walmsley, C. M.; Winnewisser, G.; Toelle, F. (1990), "Cyanoacetylene and cyanodiacetylene in interstellar clouds", *Astronomy and Astrophysics* **81** (1–2): 245–250, Bibcode:1980A&A....81..245W

[135] Kawaguchi, Kentarou; et al. (1992), "Detection of isocyanoacetylene HCCNC in TMC-1", *Astrophysical Journal* **386** (2): L51–L53, Bibcode:1992ApJ...386L..51K, doi:10.1086/186290

[136] Turner, B. E.; et al. (1975), "Microwave detection of interstellar cyanamide", *Astrophysical Journal* **201**: L149–L152, Bibcode:1975ApJ...201L.149T, doi:10.1086/181963

[137] Agúndez, M.; et al. (2015), "Probing non-polar interstellar molecules through their protonated form: Detection of protonated cyanogen (NCCNH+)", *Astronomy and Astrophysics* **579**: L10, arXiv:1506.07043, Bibcode:2015A&A...579L..10A, doi:10.1051/0004-6361/201526650

[138] Remijan, Anthony J.; et al. (2008), "Detection of interstellar cyanoformaldehyde (CNCHO)", *Astrophysical Journal* **675** (2): L85–L88, Bibcode:2008ApJ...675L..85R, doi:10.1086/533529

[139] Goldhaber, D. M.; Betz, A. L. (1984), "Silane in IRC +10216", *Astrophysical Journal Letters* **279**: –L55–L58, Bibcode:1984ApJ...279L..55G, doi:10.1086/184255

[140] Hollis, J. M.; et al. (2006), "Detection of Acetamide ($CH_3CONH_2$): The Largest Interstellar Molecule with a Peptide Bond", *Astrophysical Journal* **643** (1): L25–L28, Bibcode:2006ApJ...643L..25H, doi:10.1086/505110

[141] Hollis, J. M.; et al. (2006), "Cyclopropenone (c-$H_2C_3O$): A New Interstellar Ring Molecule", *Astrophysical Journal* **642** (2): 933–939, Bibcode:2006ApJ...642..933H, doi:10.1086/501121

[142] Zaleski, D. P.; et al. (2013), "Detection of E-Cyanomethanimine toward Sagittarius B2(N) in the Green Bank Telescope PRIMOS Survey", *Astrophysical Journal Letters* **765**: L109, arXiv:1302.0909, Bibcode:2013ApJ...765L..10Z, doi:10.1088/2041-8205/765/1/L10

[143] Betz, A. L. (1981), "Ethylene in IRC +10216", *Astrophysical Journal Letters* **244**: –L105, Bibcode:1981ApJ...244L.103B, doi:10.1086/183490

[144] Remijan, Anthony J.; et al. (2005), "Interstellar Isomers: The Importance of Bonding Energy Differences", *Astrophysical Journal* **632** (1): 333–339, arXiv:astro-ph/0506502, Bibcode:2005ApJ...632..333R, doi:10.1086/432908

[145] "Complex Organic Molecules Discovered in Infant Star System". NRAO (Astrobiology Web). 8 April 2015. Retrieved 2015-04-09.

[146] First Detection of Methyl Alcohol in a Planet-forming Disc. 15 June 2016.

[147] Lambert, D. L.; Sheffer, Y.; Federman, S. R. (1979), "Interstellar methyl mercaptan", *Astrophysical Journal Letters* **234**: L139–L142, Bibcode:1979ApJ...234L.139L, doi:10.1086/183125

[148] Cernicharo, José; et al. (1997), "Infrared Space Observatory's Discovery of $C_4H_2$, $C_6H_2$, and Benzene in CRL 618", *Astrophysical Journal Letters* **546** (2): L123–L126, Bibcode:2001ApJ...546L.123C, doi:10.1086/318871

[149] Guelin, M.; Neininger, N.; Cernicharo, J. (1998), "Astronomical detection of the cyanobutadiynyl radical C_5N", *Astronomy and Astrophysics* **335**: L1–L4, arXiv:astro-ph/9805105, Bibcode:1998A&A...335L...1G

[150] Irvine, W. M.; et al. (1988), "A new interstellar polyatomic molecule - Detection of propynal in the cold cloud TMC-1", *Astrophysical Journal Letters* **335**: L89–L93, Bibcode:1988ApJ...335L..89I, doi:10.1086/185346

[151] Agúndez, M.; et al. (2014), "New molecules in IRC +10216: confirmation of $C_5S$ and tentative identification of MgCCH, NCCP, and $SiH_3CN$", *Astronomy and Astrophysics* **570**: A45, Bibcode:2014A&A...570A..45A, doi:10.1051/0004-6361/201424542

[152] *Scientists Toast the Discovery of Vinyl Alcohol in Interstellar Space*, National Radio Astronomy Observatory, October 1, 2001, retrieved 2006-12-20

[153] Dickens, J. E.; et al. (1997), "Detection of Interstellar Ethylene Oxide (c-C2H4O)", *The Astrophysical Journal* **489** (2): 753–757, Bibcode:1997ApJ...489..753D, doi:10.1086/304821, PMID 11541726

[154] Kaifu, N.; Takagi, K.; Kojima, T. (1975), "Excitation of interstellar methylamine", *Astrophysical Journal* **198**: L85–L88, Bibcode:1975ApJ...198L..85K, doi:10.1086/181818

[155] McCarthy, M. C.; et al. (2006), "Laboratory and Astronomical Identification of the Negative Molecular Ion $C_6H^-$", *Astrophysical Journal* **652** (2): L141–L144, Bibcode:2006ApJ...652L.141M, doi:10.1086/510238

[156] Halfven, D. T.; et al. (2015), "INTERSTELLAR DETECTION OF METHYL ISOCYANATE $CH_3NCO$ IN Sgr B2(N): A LINK FROM MOLECULAR CLOUDS TO COMETS", *Astrophysical Journal* **812**: L5, arXiv:1509.09305, Bibcode:2015ApJ...812L...5H, doi:10.1088/2041-8205/812/1/L5

[157] Mehringer, David M.; et al. (1997), "Detection and Confirmation of Interstellar Acetic Acid", *Astrophysical Journal Letters* **480**: L71, Bibcode:1997ApJ...480L..71M, doi:10.1086/310612

[158] Lovas, F. J.; et al. (2006), "Hyperfine Structure Identification of Interstellar Cyanoallene toward TMC-1", *Astrophysical Journal Letters* **637** (1): L37–L40, Bibcode:2006ApJ...637L..37L, doi:10.1086/500431

[159] Sincell, Mark (June 27, 2006), "The Sweet Signal of Sugar in Space", *Science* (American Association for the Advancement of Science), retrieved 2016-01-14

[160] Loomis, R. A.; et al. (2013), "The Detection of Interstellar Ethanimine $CH_3CHNH$) from Observations Taken during the GBT PRIMOS Survey", *Astrophysical Journal Letters* **765**: L9, arXiv:1302.1121, Bibcode:2013ApJ...765L...9L, doi:10.1088/2041-8205/765/1/L9

[161] Guelin, M.; et al. (1997), "Detection of a new linear carbon chain radical: $C_7H$", *Astronomy and Astrophysics* **317**: L37–L40, Bibcode:1997A&A...317L...1G

[162] Belloche, A.; et al. (2008), "Detection of amino acetonitrile in Sgr B2(N)", *Astronomy & Astrophysics* **482**: 179–196, arXiv:0801.3219, Bibcode:2008A&A...482..179B, doi:10.1051/0004-6361:20079203

[163] Remijan, Anthony J.; et al. (2014), "OBSERVATIONAL RESULTS OF A MULTI-TELESCOPE CAMPAIGN IN SEARCH OF INTERSTELLAR UREA [(NH2)$_2$CO]", *Astrophysical Journal* **783** (2): 77, arXiv:1401.4483, Bibcode:2014ApJ...783...77R, doi:10.1088/0004-637X/783/2/77

[164] Remijan, Anthony J.; et al. (2006), "Methyltriacetylene ($CH_3C_6H$) toward TMC-1: The Largest Detected Symmetric Top", *Astrophysical Journal* **643** (1): L37–L40, Bibcode:2006ApJ...643L..37R, doi:10.1086/504918

[165] Snyder, L. E.; et al. (1974), "Radio Detection of Interstellar Dimethyl Ether", *Astrophysical Journal* **191**: L79–L82, Bibcode:1974ApJ...191L..79S, doi:10.1086/181554

[166] Zuckerman, B.; et al. (1975), "Detection of interstellar trans-ethyl alcohol", *Astrophysical Journal* **196** (2): L99–L102, Bibcode:1975ApJ...196L..99Z, doi:10.1086/181753

## 18.6. REFERENCES

[167] Cernicharo, J.; Guelin, M. (1996), "Discovery of the $C_8H$ radical", *Astronomy and Astrophysics* **309**: L26–L30, Bibcode:1996A&A...309L..27C

[168] Brünken, S.; et al. (2007), "Detection of the Carbon Chain Negative Ion $C_8H^-$ in TMC-1", *Astrophysical Journal* **664** (1): L43–L46, Bibcode:2007ApJ...664L..43B, doi:10.1086/520703

[169] Remijan, Anthony J.; et al. (2007), "Detection of $C_8H^-$ and Comparison with $C_8H$ toward IRC +10 216", *Astrophysical Journal* **664** (1): L47–L50, Bibcode:2007ApJ...664L..47R, doi:10.1086/520704

[170] Bell, M. B.; et al. (1997), "Detection of $HC_{11}N$ in the Cold Dust Cloud TMC-1", *Astrophysical Journal Letters* **483** (1): L61–L64, arXiv:astro-ph/9704233, Bibcode:1997ApJ...483L..61B, doi:10.1086/310732

[171] Kroto, H. W.; et al. (1978), "The detection of cyanohexatriyne, $H(C\equiv C)_3CN$, in Heiles's cloud 2", *The Astrophysical Journal* **219**: L133–L137, Bibcode:1978ApJ...219L.133K, doi:10.1086/182623

[172] Marcelino, N.; et al. (2007), "Discovery of Interstellar Propylene ($CH_2CHCH_3$): Missing Links in Interstellar Gas-Phase Chemistry", *Astrophysical Journal* **665** (2): L127–L130, arXiv:0707.1308, Bibcode:2007ApJ...665L.127M, doi:10.1086/521398

[173] Kolesniková, L.; et al. (2014), "Spectroscopic Characterization and Detection of Ethyl Mercaptan in Orion", *Astrophysical Journal Letters* **784** (1): L7, arXiv:1401.7810, Bibcode:2014ApJ...784L...7K, doi:10.1088/2041-8205/784/1/L7

[174] Snyder, Lewis E.; et al. (2002), "Confirmation of Interstellar Acetone", *The Astrophysical Journal* **578** (1): 245–255, Bibcode:2002ApJ...578..245S, doi:10.1086/342273

[175] Hollis, J. M.; et al. (2002), "Interstellar Antifreeze: Ethylene Glycol", *Astrophysical Journal* **571** (1): L59–L62, Bibcode:2002ApJ...571L..59H, doi:10.1086/341148, retrieved 2010-07-18

[176] Hollis, J. M. (2005), "Complex Molecules and the GBT: Is Isomerism the Key?" (PDF), *Complex Molecules and the GBT: Is Isomerism the Key?*, Proceedings of the IAU Symposium 231, Astrochemistry throughout the Universe, Asilomar, CA, pp. 119–127

[177] Discovery of the Interstellar Chiral Molecule Propylene Oxide (CH3CHCH2O), 27 June 2016.

[178] Belloche, A.; et al. (May 2009), "Increased complexity in interstellar chemistry: Detection and chemical modeling of ethyl formate and n-propyl cyanide in Sgr B2(N)", *Astronomy and Astrophysics* **499** (1): 215–232, arXiv:0902.4694, Bibcode:2009A&A...499..215B, doi:10.1051/0004-6361/200811550

[179] Tercero, B.; et al. (2013), "Discovery of Methyl Acetate and Gauche Ethyl Formate in Orion", *Astrophysical Journal Letters* **770**: L13, arXiv:1305.1135, Bibcode:2013ApJ...770L..13T, doi:10.1088/2041-8205/770/1/L13

[180] Eyre, Michael (26 September 2014). "Complex organic molecule found in interstellar space". *BBC News*. Retrieved 2014-09-26.

[181] Belloche, Arnaud; Garrod, Robin T.; Müller, Holger S. P.; Menten, Karl M. (26 September 2014). "Detection of a branched alkyl molecule in the interstellar medium: iso-propyl cyanide". *Science* **345** (6204): 1584–1587. arXiv:1410.2607. Bibcode:2014Sci...345.1584B. doi:10.1126/science.1256678. Retrieved 2014-09-26.

[182] Cami, Jan; et al. (July 22, 2010), "Detection of $C_{60}$ and $C_{70}$ in a Young Planetary Nebula", *Science* **329** (5996): 1180–2, Bibcode:2010Sci...329.1180C, doi:10.1126/science.1192035, PMID 20651118

[183] Foing, B. H.; Ehrenfreund, P. (1994), "Detection of two interstellar absorption bands coincident with spectral features of C60+", *Nature* **369** (6478): 296–298, Bibcode:1994Natur.369..296F, doi:10.1038/369296a0.

[184] Berné, Olivier; Mulas, Giacomo; Joblin, Christine (2013), "Interstellar $C_{60}^+$", *Astronomy & Astrophysics* **550**: L4, arXiv:1211.7252, Bibcode:2013A&A...550L...4B, doi:10.1051/0004-6361/201220730

[185] Lacour, S.; et al. (2005), "Deuterated molecular hydrogen in the Galactic ISM. New observations along seven translucent sightlines", *Astronomy and Astrophysics* **430** (3): 967–977, arXiv:astro-ph/0410033, Bibcode:2005A&A...430..967L, doi:10.1051/0004-6361:20041589

[186] Ceccarelli, Cecilia (2002), "Millimeter and infrared observations of deuterated molecules", *Planetary and Space Science* **50** (12–13): 1267–1273, Bibcode:2002P&SS...50.1267C, doi:10.1016/S0032-0633(02)00093-4

[187] Green, Sheldon (1989), "Collisional excitation of interstellar molecules - Deuterated water, HDO", *Astrophysical Journal Supplement Series* **70**: 813–831, Bibcode:1989ApJS...70..813G, doi:10.1086/191358

[188] Butner, H. M.; et al. (2007), "Discovery of interstellar heavy water", *Astrophysical Journal* **659** (2): L137–L140, Bibcode:2007ApJ...659L.137B, doi:10.1086/517883

[189] Turner, B. E.; Zuckerman, B. (1978), "Observations of strongly deuterated molecules - Implications for interstellar chemistry", *Astrophysical Journal Letters* **225**: L75–L79, Bibcode:1978ApJ...225L..75T, doi:10.1086/182797

[190] Lis, D. C.; et al. (2002), "Detection of Triply Deuterated Ammonia in the Barnard 1 Cloud", *Astrophysical Journal* **571** (1): L55–L58, Bibcode:2002ApJ...571L..55L, doi:10.1086/341132.

[191] Hatchell, J. (2003), "High NH$_2$D/NH$_3$ ratios in protostellar cores", *Astronomy and Astrophysics* **403** (2): L25–L28, arXiv:astro-ph/0302564, Bibcode:2003A&A...403L..25H, doi:10.1051/0004-6361:20030297.

[192] Turner, B. E. (1990), "Detection of doubly deuterated interstellar formaldehyde (D2CO) - an indicator of active grain surface chemistry", *Astrophysical Journal Letters* **362**: L29–L33, Bibcode:1990ApJ...362L..29T, doi:10.1086/185840.

[193] Coutens, A.; et al. (9 May 2016). "The ALMA-PILS survey: First detections of deuterated formamide and deuterated isocyanic acid in the interstellar medium". arXiv:1605.02562.

[194] Cernicharo, J.; et al. (2013), "Detection of the Ammonium ion in space", *Astrophysical Journal Letters* **771**: L10, arXiv:1306.3364, Bibcode:2013ApJ...771L..10C, doi:10.1088/2041-8205/771/1/L10

[195] Doménech, J. L.; et al. (2013), "Improved Deterministic of the $1_0-0_0$ Rotational Frequency of NH$_3$D$^+$ from the High-Resolution Spectrum of the $\nu_4$ Infrared Band", *Astrophysical Journal Letters* **771**: L11, arXiv:1306.3792, Bibcode:2013ApJ...771L..11D, doi:10.1088/2041-8205/771/1/L10

[196] Gerin, M.; et al. (1992), "Interstellar detection of deuterated methyl acetylene", *Astronomy and Astrophysics* **253** (2): L29–L32, Bibcode:1992A&A...253L..29G.

[197] Markwick, A. J.; Charnley, S. B.; Butner, H. M.; Millar, T. J. (2005), "Interstellar CH3CCD", *The Astrophysical Journal* **627** (2): L117–L120, Bibcode:2005ApJ...627L.117M, doi:10.1086/432415.

[198] Agúndez, M.; et al. (2008-06-04), "Tentative detection of phosphine in IRC +10216", *Astronomy & Astrophysics* **485** (3): L33, arXiv:0805.4297, Bibcode:2008A&A...485L..33A, doi:10.1051/0004-6361:200810193

[199] Gupta, H.; et al. (2013), "Laboratory Measurements and Tentative Astronomical Identification of H$_2$NCO$^+$", *Astrophysical Journal Letters* **778**: L1, Bibcode:2013ApJ...778L...1G, doi:10.1088/2041-8205/778/1/L1

[200] Snyder, L. E.; et al. (2005), "A Rigorous Attempt to Verify Interstellar Glycine", *Astrophysical Journal* **619** (2): 914–930, arXiv:astro-ph/0410335, Bibcode:2005ApJ...619..914S, doi:10.1086/426677.

[201] Kuan, Y. J.; et al. (2003), "Interstellar Glycine", *Astrophysical Journal* **593** (2): 848–867, Bibcode:2003ApJ...593..848K, doi:10.1086/375637.

[202] Widicus Weaver, S. L.; Blake, G. A. (2005), "1,3-Dihydroxyacetone in Sagittarius B2(N-LMH): The First Interstellar Ketose", *Astrophysical Journal Letters* **624** (1): L33–L36, Bibcode:2005ApJ...624L..33W, doi:10.1086/430407

[203] Fuchs, G. W.; et al. (2005), "Trans-Ethyl Methyl Ether in Space: A new Look at a Complex Molecule in Selected Hot Core Regions", *Astronomy & Astrophysics* **444** (2): 521–530, arXiv:astro-ph/0508395, Bibcode:2005A&A...444..521F, doi:10.1051/0004-6361:20053599, retrieved 2010-07-18

[204] Iglesias-Groth, S.; et al. (2008-09-20), "Evidence for the Naphthalene Cation in a Region of the Interstellar Medium with Anomalous Microwave Emission", *The Astrophysical Journal Letters* **685**: L55–L58, arXiv:0809.0778, Bibcode:2008ApJ...685L..55I, doi:10.1086/592349 - This spectral assignment has not been independently confirmed, and is described by the authors as "tentative" (page L58).

[205] García-Hernández, D. A.; et al. (2011), "The Formation of Fullerenes: Clues from New C$_{60}$, C$_{70}$, and (Possible) Planar C$_{24}$ Detections in Magellanic Cloud Planetary Nebulae", *Astrophysical Journal Letters* **737** (2): L30, arXiv:1107.2595, Bibcode:2011ApJ...737L..30G, doi:10.1088/2041-8205/737/2/L30, retrieved 2011-08-12.

[206] Iglesias-Groth, S.; et al. (May 2010), "A search for interstellar anthracene toward the Perseus anomalous microwave emission region", *Monthly Notices of the Royal Astronomical Society* **407** (4): 2157–2165, arXiv:1005.4388, Bibcode:2010MNRAS.407.2157I, doi:10.1111/j.1365-2966.2010.17075.x

## 18.7 External links

- Woon, David E. (October 1, 2010). "Interstellar and Circumstellar Molecules". Retrieved 2010-10-04.

- "Molecules in Space". Universität zu Köln. August 2010. Retrieved 2010-10-04.

- Dworkin, Jason P. (February 1, 2007). "Interstellar Molecules". NASA's Cosmic Ice Lab. Retrieved 2010-12-23.

- Wootten, Al (November 2005). "The 129 reported interstellar and circumstellar molecules". National Radio Astronomy Observatory. Retrieved 2007-02-13.

- Lovas, F. J.; Dragoset, R. A. (February 2004). "NIST Recommended Rest Frequencies for Observed Interstellar Molecular Microwave Transitions, 2002 Revision". National Institute of Standards and Technology. Retrieved 2007-02-13.

# Chapter 19

# Gard model

In evolutionary biology, the **GARD (Graded Autocatalysis Replication Domain) model** is a general kinetic model for homeostatic-growth and fission of compositional-assemblies, with specific application towards lipids.[1]

In the context of abiogenesis, the lipid-world [2] suggests assemblies of simple molecules, such as lipids, can store and propagate information, thus undergo evolution.

These 'compositional assemblies' have been suggested to play a role in the origin of life. The idea is the information being transferred throughout the generations is *compositional information* – the different types and quantities of molecules within an assembly. This is different from the information encoded in RNA or DNA, which is the specific sequence of bases in such molecule. Thus, the model is viewed as an alternative or an ancestor to the RNA world hypothesis.

## 19.1 The GARD model

The composition vector of an assembly is written as: $v = n_1 \cdots n_{N_G}$. Where $n_1 \cdots n_{N_G}$ are the molecular counts of lipid type $i$ within the assembly, and NG is how many different lipid types exist (*repertiore size*).

The change in the count of molecule type $i$ is described by:

$$\frac{dn_i}{dt} = (k_f \rho_i N - k_b n_i)\left(1 + \sum_{j=1}^{N_G} \beta_{ij}\frac{n_j}{N}\right)$$

$k_f$ and $k_b$ are the basel forward (joining) and backward (leaving) rate constants, $\beta{ij}$ is a non-negative rate enhancement exerted by molecule type $j$ within the assembly on type $i$ from the environment, and $\rho$ is the environmental concentration of each molecule type.

The assembly current size is $N = \sum_{i=1}^{N_G} n_i$. The system is kept away from equilibrium by imposing a fission action once the assembly reaches a maximal size, Nmax, usually in the order of NG. This splitting action produces two progeny of same size, and one of which is grown again.

The model is subjected to a Monte Carlo algorithm based simulations, using Gillespie algorithm.

## 19.2 Selection in GARD

In 2010, Eors Szathmary and collaborators have chosen GARD as an archetypal metabolism-first realization.[3] They have introduced selection coefficient into the model, which increase or decrease the growth rate of assemblies, depending on how similar or dis-similar they are to a given target. They found that the ranking of the assemblies are un-affected by the selection pressure, and concluded that GARD does not exhibit Darwinian evolution.

In 2012 it was shown that this criticism is errorneous and was refuted.[4] Two major drawbacks of the 2010 paper were: (1) they have focused on a general assembly and not on a composome or compotype (faithfully replicating and quasispecies, respectively); (2) they have performed only a single, random, simulation to test the selectability.

## 19.3 GARD and Quasispecies

The quasispecies model describes a population of replicators that replicate with relatively high mutations. Due to mutations and back mutations the population eventually centres around a master-replicator (master sequence). GARD's populations were shown to form a quasispecies around a master-compotype and to exhibit an error threshold (evolution), similarly to classical quasispecis such as RNA viruses. [5]

## 19.4 See also

- Abiogenesis
- Protocell

## 19.5 References

[1] Segre, D.; Ben-Eli, D.; Lancet, D. (2001). "Compositional genomes: Prebiotic information transfer in mutually catalytic noncovalent assemblies". *Proc Natl Acad Sci U S A* (Elsevier): 219–230. doi:10.1073/pnas.97.8.4112.

[2] Segre, D.; Ben-Eli, D.; Deamer, D.; Lancet, D. (2001). "The lipid world". *Orig Life Evol Biosph* **31**: 119–145. doi:10.1023/A:1006746807104. PMID 11296516.

[3] Vasas, V.; Szathmary, E.; Santos, M. (2010). "Lack of evolvability in self-sustaining autocatalytic networks constraints metabolism-first scenarios for the origin of life". *Proc Natl Acad Sci U S A* **107** (4): 1470–1475. doi:10.1073/pnas.0912628107.

[4] Markovitch, O.; Lancet, D. (2012). "Excess Mutual Catalysis Is Required for Effective Evolvability". *Artificial Life* **18** (3): 243–266. doi:10.1162/artl_a_00064. PMID 22662913.

[5] Gross, R.; Fouxon, I.; Lancet, D.; Markovitch, O. (2014). "Quasispecies in population of compositional assemblies". *BMC Evolutionary Biology* **14**: 2623. doi:10.1186/s12862-014-0265-1.

## 19.6 External links

- GARD10 MATLAB code (see Markovitch and Lancet, 2012): https://github.com/ModelingOriginsofLife/GARD

- Doron Lancet homepage at Weizmann Institute of Science, who is the inventor of GARD.

- Origin of life (OOL) at the Weizmann Institute.

# Chapter 20

# PAH world hypothesis

The **PAH world hypothesis** is a speculative hypothesis that proposes that polycyclic aromatic hydrocarbons (PAH), known to be abundant in the universe,[1][2][3] including in comets,[4] and, as well, assumed to be abundant in the primordial soup of the early Earth, played a major role in the origin of life by mediating the synthesis of RNA molecules, leading into the RNA world. However, as yet, the hypothesis is untested.[5]

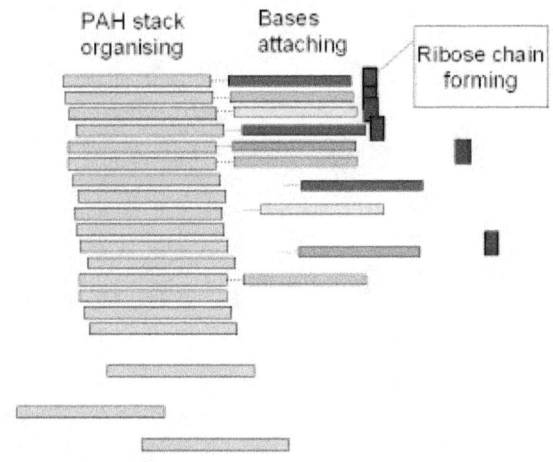

*A PAH stack assembling*

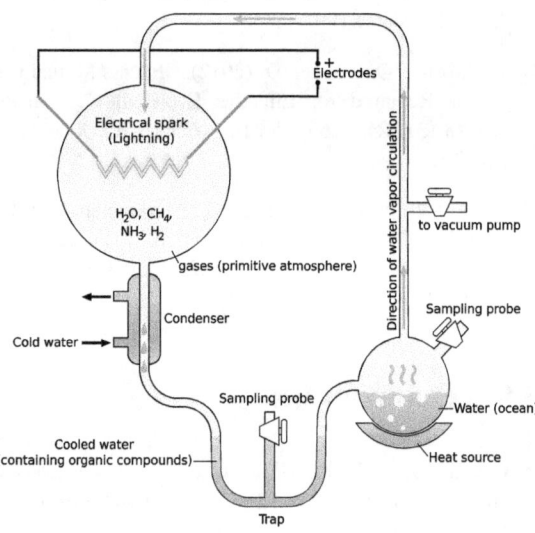

*The Miller–Urey experiment showed that organic compounds can be readily produced under the presumed conditions of the early Earth*

## 20.1 Background

The Miller–Urey experiment in 1952, and others since, demonstrated the synthesis of organic compounds, such as amino acids, formaldehyde and sugars, from the original inorganic precursors presumed to have been present in the primordial soup. This experiment inspired many others. In 1961, Joan Oró found that the nucleotide base adenine could be made from hydrogen cyanide (HCN) and ammonia in a water solution.[6] Experiments conducted later showed that the other RNA and DNA nucleobases could be obtained through simulated prebiotic chemistry with a reducing atmosphere.[7]

The RNA world hypothesis shows how RNA can become its own catalyst (a ribozyme). In between there are some missing steps such as how the first RNA molecules could be formed. The PAH world hypothesis was proposed by Simon Nicholas Platts in May 2004 to try to fill in this missing step.[8] A more thoroughly elaborated idea has been published by Ehrenfreund *et al.*.[9]

## 20.2 Polycyclic aromatic hydrocarbons

Main article: Polycyclic aromatic hydrocarbon

Polycyclic aromatic hydrocarbons are the most common

and abundant of the known polyatomic molecules in the visible universe, and are considered a likely constituent of the primordial sea.[1][2][3] PAHs, along with fullerenes (or "buckyballs"), have been recently detected in nebulae.[10] (Fullerenes are also implicated in the origin of life; according to astronomer Letizia Stanghellini, "It's possible that buckyballs from outer space provided seeds for life on Earth."[11]) In September 2012, NASA scientists reported that PAHs, subjected to interstellar medium (ISM) conditions, are transformed, through hydrogenation, oxygenation and hydroxylation, to more complex organics — "a step along the path toward amino acids and nucleotides, the raw materials of proteins and DNA, respectively".[12][13] Further, as a result of these transformations, the PAHs lose their spectroscopic signature which could be one of the reasons "for the lack of PAH detection in interstellar ice grains, particularly the outer regions of cold, dense clouds or the upper molecular layers of protoplanetary disks."[12][13]

On June 6, 2013, scientists at the IAA-CSIC reported the detection of polycyclic aromatic hydrocarbons in the upper atmosphere of Titan, the largest moon of the planet Saturn.[14]

PAHs are not normally very soluble in sea water, but when subject to ionizing radiation such as solar UV light, the outer hydrogen atoms can be stripped off and replaced with a hydroxyl group, rendering the PAHs far more soluble in water.

These modified PAHs are amphiphilic, which means that they have parts that are both hydrophilic and hydrophobic. When in solution, they assemble in discotic mesogenic (liquid crystal) stacks which, like lipids, tend to organize with their hydrophobic parts protected.

On February 21, 2014, NASA announced a greatly upgraded database[15] for tracking polycyclic aromatic hydrocarbons (PAHs) in the universe. More than 20% of the carbon in the universe may be associated with PAHs,[15][16] possible starting materials for the formation of life. PAHs seem to have been formed as early as a couple of billion years after the Big Bang, are abundant in the universe,[1][2][3] and are associated with new stars and exoplanets.[15]

## 20.3 Attachment of nucleobases to PAH scaffolding

In the self-ordering PAH stack, the separation between adjacent rings is 0.34 nm. This is the same separation found between adjacent nucleotides of RNA and DNA. Smaller molecules will naturally attach themselves to the PAH rings. However PAH rings, while forming, tend to swivel around on one another, which will tend to dislodge attached compounds that would collide with those attached to those above and below. Therefore it encourages preferential attachment of flat molecules such as pyrimidine and purine nucleobases, the key constituents (and information carriers) of RNA and DNA. These bases are similarly amphiphilic and so also tend to line up in similar stacks.

## 20.4 Attachment of oligomeric backbone

According to the hypothesis, once the nucleobases are attached (via hydrogen bonds) to the PAH scaffolding, the inter-base distance would select for "linker" molecules of a specific size, such as small formaldehyde (methanal) oligomers, also taken from the prebiotic "soup", which will bind (via covalent bonds) to the nucleobases as well as each other to add a flexible structural backbone.[5][8]

## 20.5 Detachment of the RNA-like strands

A subsequent transient drop in the ambient pH (increase in acidity), for example as a result of a volcanic discharge of acidic gases such as sulfur dioxide or carbon dioxide, would allow the bases to break off from their PAH scaffolding, forming RNA-like molecules (with the formaldehyde backbone instead of the ribose-phosphate backbone used by "modern" RNA, but the same 0.34 nm pitch).[5]

## 20.6 Formation of ribozyme-like structures

The hypothesis further speculates that once long RNA-like single strands are detached from the PAH stacks, and after ambient pH levels became less acidic, they would tend to fold back on themselves, with complementary sequences of nucleobases preferentially seeking out each other and forming hydrogen bonds, creating stable, at least partially double-stranded RNA-like structures, similar to ribozymes. The formaldehyde oligomers would eventually be replaced with more stable ribose-phosphate molecules for the backbone material, resulting in a starting milestone for the RNA world hypothesis, which speculates about further evolutionary developments from that point.[5][8][17]

## 20.7 See also

- Astrochemistry
- Atomic and molecular astrophysics
- Cosmochemistry
- Extragalactic astronomy
- Extraterrestrial materials
- History of the Earth
- Intergalactic space
- Intergalactic medium
- Intergalactic star
- Interstellar medium
- Iron-sulfur world theory
- List of interstellar and circumstellar molecules
- Thermosynthesis
- Tholin
- Other possible RNA precursors:
  - Glycol nucleic acid (GNA)
  - Peptide nucleic acid (PNA)
  - Threose nucleic acid (TNA)

## 20.8 References

[1] Carey, Bjorn (October 18, 2005). "Life's Building Blocks 'Abundant in Space'". *Space.com*. Retrieved March 3, 2014.

[2] Hudgins, Douglas M.; Bauschlicher,Jr, Charles W.; Allamandola, L. J. (October 10, 2005). "Variations in the Peak Position of the 6.2 µm Interstellar Emission Feature: A Tracer of N in the Interstellar Polycyclic Aromatic Hydrocarbon Population". *Astrophysical Journal* **632** (1): 316–332. Bibcode:2005ApJ...632..316H. doi:10.1086/432495. Retrieved March 3, 2014.

[3] Allamandola, Louis et al. (April 13, 2011). "Cosmic Distribution of Chemical Complexity". *NASA*. Retrieved March 3, 2014.

[4] Clavin, Whitney (February 10, 2015). "Why Comets Are Like Deep Fried Ice Cream". *NASA*. Retrieved February 10, 2015.

[5] Platts, Simon Nicholas, "The PAH World - Discotic polynuclear aromatic compounds as a mesophase scaffolding at the origin of life"

[6] Oró J, Kimball AP (August 1961). "Synthesis of purines under possible primitive earth conditions. I. Adenine from hydrogen cyanide". *Archives of Biochemistry and Biophysics* **94**: 217–27. doi:10.1016/0003-9861(61)90033-9. PMID 13731263.

[7] Oró J (1967). Fox SW, ed. *Origins of Prebiological Systems and of Their Molecular Matrices*. New York Academic Press. p. 137.

[8] "Prebiotic Molecular Selection and Organization", NASA's Astrobiology website

[9] Ehrenfreund, P; Rasmussen, S; Cleaves, J; Chen, L (2006). "Experimentally tracing the key steps in the origin of life: The aromatic world". *Astrobiology* **6** (3): 490–520. Bibcode:2006AsBio...6..490E. doi:10.1089/ast.2006.6.490.

[10] García-Hernández, D. A.; Manchado, A.; García-Lario, P.; Stanghellini, L.; Villaver, E.; Shaw, R. A.; Szczerba, R.; Perea-Calderón, J. V. (2010-10-28). "Formation Of Fullerenes In H-Containing Planetary Nebulae". *The Astrophysical Journal Letters* **724**: L39–L43. arXiv:1009.4357. Bibcode:2010ApJ...724L..39G. doi:10.1088/2041-8205/724/1/L39.

[11] Atkinson, Nancy (October 27, 2010). "Buckyballs Could Be Plentiful in the Universe". Universe Today. Retrieved October 28, 2010.

[12] Staff (September 20, 2012). "NASA Cooks Up Icy Organics to Mimic Life's Origins". Space.com. Retrieved September 22, 2012.

[13] Gudipati, Murthy S.; Yang, Rui (September 1, 2012). "In-Situ Probing Of Radiation-Induced Processing Of Organics In Astrophysical Ice Analogs—Novel Laser Desorption Laser Ionization Time-Of-Flight Mass Spectroscopic Studies". *The Astrophysical Journal Letters* **756** (1): L24. Bibcode:2012ApJ...756L..24G. doi:10.1088/2041-8205/756/1/L24. Retrieved September 22, 2012.

[14] López-Puertas, Manuel (June 6, 2013). "PAH's in Titan's Upper Atmosphere". *CSIC*. Retrieved June 6, 2013.

[15] Hoover, Rachel (February 21, 2014). "Need to Track Organic Nano-Particles Across the Universe? NASA's Got an App for That". *NASA*. Retrieved February 22, 2014.

[16] Hoover, Rachel (February 24, 2014). "Online Database Tracks Organic Nano-Particles Across the Universe". *Sci Tech Daily*. Retrieved March 10, 2015.

[17] Lincoln, Tracey A.; Joyce, Gerald F. (January 8, 2009). "Self-Sustained Replication of an RNA Enzyme". *Science* (New York: American Association for the Advancement of Science) **323** (5918): 1229–32. Bibcode:2009Sci...323.1229L. doi:10.1126/science.1167856. PMC 2652413. PMID 19131595. Retrieved 2009-01-13. Lay summary – *Medical News Today* (January 12, 2009).

## 20.9 External links

- The 'PAH World'

- Astrobiology magazine *Aromatic World* An interview with Pascale Ehrenfreund on PAH origin of life. - Accessed June 2006

- Life's ingredients found in early universe New Scientist Magazine 14:49 July 29, 2005

- RNA-directed amino acid homochirality

# Chapter 21

# Albert von Kölliker

**Albert von Kölliker** (born *Rudolf Albert Kölliker;* 6 July 1817 – 2 November 1905) was a Swiss anatomist, physiologist, and histologist.

## 21.1 Biography

Albert Kölliker was born in Zurich, Switzerland. His early education was carried on in Zurich, and he entered the university there in 1836. After two years, however, he moved to the University of Bonn, and later to that of Berlin, becoming a pupil of noted physiologists Johannes Peter Müller and of Friedrich Gustav Jakob Henle. He graduated in philosophy at Zurich in 1841, and in medicine at Heidelberg in 1842. The first academic post which he held was that of prosector of anatomy under Henle, but his tenure of this office was brief – in 1844 he returned to Zurich University to occupy a chair as professor extraordinary of physiology and comparative anatomy. His stay here was also brief; in 1847 the University of Würzburg, attracted by his rising fame, offered him the post of professor of physiology and of microscopical and comparative anatomy. He accepted the appointment, and at Würzburg he remained thenceforth, refusing all offers tempting him to leave the quiet academic life of the Bavarian town, where he died.

At Zurich, and afterwards at Würzburg, the title of the chair which Kölliker held laid upon him the duty of teaching comparative anatomy. Many of the numerous memoirs which he published, (including the very first paper he wrote) and which appeared in 1841, before he graduated, were on the structure of animals of the most varied kinds. Notable among these were his papers on the *Medusae* and allied creatures. His activity in this direction led him to make zoological excursions to the Mediterranean Sea and to the coasts of Scotland, as well as to undertake, conjointly with his friend Carl Theodor Ernst von Siebold, the editorship of the *Zeitschrift fur Wissenschaftliche Zoologie*, which, founded in 1848, continued under his hands to be one of the most important zoological periodicals.

His hand was one of the first to be x-rayed, by his friend Wilhelm Roentgen.[1]

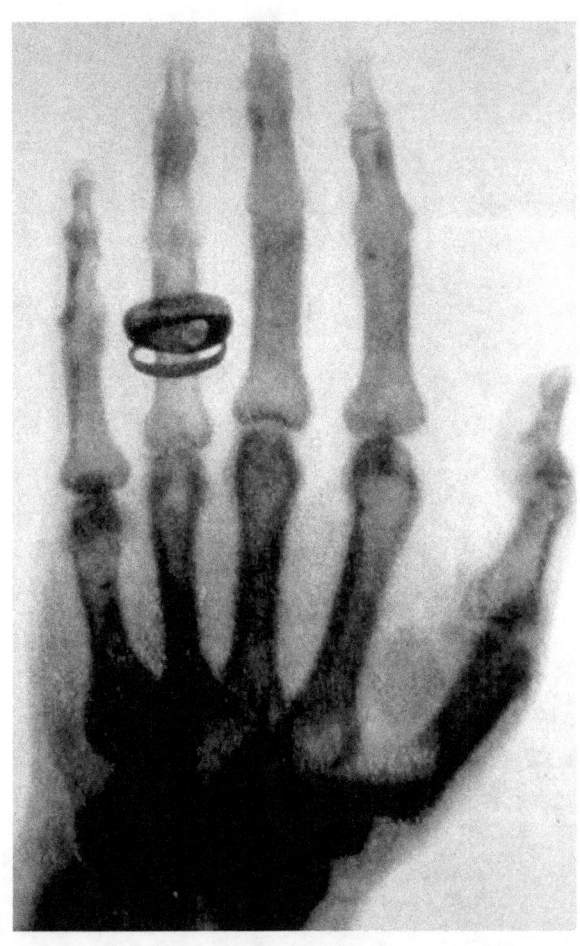

*X-ray of Kölliker's hand, made by Röntgen on 23 Jan 1896*

## 21.2 Works

Kölliker made contributions to the study of zoology. His earlier efforts were directed to the invertebrates, and his memoir on the development of cephalopods (which appeared in 1844) is considered a classical work. He soon passed on to the vertebrates, and studied the amphibians and mammalian embryos. He was among the first, if not the very first, to introduce into this branch of biological inquiry the newer microscopic technique – the methods of hardening, sectioning and staining. By doing so, not only was he enabled to make rapid progress himself, but he also placed in the hands of others the means of a similar advancement. The remarkable strides forward which embryology made during the middle and latter half of the 19th century will always be associated with his name. His *Lectures on Development*, published in 1861, at once became a standard work.

But neither zoology nor embryology furnished Kölliker's chief claim to fame. If he did much for these branches of science, he did still more for histology, the knowledge of the minute structure of the animal tissues. Among his earlier results was the demonstration in 1847 that smooth or unstriated muscle is made up of distinct units, of nucleated muscle cells. In this work, he followed in the footsteps of his master Henle. A few years before this, there was doubt whether arteries had muscle in their walls – in addition, no solid histological basis as yet existed for those views as to the action of the nervous system on the circulation, which were soon to be put forward, and which had such a great influence on the progress of physiology.

Kölliker's contributions to histology were widespread; smooth muscle, striated muscle, skin, bone, teeth, blood vessels and viscera were all investigated by Kölliker, and he touched none of them without discovering new truths. The results at which he arrived were recorded partly in separate memoirs, partly in his great textbook on microscopical anatomy, which first saw the light in 1850, and by which he advanced histology no less than by his own researches.

Lehninger asserts that Kölliker was among the first to notice the arrangement of granules in the sarcoplasm of striated muscle over a period of years beginning around 1850. These granules were later called sarcosomes by Retzius in 1890. These sarcosomes have come to be known as the mitochondria-the power houses of the cell.

In the words of Lehninger, "Kölliker should also be credited with the first separation of mitochondria from cell structure. In 1888 he teased these granules from insect muscle, in which they are very profuse, found them to swell in water, and showed them to possess a membrane."

In the case of almost every tissue, our present knowledge contains information first discovered by Kölliker – it is for his work on the nervous system that his name is most remembered. As early as 1845, while still at Zurich, he supplied the clear proof that nerve fibers are continuous with nerve cells, and so furnished the absolutely necessary basis for all sound speculations as to the actions of the central nervous system.

From that time onward he continually laboured at the histology of the nervous system, and more especially at the difficult problems presented by the intricate patterns in which nerve fibers and neurons are woven. together in the brain and spinal cord. From his early days a master of method, he saw at a glance the value of the new Golgi staining method for the investigation of the central nervous system, and, to the great benefit of science, took up once more in his old age, with the aid of a new means, the studies for which he had done so much in his youth. Kölliker contributed greatly to knowledge of the inner structure of the brain.[2]

Naturally a man of so much accomplishment was not left without honours. Kölliker was ennobled by Prince Regent Luitpold of Bavaria in 1897 and thus permitted to add the predicate "von" to his surname. He was made a member of the learned societies of many countries; in England, which he visited more than once, and where he became well known, the Royal Society made him a fellow in 1860, and in 1897 gave him its highest token of esteem, the Copley medal.

A species of lizard, *Hyalosaurus koellikeri*, is named in his honor.[3]

## 21.3 Heterogenesis

In 1864 Kölliker revived Étienne Geoffroy Saint-Hilaire's theory that evolution proceeds by large steps, under the name of heterogenesis.[5] Kölliker was a critic of Darwinism and rejected a universal common ancestor, instead he supported a theory of common descent along separate lines.[6] According to Vucinich(1988), the non-Darwinian evolution theory of Kölliker tied "organic transformism to three general ideas, all contrary to Darwin's view: the multiple origin of living forms, the internal causes of variation, and "sudden leaps" (heterogenesis) in the evolutionary process."[7]

The theory of heterogenesis was a form of saltational evolution which Kölliker claimed functioned according to a general developmental law.[8]

## 21.4 Notes

[1] http://www.christies.com/lotfinder/books-manuscripts/rontgen-wilhelm-conrad-ueber-eine-neue-5084328-details.aspx

[2] *Handbuch der Gewebelehre des Menschen*, t. 2, Leipzig, 1896.

[3] Beolens B, Watkins M, Grayson M. (2011). *The Eponym Dictionary of Reptiles*. Baltimore: Johns Hopkins University

Press. xiii + 296 pp. ISBN 978-1-4214-0135-5. ("Koelliker", p. 144).

[4] "Author Query for 'Koell.'". *International Plant Names Index*.

[5] Wright, Sewall (1984). *Evolution and the Genetics of Populations: Genetics and Biometric Foundations* **1**. University of Chicago Press. p. 10. ISBN 0226910490.

[6] Di Gregorio, Mario A. (2005). *From Here to Eternity: Ernst Haeckel and Scientific Faith*. Vandenhoeck & Ruprecht. p. 303. ISBN 3525569726.

[7] Vucinich, Alexander (1988). *Darwin in Russian Thought*. University of California Press. p. 137. ISBN 0520062833.

[8] Glick, Thomas F. (1988). *The Comparative Reception of Darwinism*. University of Chicago Press. p. 99. ISBN 0226299775.

## 21.5 References

- This article incorporates text from a publication now in the public domain: Chisholm, Hugh, ed. (1911). "article name needed". *Encyclopædia Britannica* (11th ed.). Cambridge University Press.

## 21.6 Further reading

- "Albert von Kölliker (1817–1905) Würzburger histologist". *JAMA* **206** (9): 2111–2. 1968. doi:10.1001/jama.206.9.2111. PMID 4880509.

- Albert L. Lehninger (1964). "The Mitochondrion".

## 21.7 Text and image sources, contributors, and licenses

### 21.7.1 Text

- **Abiogenesis** *Source:* https://en.wikipedia.org/wiki/Abiogenesis?oldid=731606622 *Contributors:* Damian Yerrick, AxelBoldt, Joao, Bryan Derksen, The Anome, Sjc, -- April, Ed Poor, SimonP, Maury Markowitz, AdamRetchless, Zadcat, Mjb, Heron, Someone else, Lexor, Gabbe, Martin BENOIT~enwiki, Bobby D. Bryant, Ixfd64, Cyde, Sannse, Mcarling, Ihcoyc, Ellywa, Mdebets, Cyp, JWSchmidt, Julesd, Raven in Orbit, Norwikian, Ec5618, Charles Matthews, Timwi, Steinsky, Foodman, Maximus Rex, David Shay, Populus, Omegatron, Samsara, Jackson~enwiki, Raul654, Johnleemk, Finlay McWalter, Skaffman, Twang, Jason Potter, Robbot, Fredrik, Goethean, Altenmann, Nurg, Rursus, Rebrane, Sheridan, Hadal, Wereon, Raeky, Xanzzibar, Xyzzyva, Giftlite, Mshonle~enwiki, Polsmeth, Pretzelpaws, Everyking, Dratman, Curps, Solipsist, Bobblewik, Pgan002, Andycjp, Keith Edkins, J. 'mach' wust, Sonjaaa, Quadell, Beland, Onco p53, Nograpes, Savant1984, John-Armagh, Deglr6328, Flex, Lacrimosus, Mike Rosoft, Ta bu shi da yu, Rfl, Discospinster, Rich Farmbrough, Vsmith, ArnoldReinhold, Dave souza, Paul August, Bender235, ESkog, Srbauer, RJHall, Mr. Billion, Crunchy Frog, José Gnudista, Lycurgus, Kwamikagami, Liberatus, Sietse Snel, Art LaPella, RoyBoy, Fufthmin, Guettarda, Causa sui, Bobo192, John Vandenberg, Enric Naval, Viriditas, .:Ajvol:., ZayZayEM, I9Q79oL78KiL0QTFHgyc, VBGFscJUn3, Sulai~enwiki, Hob Gadling, A Karley, Orangemarlin, Marwood, DanielVallstrom, Darrelljon, Psychofox, SlimVirgin, Ferrierd, Kocio, InShaneee, Wtmitchell, Velella, Darco, XB-70, Knowledge Seeker, Pauli133, Tainter, BerndH, Bdrasin, Linas, Mindmatrix, Anilocra, LOL, Rocastelo, Schultz.Ryan, Tabletop, Grace Note, Sadettin, GregorB, CharlesC, Wdanwatts, Essjay, Palica, Gerbrant, GSlicer, RichardWeiss, Alienus, V8rik, BD2412, Rkevins, Sjö, Drbogdan, Rjwilmsi, Mayumashu, Nightscream, Koavf, Zbxgscqf, OneWeirdDude, Bob A, XP1, Crazynas, Mikedelsol, Bfigura, SLi, Duagloth, Margosbot~enwiki, Nihiltres, Alhutch, Geologist~enwiki, Vanished user psdfiwnef3niurunfiuh234ruhfwdb7, WhyBeNormal, Knoma Tsujmai, Truthteller, Chobot, DVdm, Bgwhite, Poorsod, YurikBot, Spacepotato, RadioFan2 (usurped), GPS Pilot, The Hokkaido Crow, NawlinWiki, Rick Norwood, DragonHawk, Dysmorodrepanis~enwiki, Uberisaac, Dtrebbien, Seirscius, Zarel, SAE1962, RecSpecz, Apokryltaros, Nick, Kdbuffalo, E rulez, Crasshopper, Kortoso, Stefan Udrea, WAS 4.250, 2over0, Encephalon, Bhumiya, Smoggyrob, Davril2020, Petri Krohn, Red Jay, Fram, DisambigBot, JDspeeder1, NeilN, CIreland, Victor falk, NetRolller 3D, Quadpus, KnightRider~enwiki, SmackBot, Eperotao, PiCo, John Croft, Rtc, TestPilot, David Shear, Lankenau, Bmearns, BiT, Edgar181, Yamaguchi??, Macintosh User, Gilliam, Portillo, Betacommand, Skizzik, Eloy, Chris the speller, Kaylus, RDBrown, Davep.org, Jprg1966, Thumperward, Silly rabbit, Hibernian, Complexica, Jeff5102, Scwlong, John Hyams, JoelWhy, Jefffire, Viperphantom, Vanished User 0001, Avb, Cfassett, Ines it, Khukri, John D. Croft, Richard001, Archgoon, Smokefoot, Greg.collver, The PIPE, DMacks, Just plain Bill, Sammy1339, Daniel.Cardenas, Denise from the Cosby Show, Alan G. Archer, Ohconfucius, SashatoBot, Danielrcote, Technocratic, Gloriamarie, Attys, Atkinson 291, Khazar, John, Writtenonsand, J 1982, Butko, JoshuaZ, JorisvS, Robert Stevens, Mgiganteus1, Olin, Scetoaux, Fig wright, Extremophile, 041744, Robbins, A. Parrot, Tarcieri, Smith609, Makyen, Stevebritgimp, Tac2z, Mr Stephen, Xiaphias, Larrymcp, NJA, Novangelis, LenW, Dan Gluck, Nehrams2020, Clarityfiend, ShyK, Twas Now, Lent, The Letter J, George100, Chris55, VinnieCool, DangerousPanda, CRGreathouse, Ale jrb, Memetics, BeenAroundAWhile, Runningonbrains, RoliSoft, ButFli, WeggeBot, Moreschi, Richard Keatinge, Nnp, Myasuda, Ciyean, Abeg92, Peterdjones, Cyhawk, Hughgr, Michael C Price, Doug Weller, DumbBOT, Narayanese, DnimrevO, Ebyabe, Crum375, PKT, Thijs!bot, Barticus88, Ryansca, Pstanton, Mojo Hand, Mungomba, Headbomb, James086, Astrobiologist, Davidhorman, Chandler, Gossamers, AntiVandalBot, Luna Santin, Guy Macon, Dbrodbeck, Gnixon, TimVickers, Cstreet, Smartse, Fluffy654, Danny lost, Princeofexcess, JAnDbot, XyBot, GromXXVII, MER-C, The Transhumanist, Matthew Fennell, Mildly Mad, Andonic, Xeno, Panarjedde, TAnthony, Tstrobaugh, Rothorpe, Kornbelt888, Magioladitis, WolfmanSF, Carlwev, Sushant gupta, JNW, CattleGirl, Harelx, Trishm, Hubbardaie, Mark PEA, Recurring dreams, Zephyr2k~enwiki, Theroadislong, Cgingold, BatteryIncluded, Allstarecho, Lyonscc, Der-Hexer, Edward321, Urco, JohanViklund, Mdsats, Robin S, Drm310, Tsinoyboi, Keith D, R'n'B, CommonsDelinker, Verdatum, Leyo, Mzaki, Player 03, PhageRules1, Ulisse0, Ifomichev~enwiki, AstroHurricane001, Rlsheehan, Hans Dunkelberg, Sidhekin, AmagicalFishy, Dispenser, It Is Me Here, Enuja, McSly, Tarotcards, Janet1983, Davy p, RobinGrant, Lbeaumont, Jorfer, Cmichael, KylieTastic, AzureCitizen, IceDragon64, Funandtrvl, Novernae, Jamiejoseph, Speaker to wolves, Philip Trueman, Sub-life, Vipinhari, GcSwRhIc, Charlesdrakew, Matthewrossing, Littlealien182, Steven J. Anderson, Awl, AllGloryToTheHypnotoad, Noformation, MacFodder, Mannafredo, Mishlai, Gibson Flying V, Wiae, Maxim, Shanata, WinTakeAll, Distinguisher, SheffieldSteel, Wolfrock, Lamro, Synthebot, Zarcoen, Omermar, Northfox, Rep07, Planet-man828, Hrafn, Nachohosking, EGMAG, Tczuel, Macdonald-ross, Gnocchi, Carny, KatieandHandy, Nihil novi, ToePeu.bot, Meldor, Dawn Bard, ConfuciusOrnis, Odd nature, Yintan, 0xFFFF, Abhishikt, Chhandama, Oda Mari, Jc-S0CO, Oxymoron83, Lightmouse, Helikophis, RW Marloe, Manifolds~enwiki, Jruderman, RyanParis, Sunrise, Diego Grez-Cañete, Skeptical scientist, StaticGull, Mos bratrud, Tesi1700, Hamiltondaniel, Driftwood87, Kalidasa 777, Marmenta, Lucius Sempronius Turpio, Twinsday, Sfan00 IMG, ClueBot, Tmol42, Fyyer, The Thing That Should Not Be, AstroMark, Sexiestjen4u, Desoto10, Pi zero, Unbuttered Parsnip, Jumacdon, Canopus1, Polyamorph, Timberframe, Tfpsly, Niceguyedc, Baegis, Alexis Brooke M, Rotational, Jandew, Paulcmnt, Excirial, Gustavocarra, Winston365, Vital Forces, Shinkolobwe, Abeo iniuria, Sun Creator, Eznight, Coinmanj, NuclearWarfare, SchreiberBike, Audaciter, BOTarate, Truth is relative, understanding is limited, Thusled, Thingg, Aitias, AC+79 3888, Johnuniq, Egmontaz, Editor2020, Goodvac, Bentheadvocate, Darkicebot, CaptainVideo890, XLinkBot, Roxy the dog, Jytdog, Jovianeye, Rror, Bradv, Elfgeek, Ost316, Jungfruchallan, Aloboof123, Opaq87, Aunt Entropy, Virajelix, Thatguyflint, Janisterzaj, Addbot, Roentgenium111, DOI bot, Landon1980, Swissmeister, Ronhjones, CanadianLinuxUser, Dsmith77, Lindert, Download, Proxima Centauri, Redheylin, Bernstein0275, Camedit, Blade13125, Wildreceleste, Polyp2, LinkFA-Bot, Quietmarc, Partofwhole, U3190, Tide rolls, TL782, Romaioi, Nase, Yobot, StarTroll, Scepticus2, Yngvadottir, The Earwig, Punu, 489thCorsica, Cseppala, CinchBug, Dr.Buttons, AnomieBOT, Brroga, Mike Hayes, Kerfuffler, JWSurf, Trabucogold, Csigabi, Mann jess, Materialscientist, Citation bot, Quebec99, Romandoggie, LilHelpa, FreeRangeFrog, Xqbot, Sventington the Second, Blorblowthno, JimVC3, Wapondaponda, Mnnlaxer, Δζ, Nasnema, Mononomic, Turk oğlan, JJMesserly, Crzer07, 7h3 3L173, DerryTaylor, ProtectionTaggingBot, Gui le Roi, Conquistador, Sophus Bie, Ramssiss, Shadowjams, Methcub, Eugene-elgato, Joaquin008, Biem, FrescoBot, Finstergeist, Yanima, Hoffmannrungethailand, Krj373, Riventree, Nahuk82, BKMBC3, Machine Elf 1735, Trkiehl, PJsg1011, Citation bot 1, Redrose64, ANDROBETA, DrilBot, Winterst, Gravityguy, WaveRunner85, Gamocamo, Jonesey95, Helzrule19, Tom.Reding, Deleteduser2015, Hoo man, SpaceFlight89, FormerIP, Tanzania, Jerrywickey, Mikespedia, Jandalhandler, Kibi78704, MichaelExe, Fartherred, SkyMachine, IVAN3MAN, Trappist the monk, Silenceisgod, MEPK, Fama Clamosa, Comet Tuttle, Mcfl16, Vrenator, Victorfrogg, Jimmetry, Clarkcj12, Bcoolsdad, Diannaa, 564dude, Jynto, Gregrutz, Myrmidon1, DARTH SIDIOUS 2, Tor1714, Onel5969, RjwilmsiBot, Apotheosa, Hppa, Plommespiser, WildBot, Tesseract2, I belong to Jesus Christ, EmausBot, JeffHughes22, Immunize, Dominus Vobisdu, Niluop, Dewritech, Ibbn, RespoonsibileSQ, Tamtrible, Jmv2009, Pboehnke, Slightsmile, Tommy2010, Kiran Gopi, Mmeijeri, Dcirovic, Solomonfromfinland, Ofekalef, Gershake, Kiwi128, H3llBot, Wayne Slam, David

## 21.7. TEXT AND IMAGE SOURCES, CONTRIBUTORS, AND LICENSES

J Johnson, Ksarasofi, Korztin, Jesanj, Brandmeister, L Kensington, Jess, Scientific29, Ego White Tray, Tanoan, Renji911, SemanticMantis, Dr. Hipopotamo, JanetteDoe, Sven Manguard, JonRichfield, Ldvhl, Zuky79, Gary Dee, AUN4, ClueBot NG, Don Para, E3cubestore, Afterrock81, Colin Fredericks, Rainbowwrasse, Jorge 2701, Joefromrandb, Sketchup123, DonaldRichardSands, DS Belgium, Sjmantyl, Asukite, Widr, Telpardec, Keenedged, Wikiwiki180, MerlIwBot, Lotterox, Michaeltdeans, Helpful Pixie Bot, Elefnose, Cinnaplum, Anentiresleeve, Curb Chain, Bibcode Bot, Mwregehr, BG19bot, Lebs27, Expewikiwriter, Vevanpelt, Karmstrong909, Knowledge Examiner, Halstedcw, Mark Arsten, IraChesterfield, Drewrainey, Dkspartan1, Cauhtcoatl, Գարիբելյան Սչուն, MLearry, Cadiomals, Gorthian, Mthoodhood, Ghostsarememories, Harizotoh9, Blackstar167, HMman, Dontreader, Zedshort, Zetazeros, Dontshootimgay, Benyboy2, BattyBot, Decruft, Sfarney, Hghyux, Marc Tessera, Jimw338, SkepticalRaptor, David B Stephens, Soulbust, TheJJJunk, Garamond Lethe, Tanookiinashu, Khazar2, Ekren, Nathanielfirst, Elfinanciero222, Cmw255, Соляриcт, Pterodactyloid, RGA1980, Dexbot, Webclient101, Jinx69, Cerabot~enwiki, TippyGoomba, Mbreht, TheTahoeNatrLuvnYaho, CuriousMind01, Saehry, Leptus Froggi, 93, TruthOrTruthy, Corinne, Frivolous Consultant, HerbertHuey, Flavius-Ferry, Tjmiler, Reatlas, Anastronomer, Bret palmer, Faizan, ICameHereToEdit, Surfer43, KnowledgeIncreases07, Analiticus, StewartGriffiths, Nirendeka, DavidLeighEllis, Ronaldo Laranja, Nigellwh, AbioScientistGenesis, SzostakJack, Mj12hoaxwriter, MDPub13, PubMed2015, Andreas.Geisler, EunuchRU, PrivateMasterHD, SpazAbiogenesis, Leptinresistinadiponectin, NottNott, SuperFreakCell, Anrnusna, Stamptrader, Sstur, Dodi 8238, Suelru, Chaya5260, *thing goes, Inphynite, Kkosman, Baltazorgue, Johngraybosch, Monkbot, BethNaught, Acagastya, Garfield Garfield, Shandck, Signedzzz, Brianbleakley, Pombrand, Ruwdaman, Fried Vegetables, BicelPhD, Yazan atheos, BlueFenixReborn, Strongjam, Sarr Cat, Imradinmyownway, Washington Charter, Chemistryorigin, Michaelo1019, One sanguin, KasparBot, Fernando orrego, MusikBot, Kanashimi, Atchoum, Ktns, Paula NK, Joholub123, Shadowblade001, Bik0ser, GoldCar, Dylangenetic, Dkspartan1835, Allthefoxes, Gongwool, Oucherowl and Anonymous: 750

- **Spontaneous generation** *Source:* https://en.wikipedia.org/wiki/Spontaneous_generation?oldid=723904499 *Contributors:* Ed Poor, Heron, Olivier, Sannse, Ihcoyc, Charles Matthews, Choster, Dysprosia, Jackson~enwiki, Jonzim, Everyking, Kennethduncan, Discospinster, Bender235, Calamarain, Carlon, Causa sui, Deathawk, Hob Gadling, Alansohn, Gary, Wtmitchell, Saga City, Camw, Steinbach, BD2412, Dr-bogdan, Rjwilmsi, XP1, King of Hearts, Darker Dreams, Apokryltaros, Ospalh, Th1rt3en, Caballero1967, Zanoni, BiT, Gilliam, Egsan Bacon, Richard001, Minaker, Euchiasmus, Scientizzle, A. Parrot, Novangelis, Beefyt, Keahapana, Jetman, ChemicalBit, Chris55, Epbr123, Marek69, TimVickers, Danny lost, JAnDbot, Omeganian, Sophie means wisdom, Tstrobaugh, JPG-GR, Catgut, R'n'B, Verdatum, Mzaki, Trusilver, Hans Dunkelberg, Uncle Dick, TomS TDotO, Belovedfreak, Jpkole, Pdcook, IceDragon64, VolkovBot, Philip Trueman, TXiKiBoT, Technopat, LeaveSleaves, Wykypydya, Synthebot, Fratrep, Sunrise, Denisarona, Escape Orbit, Dlrohrer2003, ClueBot, Tmol42, GorillaWarfare, Magneticstockbrokingpetdetective, Mx3, Excirial, Jusdafax, SpikeToronto, SchreiberBike, Johnuniq, Gnowor, Conor fitzpatrick, Aunt Entropy, Thatguyflint, Addbot, Proofreader77, Leszek Jańczuk, Download, LaaknorBot, PranksterTurtle, Tide rolls, Yobot, Redirect fixer, Armchair info guy, AnomieBOT, Rubinbot, JackieBot, Bluerasberry, Materialscientist, Citation bot, LilHelpa, Xqbot, Blorblowthno, JimVC3, British-Watcher, HannesP, FrescoBot, D'ohBot, OgreBot, Citation bot 1, Intelligentsium, I dream of horses, ChristianD35, RedBot, Jeffrd10, Rjwilmsi Bot, EmausBot, Atwarwiththem, Wikipelli, Dcirovic, Solomonfromfinland, Thecheesykid, Fæ, Midas02, H3llBot, JonRichfield, ClueBot NG, E3cubestore, PedR, Dreth, Widr, MerlIwBot, Helpful Pixie Bot, Bibcode Bot, Jade Green Eyes, BG19bot, Ramos1990, Harizotoh9, Oleg-ch, BattyBot, DarafshBot, SkepticalRaptor, Cyberbot II, JYBot, Mogism, Jinx69, Unchartered, Jamesx12345, The Anonymouse, أسامة الساعدي, Eyesnore, Phdpeter, JaconaFrere, Monkbot, Spontaneoustaco, Vieque, Crystallizedcarbon, -lolololololololololoolol, GeneralizationsAreBad, Zaki1996 and Anonymous: 165

- **Self-organization** *Source:* https://en.wikipedia.org/wiki/Self-organization?oldid=728903431 *Contributors:* The Anome, Miguel~enwiki, Tedernst, Edward, Michael Hardy, Lexor, Kku, MartinHarper, EntmootsOfTrolls, Charles Matthews, Dysprosia, Nickg, Robbot, Fredrik, Rursus, Moink, Michael Snow, Mu6, Dina, Snobot, Ancheta Wis, Alensha, Pcarbonn, Margana, Karol Langner, The Land, Elektron, Pgreenfinch, Robin klein, Andreas Kaufmann, RevRagnarok, Chris Howard, Jwdietrich2, Ronaldo~enwiki, MiddleOfNowhere, Rich Farmbrough, Avriette, Vsmith, Wk muriithi, Smyth, Dave souza, JimR, Dmr2, Bender235, FirstPrinciples, Shrike, Zenohockey, Alex Kosorukoff, RoyBoy, Cretog8, Smalljim, Viriditas, .:Ajvol:., Physicistjedi, Ire and curses, Mdd, HasharBot~enwiki, Jheald, RJII, DV8 2XL, Sylvainremy, Rvanschaik, BryanKaplan, Grammarbot, Rjwilmsi, KYPark, Pleiotrop3, ElKevbo, The wub, Jeffmcneill, Mathbot, Diza, Hamidifar, YurikBot, Wavelength, Mukkakukaku, Duracell~enwiki, Pseudomonas, CLW, Curpsbot-unicodify, KnightRider~enwiki, SmackBot, Stpalli, WebDrake, Vald, Pokipsy76, M stone, Skizzik, Mobius27, Thumperward, Complexica, Colonies Chris, Royboycrashfan, Fotoguzzi, Cícero, Ericbritton, Will Beback, Eliyak, Nick Green, JoseREMY, Camazine, Kerbii, Dave Runger, Mr3641, Zarex, N2e, Pfhenshaw, Cydebot, Krauss, Gmusser, Skittleys, Miguel de Servet, Oszillodrom, Letranova, Kilva, Noclevername, Luna Santin, Rudick.JG, Davedrh, Smartse, Phanerozoic, JAnDbot, Narssarssuaq, Athkalani~enwiki, Gerculanum, Freshacconci, GrahameKing, Vernanimalcula, Economizer, Snowded, KConWiki, Dirac66, David Eppstein, User A1, Rvsole, Masaki K, Jim.henderson, Emathematica, Pilgaard, Keesiewonder, Grosscha, Crakkpot, 1000Faces, Korotkikh, Elizabeth McMillan, Pleasantville, Dggreen, Crscrs, Rollo44, Vipinhari, AllGloryToTheHypnotoad, Ordermaven, Northfox, Gbawden, SieBot, Thehotelambush, GeneCallahan, Adelanwar, Der Golem, Techdoer, Synergier, Gulmammad, Rhododendrites, Sun Creator, EhJJ, Bracton, SchreiberBike, Adriansrfr, Life of Riley, Koumz, Xiaoju zheng, Dthomsen8, Cyberoo, Fd42, WikHead, Thomas h ray, Addbot, USchick, Unesn6iduja, MrOllie, LarryJeff, Lightbot, Mcamus, Jarble, سعى, Luckas-bot, Yobot, II MusLiM HyBRiD II, Azcolvin429, AnomieBOT, Jim1138, Phantom Hoover, Materialscientist, Citation bot, LilHelpa, The Banner, Omnipaedista, Sahehco, Chjoaygame, FrescoBot, TheSen, Citation bot 1, Winterst, Gray1, Charbee, Regular Polyhedron, Jandalhandler, Ambarsande, Trappist the monk, Reflexinio, Barryclemson, We system, Blueshifting, Noresponse, Lithistman, Hhhippo, Quickmute, JuanCano, Cymru.lass, Carl Wivagg, Allanwik, Robbiemorrison, Ems2715, NinjaQuick, TuxFighter, Jrichardliston, ClueBot NG, Fgunnars, Panleek, Joel B. Lewis, MerlIwBot, Helpful Pixie Bot, Richardjb25, Revisor2011, RogerBF, BG19bot, GlaedrH, DPL bot, Terrykel, Kfriston, Soler99, Elizah379, Khazar2, Nathanielfirst, IjonTichyIjonTichy, Dexbot, Makecat-bot, BurritoBazooka, Mre env, Samotny Wędrowiec, Iztwoz, Andy Quarry, Duchifat, Otherocketman, FrB.TG, Monkbot, 卷卷卷, Mit0126, Asuscreative, Isambard Kingdom, KasparBot, Jman9058, Sangqiu5, Robcduk, Shahbazbegian and Anonymous: 136

- **Self-replication** *Source:* https://en.wikipedia.org/wiki/Self-replication?oldid=731454323 *Contributors:* Damian Yerrick, LC~enwiki, Lee Daniel Crocker, Mav, Bryan Derksen, Andre Engels, Ray Van De Walker, AdamRetchless, AdSR, Lexor, Dominus, Kku, BigFatBuddha, Kimiko, Gamma~enwiki, Timwi, Hyacinth, Omegatron, Finlay McWalter, Robbot, Altenmann, Hadal, Tea2min, Enochlau, DavidCary, Spottedowl, TheObtuseAngleOfDoom, DanielCristofani, Eric Shalov, RJHall, Liberatus, Sietse Snel, Rebroad, Woohookitty, David Haslam, Barrylb, Apokrif, Zbxgscqf, RexNL, Pete.Hurd, Diza, SteveBaker, Gwernol, RussBot, Gaius Cornelius, Grafen, BirgitteSB, Mosquitopsu, William R. Buckley, Curpsbot-unicodity, Kiv, SmackBot, Bbewsdirector, DCDuring, Chris the speller, D-Rock, Dacoutts, Sagaciousuk, Stannered, Gioto, Just Chilling, MER-C, Avaya1, VoABot II, BatteryIncluded, David Eppstein, Gwern, R'n'B, Tarotcards, Dxhtml, Moshesipper, Jamelan, Sunrise, JL-Bot, Arnos78, Tweetlebeetle367, DumZiBoT, XLinkBot, Ost316, Addbot, Hakan Kayı, CarsracBot, Lightbot, Guyonthesub-

- way, PMLawrence, 1exec1, ChildofMidnight, Plasmon1248, J04n, Solphusion~enwiki, Brunonar, Miyagawa, CES1596, FrescoBot, Foobarnix, RjwilmsiBot, Mehdiabbasi, Lopifalko, Racerx11, Faolin42, Starcheerspeaksnewslostwars, Mussermaster, Ontyx, Mikhail Ryazanov, Knowledge Examiner, OakRunner, Comp.arch, Fixature and Anonymous: 72

- **Orthogenesis** *Source:* https://en.wikipedia.org/wiki/Orthogenesis?oldid=704185738 *Contributors:* Manning Bartlett, Michael Hardy, Lexor, Steinsky, MSchnitzler2000, Alba, Jacob1207, Bensaccount, Superborsuk, Sam Hocevar, J0m1eisler, Rich Farmbrough, Sietse Snel, Viriditas, Eras-mus, RichardWeiss, Opie, Drbogdan, Fred Hsu, Entropix, Vclaw, Ewlyahoocom, Gdrbot, YurikBot, Zafiroblue05, Smaines, SmackBot, Vanished User 0001, StN, Bcasterline, Extremophile, Ceramufary, Meco, Panaceus, Dcflyer, Chris55, Headbomb, Lauranrg, Infophile, Magioladitis, Rusty Cashman, Filll, Memestream, Nwbeeson, 3tcetera, Dgri, EoGuy, Mild Bill Hiccup, Niceguyedc, GoEThe, Alexbot, Muhandes, Johnuniq, Ano-User, Addbot, Aceofhearts1968, Lightbot, Luckas-bot, Legobot II, Armchair info guy, Richardbrucebaxter, Sz-iwbot, Citation bot, Xqbot, FanCollector, Nosson77, GrouchoBot, FrescoBot, Stephen Morley, 564dude, Brichard37, TeleComNasSprVen, AManWithNoPlan, Hypercephalic, Liveintheforests, Savantas83, Randomneologism, Vagobot, Earthisalive, MrBill3, Khazar2, Makecat-bot, A little angry and Anonymous: 31

- **Primordial soup** *Source:* https://en.wikipedia.org/wiki/Primordial_soup?oldid=723778269 *Contributors:* Julesd, Decumanus, Discospinster, Vsmith, LindsayH, BD2412, Biwhite2, JAnDbot, Magioladitis, DaWarMage, PhageRules1, Jorfer, Hrafn, Sunrise, Serge020, Addbot, Michaelwuzthere, Biem, I dream of horses, Dcs002, Dcirovic, Donner60, ClueBot NG, Klilidiplomus, CogitoErgoSum14 and Anonymous: 21

- **Miller–Urey experiment** *Source:* https://en.wikipedia.org/wiki/Miller%E2%80%93Urey_experiment?oldid=725029996 *Contributors:* Bryan Derksen, Zundark, The Anome, Hannes Hirzel, Caltrop, D, Michael Hardy, Lexor, Gabbe, Skysmith, Julesd, Cyan, HolIgor, Ec5618, Tpbradbury, Omegatron, Jeffq, Phil Boswell, Robbot, 1984, Nurg, Postdlf, Rursus, Rhombus, Bkell, Sternthinker, Pengo, Alan Liefting, Leighxucl, Capitalistroadster, Michael Devore, Stern~enwiki, Solipsist, Auximines, Antandrus, Beland, Maneesh, Deglr6328, Chris Howard, Spiffy sperry, Inkypaws, Vsmith, Moonrocks, Dave souza, Bender235, Art LaPella, RoyBoy, WorldDownInFire, Gyll, Bobo192, Liffey, Mdhowe, Haham hanuka, Yalbik, Merope, Orangemarlin, Eric Kvaalen, Keenan Pepper, Wouterstomp, Riana, GabrielF, Sleigh, Dan100, Mu301, BenWilson, MONGO, Wikiklrsc, Hard Raspy Sci, Wdanwatts, Gimboid13, GSlicer, Deltabeignet, Jan van Male, V8rik, Athelwulf, Xerxes2k~enwiki, Drbogdan, Rjwilmsi, Jweiss11, XP1, Rillian, Ronocdh, Nneonneo, TheIncredibleEdibleOompaLoompa, DoubleBlue, Thecurran, Jsdratm, Jaraalbe, Whosasking, Huw Powell, Chris Capoccia, Bergsten, SpuriousQ, Hellbus, Bolinstephen, Shell Kinney, Tungsten, GeeJo, Thane, Draeco, NawlinWiki, Ragesoss, Peter Delmonte, Tony1, Kortoso, Zzuuzz, Closedmouth, Reyk, Wootini, JDspeeder1, Scolaire, True Pagan Warrior, SmackBot, Loukinho, ScaldingHotSoup, Stepa, Verne Equinox, Canthusus, Gaff, Andy M. Wang, Tyciol, Thumperward, Starless~enwiki, Jeff5102, Kostmo, DHN-bot~enwiki, Danielkueh, Vanished User 0001, Rrburke, RedHillian, Wen D House, Jgrahamc, Bmgoau, Twir, Kukini, Clicketyclack, ArglebargleIV, JorisvS, Extremophile, Waggers, Joseph Solis in Australia, Tawkerbot2, Merryjman, Xcentaur, Leujohn, Double Think, Rifleman 82, Tawkerbot4, Narayanese, Epbr123, Lord Hawk, Wikid77, Kablammo, Anupam, Nonagonal Spider, Headbomb, Sonicblade128, I do not exist, Tellyaddict, LeoTrottier, Cooljuno411, AntiVandalBot, Fedayee, QuiteUnusual, Dvunkannon, DarkAudit, Earlysda, Myanw, Res2216firestar, .anacondabot, Anþony, Narinukositkul, AuburnPilot, Transcendence, JamesBWatson, Shorty1357, Tekn04, WeeWillieWiki, BatteryIncluded, Adrian J. Hunter, Su-no-G, Edward321, Squidonius, Urco, MartinBot, Bbi5291, Rettetast, Zouavman Le Zouave, CommonsDelinker, Nelsonkcheng, Nono64, J.delanoy, Hans Dunkelberg, Extransit, TomS TDotO, AdderUser, McSly, Plasticup, NewEnglandYankee, Hugo999, Jlaramee, Jeff G., AlnoktaBOT, Shuvaev, Rei-bot, DianaGaleM, DennyColt, Don4of4, Lamro, Zarcoen, Enviroboy, Northfox, SieBot, Randommelon, Sakkura, Matthew Yeager, Chhandama, Bananastalktome, Aaronomus, Oxymoron83, PhilMacD, Lizardabuff, Sunrise, Sphilbrick, Amy--n--ash, Atif.t2, Dongkim, ClueBot, YassineMrabet~enwiki, Mild Bill Hiccup, Wikilost, Auntof6, RoundemUpJeff, Rg006, Abrech, Vital Forces, NuclearWarfare, Percentyield, Dokuhebi, Acabashi, Johnuniq, DumZiBoT, Crazy Boris with a red beard, Bearsona, SilvonenBot, Addbot, Jacopo Werther, Substar, DOI bot, Yobmod, Cst17, Numbo3-bot, Tide rolls, Alfredtheweird, Zorrobot, David0811, Relientk388, Luckas-bot, Yobot, Ptbotgourou, Niklo sv, II MusLiM HyBRiD II, Amirobot, Nallimbot, KamikazeBot, AnomieBOT, Krasss, Ulric1313, Citation bot, Patsboy1212, Psrq, Lesatdrew, Emon86, Millelacs, GrouchoBot, Leor klier, A little insignificant, Kemisulis, RedBot, Serols, HarmonicSeries, IiiTheMARTYRiii, Microboy, IVAN3MAN, Rach falk, Gwyneth99, BoracayOcean69, Updatehelper, RjwilmsiBot, EmausBot, Dcirovic, ZéroBot, H3llBot, Gamma Spelazi, Factchecker40, Shantycakes, Anthony1592, Orange Suede Sofa, Tanoan, JohnLloydScharf, Will Beback Auto, ClueBot NG, Smtchahal, O.Koslowski, Widr, Anonymus31, Tiberiusgaius, Bibcode Bot, MusikAnimal, Zerbu, ReliableFactz, Harizotoh9, Shaun, BattyBot, Garamond Lethe, RGA1980, Dexbot, Baihe1101, Lolgast, Jinx69, Discuss-Dubious, Project Osprey, Irecol59, Earth Formation, Everymorning, The Herald, T.slich, Stamptrader, Dodi 8238, Chaya5260, Monkbot, Vieque, BethNaught, Geybag, ToBk, Eurodyne, DalekSupreme, Aditya Sainiarya, 15bpetrovich, Alex why not?, GSS-1987, Toxichris2, Venjack and Anonymous: 347

- **Biogenic substance** *Source:* https://en.wikipedia.org/wiki/Biogenic_substance?oldid=657354514 *Contributors:* Mav, The Anome, Emperorbma, Gentgeen, Blainster, Bensaccount, Icairns, Vsmith, Walden, Jag123, Markornikov, Graham87, SmackBot, Bluebot, Clicketyclack, DabMachine, Alaibot, Avicennasis, WRJF, Lamro, ClueBot, Dragonjk96, Addbot, Doubious, AnomieBOT, Erik9bot, A.amitkumar, FrescoBot, ClueBot NG, Asoni725, CyberChook, RedLocustEntertainment, Rahul69404, Vhrithik18 and Anonymous: 11

- **Biotic material** *Source:* https://en.wikipedia.org/wiki/Biotic_material?oldid=729125250 *Contributors:* Altenmann, HaeB, Pengo, Rich Farmbrough, Paleorthid, Drbogdan, Rjwilmsi, Chris the speller, Lsjzl, Ventifact, Brewhaha@edmc.net, BatteryIncluded, MartinBot, Sheep2000, AzureIcicle, DRTllbrg, Anxietycello, Erik9bot, Look2See1, Dcirovic, Ego White Tray, Mark Arsten, NottNott, Kevt2002, Nahid monavarian and Anonymous: 8

- **Common descent** *Source:* https://en.wikipedia.org/wiki/Common_descent?oldid=731623015 *Contributors:* The Anome, Ed Poor, Rgamble, Edward, Palnatoke, Lexor, MartinHarper, Gabbe, Menchi, Cyde, 168..., Poor Yorick, Ec5618, Alexs, Samsara, Donarreiskoffer, Robbot, Peak, Naddy, Henrygb, Alan Liefting, FeloniousMonk, Andycjp, PDH, Pmanderson, Sam Hocevar, Deglr6328, Randwicked, Rich Farmbrough, Vsmith, Pavel Vozenilek, Bender235, CanisRufus, Guettarda, Stesmo, Kjkolb, Orangemarlin, Frank101, Ungtss, Harburg, Gpvos, Sven Heinicke, Ceyockey, Firsfron, Mindmatrix, RHaworth, WadeSimMiser, GSlicer, RichardWeiss, Opie, Drbogdan, Rjwilmsi, Cribbswh, Thomas Arelatensis, Tomtheman5, Burris, Dionyseus, Nihiltres, Jorgesalgueiro~enwiki, Dfinch, SpuriousQ, Stephenb, Curtis Clark, Epipelagic, DRosenbach, WAS 4.250, Jogers, Petri Krohn, Finell, Sardanaphalus, SmackBot, Jim62sch, Eskimbot, DLH, Yamaguchi??, Skizzik, Persian Poet Gal, Valich, SundarBot, Bubbamagic87, Richard001, Alan G. Archer, Bcasterline, JH-man, JoshuaZ, Robert Stevens, Ben Moore, JHunterJ, Vanished user, Petr Matas, CmdrObot, Agathman, Logicus, Michael Johnson, Headbomb, Bibliophile20, Seaphoto, Deeplogic, Doc Tropics, Zappernapper, Princeofexcess, Jondw, Magioladitis, Bongwarrior, VoABot II, Catslash, Cat Whisperer, Catgut, David Eppstein, Glen, CommonsDelinker, Jarhed, Mausy5043, J.delanoy, Filll, TomS TDotO, Pbarnes, Mccajor, Nwbeeson, DadaNeem, Martial75, VolkovBot, Ewe nik, Littlealien182, Demfranchize, Winterschlaefer, SheffieldSteel, Synthebot, Rafikgl, Northfox, Infraredeclipse, Dawn Bard, Flyer22 Reborn,

## 21.7. TEXT AND IMAGE SOURCES, CONTRIBUTORS, AND LICENSES 165

Sunrise, Capitalismojo, Webridge, ClueBot, Rumping, EoGuy, Mild Bill Hiccup, Auntof6, Christian Skeptic, Octavian history, Horselover Frost, ForestDim, Johnuniq, Aunt Entropy, Addbot, DOI bot, Mr. Wheely Guy, LaaknorBot, 84user, Numbo3-bot, Luckas-bot, Yobot, Anypodetos, Azcolvin429, Armchair info guy, Categorer, Archaeopteryx, Citation bot, Join Tile, Gilo1969, Digitwoman, GrouchoBot, Cwbm (commons), Stiepan Pietrov, Cheddarbob23, Citation bot 1, Pinethicket, Cement123, RedBot, PoolPartay, Trappist the monk, Strictscrutiny, Livingrm, Dtheobald, Keegscee, Rogue-pilot, Dominus Vobisdu, Haggiaomega, Traxs7, Medeis, Brandmeister, ClueBot NG, Wallace Kneeland, Jorge 2701, Dufusrex, Snotbot, MerlIwBot, Cadiomals, Klilidiplomus, SaudiPseudonym, Billyshiverstick, Equilibrium Allure, CsDix, Monkbot, Mama meta modal, Georgeanewman, Ilikelifesciences, GSS-1987 and Anonymous: 111

- **Last universal ancestor** *Source:* https://en.wikipedia.org/wiki/Last_universal_ancestor?oldid=731675604 *Contributors:* The Anome, Mrwojo, Lexor, Paul A, Julesd, Glenn, Schneelocke, Greenrd, Samsara, Peak, Ashley Y, Rorro, Meanos~enwiki, Xanzzibar, Unfree, David Gerard, BenFrantzDale, Pgan002, Superborsuk, Taka, FT2, Vsmith, Smyth, JimR, Dbachmann, Night Gyr, Bender235, Violetriga, Reinyday, Jumbuck, PatrickFisher, Axl, Hohum, Mindmatrix, SCEhardt, Driftwoodzebulin, RichardWeiss, Sjö, Drbogdan, Rjwilmsi, Thomas Arelatensis, Mathrick, Diza, Johnhalton, Samwaltz, YurikBot, Hairy Dude, Chris Capoccia, Motmot, Emmanuelm, Eleassar, Apokryltaros, Epipelagic, WAS 4.250, Calaschysm, Redgolpe, Pietdesomere, Petri Krohn, Bill, SmackBot, John Croft, Brya, EncycloPetey, SvGeloven, Valich, Iridescence, Bejnar, Byelf2007, Loodog, JorisvS, Mgiganteus1, Smith609, Larrymcp, Hu12, Joseph Solis in Australia, Lottamiata, New User, CmdrObot, Woudloper, Im.a.lumberjack, Article editor, Richard Keatinge, A876, Was a bee, Thijs!bot, Headbomb, Bob the Wikipedian, Noclevername, Smartse, Byeee, IanOsgood, Magioladitis, BatteryIncluded, David Eppstein, Thibbs, Squidonius, Robin S, CommonsDelinker, Petter Bøckman, Sidhekin, Justin, WindAndConfusion, Tom Schmal, Tdadamemd, HiLo48, Blueshifter, Nwbeeson, Maximusthaler, Phoenix1304, VolkovBot, Scilit, BotKung, Macdonald-ross, SieBot, StAnselm, BotMultichill, Dawn Bard, Mimihitam, Pittsburghmuggle, Sunrise, Hamiltondaniel, Ecthelion83, WurmWoode, Jackwestmore, Niceguyedc, GoEThe, Johnuniq, Fastily, Subversive.sound, Addbot, Energyequation, Non-dropframe, GPWM, Junegirl7, Luckas-bot, Ptbotgourou, Legobot II, Adi, Dmarquard, AnomieBOT, Tryptofish, SporeGames, JackieBot, InsufficientData, Citation bot, Xqbot, FrescoBot, Jewlrzeye, Moonraker, RedBot, Double sharp, RjwilmsiBot, Liamzebedee, Tesseract2, The Mysterious El Willstro, Dcirovic, Solomonfromfinland, Thecheesykid, A2soup, WeijiBaikeBianji, Vanished user fijtji34toksdcknqrjn54yoimascj, Linda 444, ClueBot NG, Dharam66, Davidddwd, Bibcode Bot, BG19bot, Videsh Ramsahai, ElphiBot, Harizotoh9, Jimw338, SkepticalRaptor, Jinx69, CuriousMind01, GrendelGreyfur, Booklaunch, EtymAesthete, Odysnes, Monkbot, Mama meta modal, Yolol25, Macofe, Geo-Science-International, Hannahquack, Mythslayer7, Saturn comes back around, Rigsofrods, Lorricotton, Andrewrobinson01070111GroupD ext2015 and Anonymous: 97

- **Proteinoid** *Source:* https://en.wikipedia.org/wiki/Proteinoid?oldid=717434725 *Contributors:* Bryan Derksen, Robbot, Eroica, Guanabot, Firsfron, Rjwilmsi, Bgwhite, Welsh, SmackBot, Lankenau, Jrosenau, Emre D., Takowl, Chrumps, BatteryIncluded, WaiteDavid137, Wiae, Sensonet, Jacopo Werther, Hakan Kayı, Download, Legobot, Yobot, AnomieBOT, Xqbot, Erik9bot, Trappist the monk, Jesse V., Aozf05, BattyBot, Koza1983, Monkbot and Anonymous: 17

- **Autocatalysis** *Source:* https://en.wikipedia.org/wiki/Autocatalysis?oldid=724193320 *Contributors:* Lexor, Shyamal, Raven in Orbit, Dino, Random832, Azra99, RedWolf, UtherSRG, Jason Quinn, D3, Xenoglossophobe, TedPavlic, Ebradsha, Ceyockey, Rjwilmsi, Misternuvistor, Butonic~enwiki, Scope creep, Cedar101, Eno-ja, Limhes, Yakudza, SmackBot, Thumperward, Complexica, Mion, Olin, Tarcieri, Knights who say ni, Imsobored, Ebaskerv, John Riemann Soong, Rifleman 82, Christian75, Thijs!bot, Dirac66, User A1, TechnoFaye, CommonsDelinker, Daniele.tampieri, STBotD, RedAndr, Ordermaven, SieBot, Liamstone, Sunrise, Alexbot, Addbot, EconoPhysicist, Lightbot, LilHelpa, GrouchoBot, DerryTaylor, Nathanielvirgo, Citation bot 1, Dinamik-bot, Dcirovic, Bollyjeff, Surya Prakash.S.A., RockMagnetist, Helpful Pixie Bot, Khazar2, Jamesx12345, Kernsters, Yahadzija, Alékos Elefthérios, ❏❏❏❏, Skylord a52 and Anonymous: 24

- **Homochirality** *Source:* https://en.wikipedia.org/wiki/Homochirality?oldid=730167618 *Contributors:* Jengod, Daniel Newby, V8rik, Drbogdan, Rjwilmsi, Ian Pitchford, Rekleov, MadMax, Kortoso, SmackBot, M stone, Chris the speller, Meaningless, DMacks, Ohconfucius, CmdrObot, Rzepa, Rifleman 82, Christian75, WVhybrid, DOSGuy, Jingxin, BatteryIncluded, Su-no-G, STBot, Speciate, Dmbius, VolkovBot, Shosukekojo~enwiki, LabFox, Wenli, Nave.notnilc, Northfox, Lesterama, Sunrise, Mnefliu, DragonBot, Addbot, Wickey-nl, Hakan Kayı, Numbo3-bot, Legobot, Yobot, AnomieBOT, Krasss, Wiki007wiki, Citation bot, Sprlzrd, Saehrimnir, Citation bot 1, Mikespedia, Artoannila, Craxyxarc, Dcirovic, JPBoyd, Karl 334, Bibcode Bot, HPBiochemie, Dartvox78, Project Osprey, I am One of Many, Blackbombchu, NFHC, Die Antworten, Crito10, Srednuas Lenoroc, RIT RAJARSHI and Anonymous: 23

- **Protocell** *Source:* https://en.wikipedia.org/wiki/Protocell?oldid=729218917 *Contributors:* IceKarma, Vsmith, Drbogdan, Rjwilmsi, NawlinWiki, ❏❏❏❏ robot, CRGreathouse, BatteryIncluded, Kalidasa 777, SallyForth123, Redrose64, Lowercase sigmabot, BG19bot, CitationCleanerBot, CuriousMind01, Monkbot, Χρυσάνθη Λυκούση, Callum radiator and Anonymous: 15

- **Iron–sulfur world hypothesis** *Source:* https://en.wikipedia.org/wiki/Iron%E2%80%93sulfur_world_hypothesis?oldid=731709085 *Contributors:* AxelBoldt, Lexor, Timwi, IceKarma, Lumos3, Donarreiskoffer, Peak, Barbara Shack, Pgan002, Julianonions, Rich Farmbrough, Cacycle, Vsmith, Bender235, Art LaPella, Viriditas, Paleorthid, Feline1, Vanished user dfvkjmet9jweflkmdkcn234, GregorB, V8rik, Drbogdan, Rjwilmsi, Koavf, Parutakupiu, TexasAndroid, Open2universe, Petri Krohn, SmackBot, Rtc, Dauto, Bluebot, Scwlong, John D. Croft, Smokefoot, Andrew c, Daniel.Cardenas, MegaHasher, Extremophile, Twas Now, Ziusudra, Yashgaroth, CmdrObot, AndrewHowse, Rgas, Rifleman 82, Ttiotsw, Pro crast in a tor, Res2216firestar, VoABot II, BatteryIncluded, Ugajin, Geekdiva, Grock2, TXiKiBoT, Richwil, OMCV, FghIJklm, General Epitaph, Sid-Vicious, Gustavocarra, AC+79 3888, Dthomsen8, SilvonenBot, Addbot, DOI bot, Favonian, Ginosbot, 84user, Luckas-bot, Yobot, Dr.Buttons, AnomieBOT, Citation bot, RibotBOT, Linakieper, Citation bot 1, Citation bot 4, Tornado00, Tom.Reding, Trappist the monk, MarkGT, Siddunitw, EmausBot, Insorak, Jasonanaggie, ZéroBot, AManWithNoPlan, EvenGreenerFish, Helpful Pixie Bot, Bibcode Bot, Farid320, NotWith, BattyBot, EagerToddler39, Monkbot and Anonymous: 38

- **Panspermia** *Source:* https://en.wikipedia.org/wiki/Panspermia?oldid=729559598 *Contributors:* Mav, Bryan Derksen, Timo Honkasalo, Taw, Andre Engels, William Avery, SimonP, Shii, Maury Markowitz, Stevertigo, Edward, Alan Peakall, Alfio, CesarB, JWSchmidt, Whkoh, Kimiko, Timwi, RickK, Rednblu, Abscissa, VeryVerily, Populus, Paul-L~enwiki, Omegatron, Samsara, Warofdreams, Proteus, Tonderai, Jeffq, Radical-Bender, Northgrove, Chris Roy, Postdlf, Sverdrup, Rholton, Rebrane, Wlievens, David Edgar, Jor, Cyberia23, Per Abrahamsen, Nagelfar, Alan Liefting, Ncox, Kbahey, Jyril, Bfinn, Dmb000006, MingMecca, Duncharris, Solipsist, SWAdair, Bobblewik, Pgan002, Alexf, Bcameron54, Piotrus, Melikamp, Csmiller, Taka, Obby~enwiki, Mzajac, Cglassey, Sam, Lacrimosus, Rich Farmbrough, Vsmith, ArnoldReinhold, YUL89YYZ, Dbachmann, Bender235, BACbKA, RJHall, CanisRufus, RoyBoy, Godfreylouis, Viriditas, Cmdrjameson, Nicke Lilltroll~enwiki, Dejitarob, L33tminion, Hob Gadling, Vicarage, Pschemp, Slipperyweasel, Calebe, Chino, Arthena, JoaoRicardo, Ahruman, Hu, Mnemo, TahitiB~enwiki,

Pauli133, Gene Nygaard, Dismas, Stephen, Gmaxwell, Mindmatrix, TotoBaggins, Pictureuploader, DanHobley, Holek, Dbutler1986, Mandarax, Marskell, Drbogdan, Rjwilmsi, Koavf, Oblivious, Miserlou, SonicSpike, Erkcan, Krash, Dionyseus, FlaBot, Ian Pitchford, John Baez, Papacha, Diza, SteveBaker, Gurubrahma, Chobot, Voodoom, Bgwhite, Wavelength, Acefox, Chris Capoccia, Akamad, Dysmorodrepanis~enwiki, Icelight, Anetode, Wangi, Emdx, DeadEyeArrow, Skepticsteve, Elkman, Pegship, Noosfractal, 2over0, Jules.LT, Chase me ladies, I'm the Cavalry, Abune, SMcCandlish, Reyk, Petri Krohn, Pádraic MacUidhir, Brentt, SmackBot, Judith.d, Melchoir, Elfsareus, Kintetsubuffalo, Portillo, Ohnoitsjamie, Hmains, Jushi, Kinhull, Qwasty, Sumthingweird, RDBrown, Thumperward, Starless~enwiki, Nemodomi, Silly rabbit, Hibernian, Zephyr707, WDGraham, Jefffire, Frap, OrphanBot, Nixeagle, Thomqi, Soosed, Wen D House, Tsop, Iamdaniel, Kismetmagic, Wizardman, Bklyce, StN, Virago, BrownHairedGirl, Eaglecros, John, Siddharth srinivasan, John Cumbers, Neodarksaver, ISoron, Extremophile, Skymist, Novangelis, Corykoski, Jason.grossman, FelisSchrödingeris, Greygirlbeast, Courcelles, Brainbark, Centered1, Banedon, Kylu, RagingR2, JFreeman, Daniel J. Leivick, Dancter, Michael C Price, DumbBOT, Iliank, Robertinventor, Arb, UberScienceNerd, Thijs!bot, JAF1970, Mpallen, Headbomb, John254, Second Quantization, Z10x, Iulius, Morgana The Argent, The Hams, NeilEvans, Gnixon, Yellowdesk, Rothorpe, LittleOldMe, Magioladitis, Professor marginalia, Yakushima, Theroadislong, Gabriel Kielland, BatteryIncluded, Joe hill, Thibbs, DerHexer, Lelek, Waninge, Urco, Jim.henderson, Rettetast, Ulisse0, AlphaEta, Trusilver, Terrek, Nsande01, Ian.thomson, ABVS1936, Skier Dude, Kukec, RenniePet, Davy p, M-le-mot-dit, EyeRmonkey, Diletante, Idioma-bot, Evolvearth, Soliloquial, Dom Kaos, Philip Trueman, JayEsJay, TXiKiBoT, Mathwhiz 29, Cloudswrest, Stickyhammer, Agmon, AlleborgoBot, Hrafn, SieBot, Coffee, Hugh16, Arbor to SJ, Tiki 92090, Mimihitam, Excutio, LSmok3, Sunrise, Anchor Link Bot, Mnmautner, Martarius, Doyee5, ClueBot, CarolSpears, Foxj, Wanderer57, DrFO.Jr.Tn~enwiki, 1111news, Niceguyedc, Nanobear~enwiki, Sid-Vicious, Sjdunn9, Deselliers, Scog, BSmith821, HRosenberg, Fitzburgh, Unmerklich, Aitias, Apparition11, Crowsnest, Mrmpsy, DumZiBoT, Hotcrocodile, Rror, Ost316, Oogaboogabooga, Fabio6043, BrucePodger, Addbot, Basilicofresco, DOI bot, Potentialten, Ronhjones, Chamberlain2007, Tdehel, MrOllie, Download, Glane23, SamatBot, Scienceislife, Beren, MuZemike, Cannizzaro S, Legobot, Luckas-bot, Yobot, Wikipedian Penguin, Untrue Believer, AnomieBOT, Taylordw, Rubinbot, Citation bot, Donkyhotay, Eumolpo, Quebec99, LilHelpa, Xqbot, Silver Spoon Sokpop, Albalma, Felipe Schenone, GrouchoBot, Knightofcydonia49, Omnipaedista, N419BH, Hdrosenberg, Unused0011, A.amitkumar, Nagualdesign, FrescoBot, LucienBOT, Styxpaint, Citation bot 1, Solarflaredigital, Pinethicket, Jonesey95, Tom.Reding, Fumitol, SkyMachine, IVAN3MAN, Trappist the monk, Jonkerz, RjwilmsiBot, Damaavand, Ansaazi, Androstachys, EmausBot, John of Reading, Clive tooth, Emmajanej, Dcirovic, Evanh2008, JacobSheehy, Kp grewal, ZéroBot, AManWithNoPlan, David J Johnson, Actsmart, Orange Suede Sofa, Spicemix, Grapple X, ClueBot NG, Njh321, Kikichugirl, Liveintheforests, RocketLauncher2, Old wombat, BrigKlyce, Russellml, Widr, Antiqueight, Secret of success, Helpful Pixie Bot, Curb Chain, Gob Lofa, Bibcode Bot, Lilman0509, BG19bot, Mariansavu, Chemistryfan, Walterfarah, Kooky2, BattyBot, Arodr451, MichaelEF71, ChrisGualtieri, Khazar2, Orestesgaolin, Rchouake, Dexbot, SummerWillow, Adam2828, Abepeace, Tomdarkblade, Melonkelon, KingSupernova, AlexeiSharov, Praemonitus, Syd Menon, Npascucci01, Someone not using his real name, HarbingerOfLunch, Monkbot, Tigercompanion25, Denny123123, Formuse, BicelPhD, Mapsfly, HiBlueSky, Cspoleta, ComicsAreJustAllRight, Dsmith125, Alexis Gervais, Nøkkenbuer, Jnav7, AusLondonder, Sir Cumference, Dutral, PiNerd3 and Anonymous: 327

- **List of interstellar and circumstellar molecules** *Source:* https://en.wikipedia.org/wiki/List_of_interstellar_and_circumstellar_molecules?oldid=728930894 *Contributors:* Bryan Derksen, CBDunkerson, Dbenbenn, Graeme Bartlett, Dratman, BrendanRyan, Jorge Stolfi, Foobar, Darrien, Karol Langner, Spiffy sperry, Tompw, RJHall, Huntster, Bnikolic, Pearle, Snarfevs, John Coupe, EagleFalconn, Shoefly, ThomasWinwood, Drbogdan, Rjwilmsi, Marasama, Mike s, Mike Peel, HappyCamper, Takometer, Bgwhite, Spacepotato, Deville, Reyk, Poulpy, Tropylium, SmackBot, Edgar181, Chris the speller, Bduke, Modest Genius, Chlewbot, Thor Dockweiler, JohnI, AstroChemist, JorisvS, RekishiEJ, CmdrObot, Cydebot, Astrochemist, Rifleman 82, Nikopoley, Headbomb, Pixelface, JAnDbot, Jingxin, Magioladitis, BatteryIncluded, Heril, Nono64, Leyo, James McBride, Schmackes, Lightmouse, Nergaal, ImageRemovalBot, NuclearWarfare, Scog, BSmith821, Ost316, Addbot, DOI bot, LinkFA-Bot, Yobot, AnomieBOT, Citation bot, Blundgr2, Hcnhplus, Citation bot 1, Jonesey95, Tom.Reding, Mikespedia, Trappist the monk, InvaderXan, Jynto, RjwilmsiBot, John of Reading, GoingBatty, TuHan-Bot, Dcirovic, H3llBot, Whoop whoop pull up, ClueBot NG, Kikichugirl, Frietjes, Polskivinnik, Helpful Pixie Bot, Bibcode Bot, BG19bot, BattyBot, Cyberbot II, Dexbot, Project Osprey, Ruby Murray, Eyesnore, Monkbot, CxHy and Anonymous: 37

- **Gard model** *Source:* https://en.wikipedia.org/wiki/Gard_model?oldid=684477784 *Contributors:* Michael Hardy, Bearcat, Rjwilmsi, Malcolma, Guy Macon, Carlwev, BatteryIncluded, Terrek, Omermar, Whmice, Fluffernutter, Yobot, Orenburg1, John of Reading, BG19bot and Jamesmcmahon0

- **PAH world hypothesis** *Source:* https://en.wikipedia.org/wiki/PAH_world_hypothesis?oldid=704792931 *Contributors:* Lumos3, Gadfium, Julianonions, Keenan Pepper, Cwgannon, Blaxthos, Drbogdan, Rjwilmsi, Loukinho, Bluebot, Sbharris, Joseph Solis in Australia, Crum375, SilentWings, BatteryIncluded, Squidonius, Philcha, AquamarineOnion, VolkovBot, TXiKiBoT, Fanra, Jdaloner, Lightmouse, Rosiestep, Martarius, Sid-Vicious, Gustavocarra, Johnuniq, DumZiBoT, XLinkBot, Ost316, Addbot, Luckas-bot, JackieBot, Trappist the monk, The Mysterious El Willstro, Just granpa, Bibcode Bot, Kmzayeem, SomeFreakOnTheInternet, AioftheStorm, Monkbot, Brianbleakley and Anonymous: 22

- **Albert von Kölliker** *Source:* https://en.wikipedia.org/wiki/Albert_von_K%C3%B6lliker?oldid=728653183 *Contributors:* Amillar, Rsabbatini, Rick Block, Creidieki, Klemen Kocjancic, D6, Rich Farmbrough, Mwng, Bender235, TheParanoidOne, Lokicarbis, FeanorStar7, SCEhardt, Eras-mus, Emerson7, Fish and karate, Yopohari~enwiki, YurikBot, A314268, Nrets, Encephalon, Јованб, SmackBot, KocjoBot~enwiki, Vanished User 0001, SashatoBot, Eliyak, Ryulong, DumbBOT, Thijs!bot, Rosarinagazo, Mnpeter, SuperGirl, KowDude, Plindenbaum, STBotD, RB972, Squids and Chips, VolkovBot, TXiKiBoT, Rei-bot, SwordSmurf, Dassiebtekreuz, GirasoleDE, MadmanBot, Niceguyedc, Dupdater, Ottawahitech, Addbot, Tide rolls, Lightbot, Yobot, TaBOT-zerem, Materialscientist, Citation bot, LilHelpa, Xqbot, Omnipaedista, Hamamelis, Plucas58, Vrenator, EmausBot, Gcastellanos, AvicBot, ZéroBot, Cocoadependent, Balm oral winds or sand ring ham, ClueBot NG, BattyBot, Adelshaus, VIAFbot, Fodor Fan, Lyttle-Wight, BethNaught, Jonarnold1985, KasparBot, Y~ruwiki, Aiujdfhuiajf and Anonymous: 14

## 21.7.2 Images

- **File:080205_Brusselator_picture.jpg** *Source:* https://upload.wikimedia.org/wikipedia/commons/a/a4/080205_Brusselator_picture.jpg *License:* Public domain *Contributors:* Own work (Original text: *self-made*) *Original artist:* Complexica at English Wikipedia

- **File:10_small_subunit.gif** *Source:* https://upload.wikimedia.org/wikipedia/commons/3/3d/10_small_subunit.gif *License:* Public domain *Contributors:* <a data-x-rel='nofollow' class='external text' href='http://www.pdb.org/pdb/static.do?p=education_discussion/molecule_of_the_

## 21.7. TEXT AND IMAGE SOURCES, CONTRIBUTORS, AND LICENSES

month/pdb10_1.html'>*Molecule of the Month*</a> at the RCSB Protein Data Bank *Original artist:* Animation by David S. Goodsell, RCSB Protein Data Bank

- **File:1990_s32_LDEF_stow.jpg** *Source:* https://upload.wikimedia.org/wikipedia/commons/0/0d/1990_s32_LDEF_stow.jpg *License:* Public domain *Contributors:* NASA *Original artist:* NASA/exploitcorporations
- **File:A_rep-tile-based_setiset_of_order_4.png** *Source:* https://upload.wikimedia.org/wikipedia/commons/4/45/A_rep-tile-based_setiset_of_order_4.png *License:* CC BY-SA 3.0 *Contributors:* http://www.leesallows.com/files/reptile_demo1a.png *Original artist:* Lee Sallows
- **File:Acetaldehyde-3D-vdW.png** *Source:* https://upload.wikimedia.org/wikipedia/commons/8/8e/Acetaldehyde-3D-vdW.png *License:* Public domain *Contributors:* ? *Original artist:* ?
- **File:Acetic-acid-3D-vdW.png** *Source:* https://upload.wikimedia.org/wikipedia/commons/d/db/Acetic-acid-3D-vdW.png *License:* Public domain *Contributors:* ? *Original artist:* ?
- **File:Aleksandr_Oparin_and_Andrei_Kursanov_in_enzymology_laboratory_1938.jpg** *Source:* https://upload.wikimedia.org/wikipedia/commons/f/f4/Aleksandr_Oparin_and_Andrei_Kursanov_in_enzymology_laboratory_1938.jpg *License:* Public domain *Contributors:* ? *Original artist:* ?
- **File:Argopecten_irradians.jpg** *Source:* https://upload.wikimedia.org/wikipedia/commons/0/09/Argopecten_irradians.jpg *License:* Public domain *Contributors:* Transferred from en.wikipedia to Commons using CommonsHelper. *Original artist:* Rachael Norris and Marina Freudzon / Mayscallop at en.wikipedia
- **File:Big_and_little_dog_1.jpg** *Source:* https://upload.wikimedia.org/wikipedia/commons/e/e9/Big_and_little_dog_1.jpg *License:* CC-BY-SA-3.0 *Contributors:* http://en.wikipedia.org/wiki/Image:IMG013biglittledogFX_wb.jpg *Original artist:* Ellen Levy Finch / en:User:Elf (uploaded by TBjornstad 14:51, 17 August 2006 (UTC))
- **File:Blacksmoker_in_Atlantic_Ocean.jpg** *Source:* https://upload.wikimedia.org/wikipedia/commons/6/6f/Blacksmoker_in_Atlantic_Ocean.jpg *License:* Public domain *Contributors:* NOAA Photo Library *Original artist:* P. Rona
- **File:Blue_morpho_butterfly.jpg** *Source:* https://upload.wikimedia.org/wikipedia/commons/6/65/Blue_morpho_butterfly.jpg *License:* CC-BY-SA-3.0 *Contributors:* Own work *Original artist:* Gregory Phillips
- **File:Branta_leucopsis.jpg** *Source:* https://upload.wikimedia.org/wikipedia/commons/5/5f/Branta_leucopsis.jpg *License:* CC BY-SA 2.5 *Contributors:* Uploaded by Linnea *Original artist:* Linnea Samila
- **File:Buckminsterfullerene-perspective-3D-balls.png** *Source:* https://upload.wikimedia.org/wikipedia/commons/0/0f/Buckminsterfullerene-perspective-3D-balls.png *License:* Public domain *Contributors:* Own work *Original artist:* Benjah-bmm27
- **File:Carbon-monoxide-3D-vdW.png** *Source:* https://upload.wikimedia.org/wikipedia/commons/a/a7/Carbon-monoxide-3D-vdW.png *License:* Public domain *Contributors:* ? *Original artist:* ?
- **File:CentralTendencyLV.jpg** *Source:* https://upload.wikimedia.org/wikipedia/en/6/60/CentralTendencyLV.jpg *License:* PD *Contributors:* Own work

  *Original artist:*

  Elb2000 (talk) (Uploads)
- **File:Champagne_vent_white_smokers.jpg** *Source:* https://upload.wikimedia.org/wikipedia/commons/a/aa/Champagne_vent_white_smokers.jpg *License:* Public domain *Contributors:* http://oceanexplorer.noaa.gov/explorations/04fire/logs/hirez/champagne_vent_hirez.jpg *Original artist:* NOAA
- **File:Coldecygne.svg** *Source:* https://upload.wikimedia.org/wikipedia/commons/6/63/Coldecygne.svg *License:* GFDL *Contributors:* Own work *Original artist:* This vector image was created with Inkscape.
- **File:CollapsedtreeLabels-simplified.svg** *Source:* https://upload.wikimedia.org/wikipedia/commons/1/12/CollapsedtreeLabels-simplified.svg *License:* Public domain *Contributors:* Own work (Original text: *Self made.*) *Original artist:* Original uploader was User:TimVickers, SVG conversion by User:User_A1
- **File:Commons-logo.svg** *Source:* https://upload.wikimedia.org/wikipedia/en/4/4a/Commons-logo.svg *License:* CC-BY-SA-3.0 *Contributors:* ? *Original artist:* ?
- **File:ConvectionCells.svg** *Source:* https://upload.wikimedia.org/wikipedia/commons/f/f5/ConvectionCells.svg *License:* CC-BY-SA-3.0 *Contributors:* Own work *Original artist:* Eyrian Con-struct
- **File:Cyanooctatetrayne-3D-vdW.png** *Source:* https://upload.wikimedia.org/wikipedia/commons/f/f7/Cyanooctatetrayne-3D-vdW.png *License:* Public domain *Contributors:* Derived from File:Acetylene-3D-vdW.png and File:Hydrogen-cyanide-3D-vdW.png. *Original artist:* Ben Mills and Jynto
- **File:DNA_chemical_structure.svg** *Source:* https://upload.wikimedia.org/wikipedia/commons/e/e4/DNA_chemical_structure.svg *License:* CC-BY-SA-3.0 *Contributors:* <a href='//validator.w3.org/' data-x-rel='nofollow'><img alt='W3C' src='https://upload.wikimedia.org/wikipedia/commons/thumb/1/1a/Valid_SVG_1.1_%28green%29.svg/88px-Valid_SVG_1.1_%28green%29.svg.png' width='88' height='30' style='vertical-align: top' srcset='https://upload.wikimedia.org/wikipedia/commons/thumb/1/1a/Valid_SVG_1.1_%28green%29.svg/132px-Valid_SVG_1.1_%28green%29.svg.png 1.5x, https://upload.wikimedia.org/wikipedia/commons/thumb/1/1a/Valid_SVG_1.1_%28green%29.svg/176px-Valid_SVG_1.1_%28green%29.svg.png 2x' data-file-width='91' data-file-height='31' /></a>iThe source code of this SVG is <a data-x-rel='nofollow' class='external text' href='//validator.w3.org/check?uri=https%3A%2F%2Fcommons.wikimedia.org%2Fwiki%2FSpecial%3AFilepath%2FDNA_chemical_structure.svg,,&,,ss=1#source'>valid</a>.

  *Original artist:* Madprime (talk · contribs)

- **File:DNA_nanostructures.png** *Source:* https://upload.wikimedia.org/wikipedia/commons/5/55/DNA_nanostructures.png *License:* CC BY 2.5 *Contributors:* Strong M: *Protein Nanomachines.* PLoS Biol 2/3/2004: e73. http://dx.doi.org/10.1371/journal.pbio.0020073 *Original artist:* (Images were kindly provided by Thomas H. LaBean and Hao Yan.)
- **File:Darwin'{}s_finches.jpeg** *Source:* https://upload.wikimedia.org/wikipedia/commons/9/97/Darwin%27s_finches.jpeg *License:* Public domain *Contributors:* From "Voyage of the Beagle" as found on [1] and [2] *Original artist:* John Gould (14.Sep.1804 - 3.Feb.1881)
- **File:Darwin_restored2.jpg** *Source:* https://upload.wikimedia.org/wikipedia/commons/b/b6/Darwin_restored2.jpg *License:* Public domain *Contributors:* Library of Congress[1] *Original artist:* Elliott & Fry
- **File:Diacetylene-3D-vdW-B.png** *Source:* https://upload.wikimedia.org/wikipedia/commons/6/64/Diacetylene-3D-vdW-B.png *License:* Public domain *Contributors:* Derived from File:Acetylene-3D-vdW.png. *Original artist:* Ben Mills and Jynto
- **File:Diagram_of_the_Monitor-Analyse-Record-Reflect-Reconstruct-Review-Spiral_algorithm.jpg** *Source:* https://upload.wikimedia.org/wikipedia/commons/b/b8/Diagram_of_the_Monitor-Analyse-Record-Reflect-Reconstruct-Review-Spiral_algorithm.jpg *License:* CC BY-SA 3.0 *Contributors:* Created by Laurie F. Thomas and published in the book "Learning Conversations" and published by Routledge, copyright has now returned to the Author(s) *Original artist:* Soler99
- **File:Drugroutemap.gif** *Source:* https://upload.wikimedia.org/wikipedia/commons/6/64/Drugroutemap.gif *License:* Public domain *Contributors:* CIA Employee *Original artist:* CIA Employee
- **File:EXPOSE_location_on_the_ISS.jpg** *Source:* https://upload.wikimedia.org/wikipedia/commons/d/da/EXPOSE_location_on_the_ISS.jpg *License:* Public domain *Contributors:* http://spaceflight.nasa.gov/gallery/images/shuttle/sts-124/html/s124e009982.html *Original artist:* NASA
- **File:Edit-clear.svg** *Source:* https://upload.wikimedia.org/wikipedia/en/f/f2/Edit-clear.svg *License:* Public domain *Contributors:* The *Tango! Desktop Project. Original artist:*

    The people from the Tango! project. And according to the meta-data in the file, specifically: "Andreas Nilsson, and Jakub Steiner (although minimally)."
- **File:Ethanol-3D-balls.png** *Source:* https://upload.wikimedia.org/wikipedia/commons/b/b0/Ethanol-3D-balls.png *License:* Public domain *Contributors:* ? *Original artist:* ?
- **File:Folder_Hexagonal_Icon.svg** *Source:* https://upload.wikimedia.org/wikipedia/en/4/48/Folder_Hexagonal_Icon.svg *License:* Cc-by-sa-3.0 *Contributors:* ? *Original artist:* ?
- **File:Formaldehyde-3D-vdW.png** *Source:* https://upload.wikimedia.org/wikipedia/commons/a/a3/Formaldehyde-3D-vdW.png *License:* Public domain *Contributors:* ? *Original artist:* ?
- **File:Formamide-3D-vdW.png** *Source:* https://upload.wikimedia.org/wikipedia/commons/c/cf/Formamide-3D-vdW.png *License:* Public domain *Contributors:* ? *Original artist:* ?
- **File:Formation_of_Glycolaldehyde_in_star_dust.png** *Source:* https://upload.wikimedia.org/wikipedia/commons/4/46/Formation_of_Glycolaldehyde_in_star_dust.png *License:* Public domain *Contributors:* NASA *Original artist:* Lara Clemence
- **File:Fugle,_ørnsø_073.jpg** *Source:* https://upload.wikimedia.org/wikipedia/commons/d/d6/Fugle%2C_%C3%B8rns%C3%B8_073.jpg *License:* Public domain *Contributors:* Own work *Original artist:* Christoffer A Rasmussen (Rasmussen29892 at da.wikipedia)
- **File:Glass_flask_used_by_Louis_Pasteur,_1860s._(9660573541).jpg** *Source:* https://upload.wikimedia.org/wikipedia/commons/5/5b/Glass_flask_used_by_Louis_Pasteur%2C_1860s._%289660573541%29.jpg *License:* CC BY-SA 2.0 *Contributors:* Glass flask used by Louis Pasteur, 1860s. *Original artist:* Science Museum London / Science and Society Picture Library
- **File:Gospers_glider_gun.gif** *Source:* https://upload.wikimedia.org/wikipedia/commons/e/e5/Gospers_glider_gun.gif *License:* CC-BY-SA-3.0 *Contributors:* Own work *Original artist:* Kieff
- **File:Henry_Fairfield_Osborn.jpg** *Source:* https://upload.wikimedia.org/wikipedia/commons/4/4a/Henry_Fairfield_Osborn.jpg *License:* Public domain *Contributors:* ? *Original artist:* ?
- **File:Homochiralproline.png** *Source:* https://upload.wikimedia.org/wikipedia/commons/2/2d/Homochiralproline.png *License:* CC-BY-SA-3.0 *Contributors:* en:Image:Homochiralproline.png *Original artist:* V8rik
- **File:HomunculusLarge.png** *Source:* https://upload.wikimedia.org/wikipedia/commons/7/7a/HomunculusLarge.png *License:* Public domain *Contributors:* ? *Original artist:* ?
- **File:Horizontal-gene-transfer.jpg** *Source:* https://upload.wikimedia.org/wikipedia/commons/1/18/Horizontal-gene-transfer.jpg *License:* Attribution *Contributors:* Barth F. Smets, Ph.D., with permission *Original artist:* Dr. Smets and perhaps others
- **File:Kölliker_Rudolph_Albert_von_1818-1902.jpg** *Source:* https://upload.wikimedia.org/wikipedia/commons/2/22/K%C3%B6lliker_Rudolph_Albert_von_1818-1902.jpg *License:* Public domain *Contributors:* ? *Original artist:* ?
- **File:Lipid_bilayer_section.gif** *Source:* https://upload.wikimedia.org/wikipedia/commons/f/f0/Lipid_bilayer_section.gif *License:* Public domain *Contributors:* http://en.wikipedia.org/wiki/Image:Lipid_bilayer_section.gif *Original artist:* Bensaccount
- **File:Methane-2D-stereo.svg** *Source:* https://upload.wikimedia.org/wikipedia/commons/9/92/Methane-2D-stereo.svg *License:* Public domain *Contributors:* Own work *Original artist:* SVG version by Patricia.fidi
- **File:Methane-3D-space-filling.svg** *Source:* https://upload.wikimedia.org/wikipedia/commons/4/4b/Methane-3D-space-filling.svg *License:* Public domain *Contributors:* Own work *Original artist:* Dbc334 (first version); Jynto (second version).
- **File:Methyldiacetylene-3D-vdW.png** *Source:* https://upload.wikimedia.org/wikipedia/commons/4/44/Methyldiacetylene-3D-vdW.png *License:* Public domain *Contributors:* Derived from File:Propyne-3D-vdW.png. *Original artist:* Ben Mills and Jynto
- **File:Micelle_scheme-en.svg** *Source:* https://upload.wikimedia.org/wikipedia/commons/4/4d/Micelle_scheme-en.svg *License:* CC BY-SA 3.0 *Contributors:* Own work *Original artist:* SuperManu

## 21.7. TEXT AND IMAGE SOURCES, CONTRIBUTORS, AND LICENSES

- **File:Miller-Urey_experiment-en.svg** *Source:* https://upload.wikimedia.org/wikipedia/commons/5/54/Miller-Urey_experiment-en.svg *License:* CC BY-SA 3.0 *Contributors:* Own work from Image:MUexperiment.png. *Original artist:* This vector image was created with Inkscape.
- **File:Miller-Urey_experiment_-_Work_by_the_C3BC_consortium,_licensed_under_CC-BY-3.0.webm** *Source:* https://upload.wikimedia.org/wikipedia/commons/a/aa/Miller-Urey_experiment_-_Work_by_the_C3BC_consortium%2C_licensed_under_CC-BY-3.0.webm *License:* CC BY 3.0 *Contributors:* Miller-Urey experiment *Original artist:* Courtney Harrington
- **File:Nb3O7(OH)_self-organization2.jpg** *Source:* https://upload.wikimedia.org/wikipedia/commons/3/3f/Nb3O7%28OH%29_self-organization2.jpg *License:* CC BY 3.0 *Contributors:* http://pubs.rsc.org/en/content/articlehtml/2014/ta/c4ta02202e *Original artist:* Sophia B. Betzler et al.
- **File:Nitrous-oxide-3D-balls.png** *Source:* https://upload.wikimedia.org/wikipedia/commons/9/93/Nitrous-oxide-3D-balls.png *License:* Public domain *Contributors:* Own work *Original artist:* Ben Mills
- **File:Office-book.svg** *Source:* https://upload.wikimedia.org/wikipedia/commons/a/a8/Office-book.svg *License:* Public domain *Contributors:* This and myself. *Original artist:* Chris Down/Tango project
- **File:Open_Access_logo_PLoS_transparent.svg** *Source:* https://upload.wikimedia.org/wikipedia/commons/7/77/Open_Access_logo_PLoS_transparent.svg *License:* CC0 *Contributors:* http://www.plos.org/ *Original artist:* art designer at PLoS, modified by Wikipedia users Nina, Beao, and JakobVoss
- **File:Open_book_nae_02.svg** *Source:* https://upload.wikimedia.org/wikipedia/commons/9/92/Open_book_nae_02.svg *License:* CC0 *Contributors:* OpenClipart *Original artist:* nae
- **File:PAHWorld.png** *Source:* https://upload.wikimedia.org/wikipedia/commons/d/d1/PAHWorld.png *License:* Public domain *Contributors:* Transferred from en.wikipedia to Commons by gustavocarra. *Original artist:* Julianonions at English Wikipedia
- **File:Panspermie.svg** *Source:* https://upload.wikimedia.org/wikipedia/commons/4/48/Panspermie.svg *License:* CC BY-SA 3.0 *Contributors:* Own work; For the proto-bacteria I used an adapted version of File:Bacteria-.svg by JrPol and for the DNA I used an adapted version of File:DNA chemical structure.svg by Madprime. Earth from File:Earth Flag.svg by Himasaram *Original artist:* Silver Spoon Sokpop
- **File:People_icon.svg** *Source:* https://upload.wikimedia.org/wikipedia/commons/3/37/People_icon.svg *License:* CC0 *Contributors:* OpenClipart *Original artist:* OpenClipart
- **File:Phospholipids_aqueous_solution_structures.svg** *Source:* https://upload.wikimedia.org/wikipedia/commons/c/c6/Phospholipids_aqueous_solution_structures.svg *License:* Public domain *Contributors:* Own work *Original artist:* Mariana Ruiz Villarreal ,LadyofHats
- **File:Phylogenetic_tree.svg** *Source:* https://upload.wikimedia.org/wikipedia/commons/7/70/Phylogenetic_tree.svg *License:* Public domain *Contributors:* NASA Astrobiology Institute, found in an article *Original artist:* This vector version: Eric Gaba (Sting - fr:Sting)
- **File:Phylogenic_Tree-en.svg** *Source:* https://upload.wikimedia.org/wikipedia/commons/5/58/Phylogenic_Tree-en.svg *License:* CC BY-SA 3.0 *Contributors:* This file was derived from Phylogenic Tree.jpg: <a href='//commons.wikimedia.org/wiki/File:Phylogenic_Tree.jpg' class='image'><img alt='Phylogenic Tree.jpg' src='https://upload.wikimedia.org/wikipedia/commons/thumb/d/de/Phylogenic_Tree.jpg/50px-Phylogenic_Tree.jpg' width='50' height='33' srcset='https://upload.wikimedia.org/wikipedia/commons/thumb/d/de/Phylogenic_Tree.jpg/75px-Phylogenic_Tree.jpg 1.5x, https://upload.wikimedia.org/wikipedia/commons/thumb/d/de/Phylogenic_Tree.jpg/100px-Phylogenic_Tree.jpg 2x' data-file-width='1069' data-file-height='713' /></a>
*Original artist:* Phylogenic Tree.jpg: John D. Croft
- **File:Pollicipes_cornucopia.jpg** *Source:* https://upload.wikimedia.org/wikipedia/commons/5/5a/Pollicipes_pollicipes.jpg *License:* CC BY-SA 4.0 *Contributors:* Own work *Original artist:* Hans Hillewaert
- **File:Polycyclic_Aromatic_Hydrocarbons.png** *Source:* https://upload.wikimedia.org/wikipedia/commons/c/c0/Polycyclic_Aromatic_Hydrocarbons.png *License:* Public domain *Contributors:* Own work by uploader, Accelrys DS Visualizer *Original artist:* Inductiveload
- **File:Pore_schematic.svg** *Source:* https://upload.wikimedia.org/wikipedia/commons/c/cc/Pore_schematic.svg *License:* Public domain *Contributors:* Own work *Original artist:* MDougM
- **File:Portal-puzzle.svg** *Source:* https://upload.wikimedia.org/wikipedia/en/f/fd/Portal-puzzle.svg *License:* Public domain *Contributors:* ? *Original artist:* ?
- **File:Question_book-new.svg** *Source:* https://upload.wikimedia.org/wikipedia/en/9/99/Question_book-new.svg *License:* Cc-by-sa-3.0 *Contributors:*
Created from scratch in Adobe Illustrator. Based on Image:Question book.png created by User:Equazcion *Original artist:* Tkgd2007
- **File:STS-46_EURECA_deployment.jpg** *Source:* https://upload.wikimedia.org/wikipedia/commons/b/bf/STS-46_EURECA_deployment.jpg *License:* Public domain *Contributors:* NASA http://images.jsc.nasa.gov/luceneweb/caption.jsp?photoId=STS046-08-010 *Original artist:* NASA
- **File:Self-replication_of_sphynx_hexidiamonds.svg** *Source:* https://upload.wikimedia.org/wikipedia/commons/f/fa/Self-replication_of_sphynx_hexidiamonds.svg *License:* Public domain *Contributors:* en:Image:Sphnxhex.png *Original artist:* en:User:Spottedowl, User:Stannered
- **File:Sigmoid_curve_for_an_autocatalytical_reaction.jpg** *Source:* https://upload.wikimedia.org/wikipedia/commons/5/55/Sigmoid_curve_for_an_autocatalytical_reaction.jpg *License:* CC-BY-SA-3.0 *Contributors:* English Wikipedia *Original artist:* Knights who say ni
- **File:Soai_autocatalysis.png** *Source:* https://upload.wikimedia.org/wikipedia/commons/9/90/Soai_autocatalysis.png *License:* CC-BY-SA-3.0 *Contributors:* en:Image:Soai_autocatalysis.png *Original artist:* V8rik
- **File:Sound-icon.svg** *Source:* https://upload.wikimedia.org/wikipedia/commons/4/47/Sound-icon.svg *License:* LGPL *Contributors:* Derivative work from Silsor's versio *Original artist:* Crystal SVG icon set

- **File:Stardust_Dust_Collector_with_aerogel.jpg** *Source:* https://upload.wikimedia.org/wikipedia/commons/1/1e/Stardust_Dust_Collector_with_aerogel.jpg *License:* Public domain *Contributors:* ? *Original artist:* ?
- **File:Stromatolites.jpg** *Source:* https://upload.wikimedia.org/wikipedia/commons/c/c0/Stromatolites.jpg *License:* Public domain *Contributors:* National Park Service - http://www.nature.nps.gov/geology/cfprojects/photodb/Photo_Detail.cfm?PhotoID=204 *Original artist:* P. Carrara, NPS
- **File:Surfactant.jpg** *Source:* https://upload.wikimedia.org/wikipedia/commons/0/03/Surfactant.jpg *License:* Public domain *Contributors:* Own work *Original artist:* Major measure
- **File:Symbol_book_class2.svg** *Source:* https://upload.wikimedia.org/wikipedia/commons/8/89/Symbol_book_class2.svg *License:* CC BY-SA 2.5 *Contributors:* Mad by Lokal_Profil by combining: *Original artist:* Lokal_Profil
- **File:The_Systems7_Diagram.jpg** *Source:* https://upload.wikimedia.org/wikipedia/en/5/59/The_Systems7_Diagram.jpg *License:* CC-BY-3.0 *Contributors:*
Created by myself (Laurie F. Thomas) during a research project and published in the book 'Learning Conversation' publishd by Routledge 1985 and the copyright has returned to myself
*Original artist:*
Soler99
- **File:Theodor_Eimer_(Professorengalerie_Universität_Tübingen).jpg** *Source:* https://upload.wikimedia.org/wikipedia/commons/b/bd/Theodor_Eimer_%28Professorengalerie_Universit%C3%A4t_T%C3%BCbingen%29.jpg *License:* Public domain *Contributors:* http://www.studion.uni-tuebingen.de/mediawiki/index.php/Professorengalerie:_Theodor_Eimer *Original artist:* Alkan Lukeroth oder Eugen Hofmeister
- **File:Tree_of_life.svg** *Source:* https://upload.wikimedia.org/wikipedia/commons/0/09/Tree_of_life.svg *License:* CC-BY-SA-3.0 *Contributors:* No machine-readable source provided. Own work assumed (based on copyright claims). *Original artist:* No machine-readable author provided. Vanished user fijtji34toksdcknqrjn54yoimascj assumed (based on copyright claims).
- **File:Tree_of_life_by_Haeckel.jpg** *Source:* https://upload.wikimedia.org/wikipedia/commons/d/de/Tree_of_life_by_Haeckel.jpg *License:* Public domain *Contributors:* First version from en.wikipedia; description page was here. Later versions derived from this scan, from the American Philosophical Society Museum. *Original artist:* Ernst Haeckel
- **File:Trihydrogen-cation-3D-vdW.png** *Source:* https://upload.wikimedia.org/wikipedia/commons/4/49/Trihydrogen-cation-3D-vdW.png *License:* Public domain *Contributors:* ? *Original artist:* ?
- **File:Visualization_of_wiki_structure_using_prefuse_visualization_package.png** *Source:* https://upload.wikimedia.org/wikipedia/commons/9/90/Visualization_of_wiki_structure_using_prefuse_visualization_package.png *License:* CC BY-SA 3.0 *Contributors:* Own work - I (Mr3641 (talk)) created this work entirely by myself.
*Original artist:* Chris Davis at en.wikipedia
- **File:Wikisource-logo.svg** *Source:* https://upload.wikimedia.org/wikipedia/commons/4/4c/Wikisource-logo.svg *License:* CC BY-SA 3.0 *Contributors:* Rei-artur *Original artist:* Nicholas Moreau
- **File:Wiktionary-logo-v2.svg** *Source:* https://upload.wikimedia.org/wikipedia/commons/0/06/Wiktionary-logo-v2.svg *License:* CC BY-SA 4.0 *Contributors:* Own work *Original artist:* Dan Polansky based on work currently attributed to Wikimedia Foundation but originally created by Smurrayinchester
- **File:X-ray_by_Wilhelm_Röntgen_of_Albert_von_Kölliker'{}s_hand_-_18960123-02.jpg** *Source:* https://upload.wikimedia.org/wikipedia/commons/f/fb/X-ray_by_Wilhelm_R%C3%B6ntgen_of_Albert_von_K%C3%B6lliker%27s_hand_-_18960123-02.jpg *License:* Public domain *Contributors:* Flipped version of File:X-ray by Wilhelm Röntgen of Albert von Kölliker's hand - 18960123-03.jpg (now deleted as a duplicate). *Original artist:* Wilhelm Röntgen; current version created by Old Moonraker.
- **File:Zoomed_view_of_carbon_nanotube.svg** *Source:* https://upload.wikimedia.org/wikipedia/commons/a/a1/Zoomed_view_of_carbon_nanotube.svg *License:* CC BY-SA 3.0 *Contributors:* This file was derived from Carbon nanotube.svg: <a href='//commons.wikimedia.org/wiki/File:Carbon_nanotube.svg' class='image'><img alt='Carbon nanotube.svg' src='https://upload.wikimedia.org/wikipedia/commons/thumb/9/9b/Carbon_nanotube.svg/50px-Carbon_nanotube.svg.png' width='50' height='187' srcset='https://upload.wikimedia.org/wikipedia/commons/thumb/9/9b/Carbon_nanotube.svg/75px-Carbon_nanotube.svg.png 1.5x, https://upload.wikimedia.org/wikipedia/commons/thumb/9/9b/Carbon_nanotube.svg/100px-Carbon_nanotube.svg.png 2x' data-file-width='159' data-file-height='594' /></a>
*Original artist:* Carbon_nanotube.svg: Guillaume Paumier (user:guillom)

### 21.7.3 Content license

- Creative Commons Attribution-Share Alike 3.0

www.ingramcontent.com/pod-product-compliance
Lightning Source LLC
Chambersburg PA
CBHW080656190526
45169CB00006B/2138